MW01144974

OAK RIDGE NATIONAL LABORATORY LIBRARIES
3 4456 0552541 0

258

# Topics in Current Chemistry

## Topics in Current Chemistry

Recently Published and Forthcoming Volumes

**Molecular Machines**
Volume Editor: Kelly, T. R.
Vol. 262, 2006

**Immobilisation of DNA on Chips II**
Volume Editor: Wittmann, C.
Vol. 261, 2005

**Immobilisation of DNA on Chips I**
Volume Editor: Wittmann, C.
Vol. 260, 2005

**Prebiotic Chemistry**
From Simple Amphiphiles to Protocell Models
Volume Editor: Walde, P.
Vol. 259, 2005

**Supramolecular Dye Chemistry**
Volume Editor: Würthner, F.
Vol. 258, 2005

**Molecular Wires**
From Design to Properties
Volume Editor: De Cola, L.
Vol. 257, 2005

**Low Molecular Mass Gelators**
Design, Self-Assembly, Function
Volume Editor: Fages, F.
Vol. 256, 2005

**Anion Sensing**
Volume Editor: Stibor, I.
Vol. 255, 2005

**Organic Solid State Reactions**
Volume Editor: Toda, F.
Vol. 254, 2005

**DNA Binders and Related Subjects**
Volume Editors:Waring, M. J., Chaires, J. B.
Vol. 253, 2005

**Contrast Agents III**
Volume Editor: Krause,W.
Vol. 252, 2005

**Chalcogenocarboxylic Acid Derivatives**
Volume Editor: Kato, S.
Vol. 251, 2005

**New Aspects in Phosphorus Chemistry V**
Volume Editor: Majoral, J.-P.
Vol. 250, 2005

**Templates in Chemistry II**
Volume Editors: Schalley, C. A.,Vögtle, F., Dötz, K. H.
Vol. 249, 2005

**Templates in Chemistry I**
Volume Editors: Schalley, C. A.,Vögtle, F., Dötz, K. H.
Vol. 248, 2004

**Collagen**
Volume Editors: Brinckmann, J., Notbohm, H., Müller, P. K.
Vol. 247, 2005

**New Techniques in Solid-State NMR**
Volume Editor: Klinowski, J.
Vol. 246, 2005

**Functional Molecular Nanostructures**
Volume Editor: Schlüter, A. D.
Vol. 245, 2005

**Natural Product Synthesis II**
Volume Editor: Mulzer, J.
Vol. 244, 2005

**Natural Product Synthesis I**
Volume Editor: Mulzer, J.
Vol. 243, 2005

# Supermolecular Dye Chemistry

Volume Editor: Frank Würthner

With contributions by

A. Ajayaghosh · T. S. Balaban · R. Dobrawa · S. De Feyter · S. J. George
A. R. Holzwarth · H. Ihmels · T. Ishi-i · V. Kriegisch · C. Lambert
D. Otto · C. R. Saha-Möller · A. P. H. J. Schenning · F. De Schryver
S. Shinkai · H. Tamiaki · F. Würthner · C.-C. You

The series *Topics in Current Chemistry* presents critical reviews of the present and future trends in polymer and biopolymer science including chemistry, physical chemistry, physics and material science. It is adressed to all scientists at universities and in industry who wish to keep abreast of advances in the topics covered.
As a rule, contributions are specially commissioned. The editors and publishers will, however, always be pleased to receive suggestions and supplementary information. Papers are accepted for *Topics in Current Chemistry* in English.
In references *Topics in Current Chemistry* is abbeviated *Top Curr Chem* and is cited as a journal.

Springer WWW home page: http://www.springeronline.com
Visit the TCC content at http://www.springerlink.com/

Library of Congress Control Number: 2005928611

ISSN 0340-1022
ISBN-10 3-540-27758-7 Springer Berlin Heidelberg New York
ISBN-13 978-3-540-27758-3 Springer Berlin Heidelberg New York
DOI 10.1007/b105136



**Springer is a part of Springer Science+Business Media**

springeronline.com


Printed in Germany



Cover design: *Design & Production* GmbH, Heidelberg
Typesetting and Production: LE-TeX Jelonek, Schmidt & Vöckler GbR, Leipzig

Printed on acid-free paper 02/3141 YL – 5 4 3 2 1 0

## Volume Editor

Prof. Dr. Frank Würthner

Institut für Organische Chemie
Universität Würzburg
Am Hubland
97074 Würzburg, Germany
*wuerthner@chemie.uni-wuerzburg.de*

## Editorial Board

## Topics in Current Chemistry Also Available Electronically

For all customers who have a standing order to Topics in Current Chemistry, we offer the electronic version via SpringerLink free of charge. Please contact your librarian who can receive a password or free access to the full articles by registering at:

springerlink.com

If you do not have a subscription, you can still view the tables of contents of the volumes and the abstract of each article by going to the SpringerLink Homepage, clicking on "Browse by Online Libraries", then "Chemical Sciences", and finally choose Topics in Current Chemistry.

You will find information about the

- Editorial Board
- Aims and Scope
- Instructions for Authors
- Sample Contribution

at springeronline.com using the search function.

# Preface

Dye chemistry was one of the initial topics of chemical research in the academic as well as industrial field. At the early stage of dye research, in the last decades of the 19th and the beginning of the 20th century, the focus was on the elucidation of structures of natural dyes aiming at the development of their chemical syntheses and to establish theoretical concepts for the understanding of the color–constitution relationship as a prerequisite for the design of new artificial colorants. The major outcome of these pioneering efforts for mankind was that color is no more a privilege of nature and, hence, multi-colored paints entered our everyday life and textiles of any desirable shade became accessible.

Nowadays most colorants have the purpose to satisfy our aesthetical needs and, thus, thousands of dyes and pigments are produced on industrial scales. Nevertheless, nearly periodically new demands arise for so-called "functional dyes" whose $\pi$-conjugated systems exhibit novel functionalities beyond aesthetical purposes. Optical brighteners or near-infrared absorbers are examples where even transparency in the visible spectrum is desired and dyes for non-linear optics, holographic optical data storage and two photon absorption are further examples where the color properties of "dyes" are insignificantly related to the functional demands.

Whereas most of these applications can still be addressed by appropriate design of the molecular properties of the $\pi$-conjugated backbone, i.e. the chromophore, nature has developed other types of functional dyes which obtain their functionality only by proper organization of dye molecules in space, typically within a protein matrix. Moreover, in most cases not a particular dye, rather a multichromophoric entity imparts the desired functionality. For example, regulation of oxygen transport requires oligomeric protein assemblies containing iron porphyrins (hemoglobin) and the photoinduced electron transfer cascade in the reaction center of photosynthesis needs a set of functional dyes arranged across the photosynthetic membrane in proper geometry. Most intriguing, to enable highly efficient conversion of sunlight into chemical energy, nature has developed light-harvesting systems which incorporate hundreds of dye molecules in well-defined spatial proximity to efficiently feed the reaction centers of photosynthesis with excitation energy.

Exploration of the structure–function relationship of such complex natural assemblies with the aim to develop efficient artificial photoactive devices constitutes, indeed, a major goal of current interdisciplinary research in the field of functional dye systems. Supramolecular chemistry contributes to this field by offering the toolbox for the synthesis of the desired architectures and their structural characterization. Time resolved spectroscopic characterization of photophysical properties of these structures allows to compare their functionality with those of the natural counterparts and to evaluate the prospects for various technical applications. It is very encouraging in this respect that in the last few years rationally designed dye assemblies proved to be successful in the fields of organic electronics and photonics where dye-dye interactions in the bulk state and at interfaces are of crucial importance for the desired functionalities of charge and energy transport.

This monograph is intended to provide coverage of a selection of different aspects of supramolecular dye chemistry with special emphasis on the elaboration of concepts for the realization of defined multichromophoric architectures in solution and at interfaces. We regret that some interesting topics, e.g. photoswitches and sensor materials, of this very rapidly growing research field could not be considered and, even not all classes of dye assemblies are covered by the given chapters. For example, the oldest class of dye assemblies, i.e. cyanine dye aggregates, is not included because our principle of organization is based on the respective noncovalent forces and not on the classes of dyes. Only in the first chapter by *T. S. Balaban*, *H. Tamiaki* and *A. R. Holzwarth* insight is provided into the structural and functional peculiarities of a special dye system, namely the chlorins, which combine hydrogen-bonding, metal-ligand coordination and $\pi$–$\pi$ stacking in the most beneficial way to accomplish nature's most successful light-harvesting machinery. The following two chapters are organized from the supramolecular point of view with a comprehensive review on metal-directed self-assembly by *C.-C. You*, *R. Dobrawa*, *C. R. Saha-Möller* and *F. Würthner* (here porphyrins and perylene bisimides constitute the most important classes of dyes) and an article dealing with hydrogen-bond directed self-assembly by *A. Ajayaghosh*, *S. J. George* and *A. P. H.J. Schenning*. Combining these noncovalent interactions (hydrogen-bonding and metal-ligand coordination) with $\pi$–$\pi$ stacking is the method of choice to create more complex materials like organogels which is the topic of the fourth chapter by *T. Ishi-i* and *S. Shinkai*.

The concept behind the other three chapters is the interaction of dye molecules with macromolecular scaffolds or surfaces. Thus, in chapter five *H. Ihmels* and *D. Otto* highlight the field of dye-based DNA intercalators. Although this field is traditionally not considered as a central topic of supramolecular chemistry, it provides a bridge towards biomedicinal applications of concepts emerging from supramolecular research. Likewise the last two chapters by *S. De Feyter* and *F. De Schryver* and by *V. Kriegisch* and *C. Lambert* constitute the interface to physics and nanotechnology covering important aspects of dye organization at surfaces.

The selection of subjects presented here is not aimed to offer a balanced compilation of "hot topics" of supramolecular dye chemistry, rather it attempts to identify concepts which hold promise for successful development of this field with tremendous prospects. I cordially thank the authors of the chapters for their efforts to provide high standard manuscripts and the publishers for giving me the opportunity to edit this volume.

Würzburg, June 2005 Frank Würthner

# Contents

## Contents of *Structure and Bonding, Vol. 108*

## Supramolecular Assembly via Hydrogen Bonds I

**Volume Editor: D. M. P. Mingos**
ISBN: 3-540-20084-3



## Contents of *Structure and Bonding, Vol. 111*

## Supramolecular Assembly via Hydrogen Bonds II

**Volume Editor: D. M. P. Mingos**
ISBN: 3-540-20086-X

Top Curr Chem (2005) 258: 1–38
DOI 10.1007/b137480

Published online: 8 August 2005

# Chlorins Programmed for Self-Assembly

Teodor Silviu Balaban[1] (✉) · Hitoshi Tamiaki[2] · Alfred R. Holzwarth[3]

[1]Institute for Nanotechnology, Forschungszentrum Karlsruhe, Postfach 3640, 76021 Karlsruhe, Germany
*silviu.balaban@int.fzk.de*

[2]Department of Bioscience and Biotechnology, Faculty of Science and Engineering, Ritsumeikan University, Kusatsu, 525-8577 Shiga, Japan
*tamiaki@se.ritsumei.ac.jp*

[3]Max-Planck-Institut für Bioanorganische Chemie (Former Max-Planck-Institut für Strahlenchemie), Stiftstr. 34–36, Postfach 101365, 45413 Mülheim a.d. Ruhr, Germany
*holzwarth@mpi-muelheim.mpg.de*



**Abstract** The supramolecular chemistry of chlorins which are the most abundant photosynthetic pigments is reviewed. In chlorophyll-protein complexes, ligation of the central magnesium atom can occur in two diastereomeric configurations. Light-harvesting complexes of purple bacteria are formed by the self-assembly of short polypeptides which bind bacteriochlorophylls into circular structures. The light-harvesting organelle of green photosynthetic bacteria, the so-called "chlorosome", is the most efficient natural antenna system and is formed by self-assembly of bacteriochlorophylls *c*, *d* or *e* without the help of a protein scaffold. Semisynthetic and fully synthetic mimics of these self-assembling bacteriochlorophylls have been prepared and their self-assemblies have been studied in detail in view of artificial light-harvesting systems. From a single crystal X-ray diffraction analysis, one could put into evidence hierarchic supramolecular interactions within such self-assembling systems. Interestingly, hydrogen bonding which all present models of bacteriochlorophyll self-assemblies contain as one of the important supramolecular interactions is absent in the fully synthetic mimics.

**Keywords** Self-assembly, chlorophyll, bacteriochlorophyll, antenna complex, chlorosome, biomimetic models, porphyrinoids.

# 1 Introduction

## 1.1 Chlorins as the Most Abundant Natural Photosynthetic Chromophores

Nature uses tetrapyrroles such as chlorophylls and bacteriochlorophylls as the main chromophores for light-harvesting in photosynthetic organisms. While porphyrins have a fully conjugated 26 $\pi$ electron system, in chlorins one of the pyrrolic double bonds is reduced and in bacteriochlorins two such double bonds are reduced (Fig. 1). In bacteriochlorins, the basic tetrapyrrole of the chlorophyllous ancestors, the single bonds are in opposite and not adjacent pyrrole rings. Corroles lack the 20-*meso* carbon atom while phthalocyanines are very robust fully synthetic pigments which have benzo-annulated pyrrole rings and nitrogen bridges instead of the four *meso*-methine units.

Chlorophylls (Chls) are chlorins which carry an additional five-membered ring having thus a phorbin skeleton and are usually encountered in cyanobac-

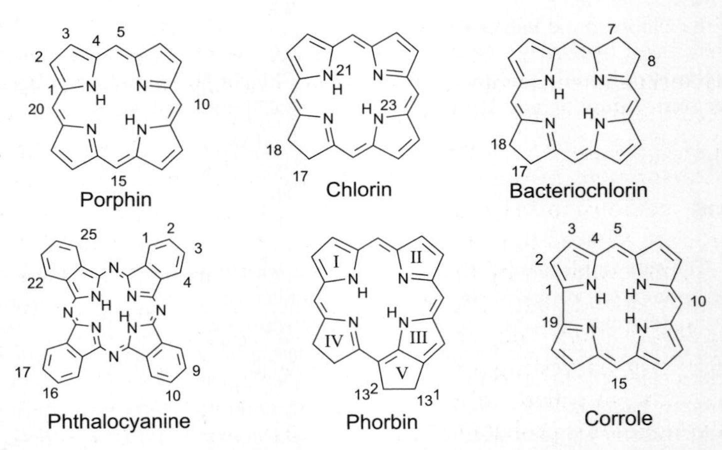


**Fig. 1** Basic cyclic tetrapyrroles shown here with the usual numbering system

teria, red algae, green algae, and higher plants. Bacteriochlorophylls (BChls) occur in photosynthetic bacteria and also possess the annulated five-membered ring. While BChls typically derive from the bacteriochlorin structure (such as BChl *a*, Fig. 2) some "bacteriochlorophylls" actually have a chlorin chemical and electronic structure (c.f. Fig. 2). These chlorin-based "bacteriochlorophylls"—due to the fact that they are present in some photosynthetic bacteria—received their trivial name before their actual chemical structure was known. This chapter focuses on the properties of special (bacterio)-chlorins which have been endowed for supramolecular self-organization. Supramolecular chemistry or the "chemistry beyond the molecule" [1] is effected via non-covalent interactions such as metal-ligation, hydrogen bonding, $\pi$-stacking and hydrophobic or dispersive interactions. All these can come into play with chlorins and often their combinations act cooperatively.

In a supramolecular system the non-covalently bound assemblies have properties that are often drastically different from those of their monomeric constituents. Thus, ensemble characteristics are dominant and novel functions emerge. Since a chlorin molecule with its peripheral substituents is a little over 1 nm in diameter, the term *functional nanostructure* is appropriate for their supramolecular assemblies.

Chls and BChls are typically found as light-harvesting pigments in the membrane-bound antenna systems of photosynthetic organisms [2]. Besides these cyclic tetrapyrroles, carotenoids are also encountered in most antenna systems. Some special photosynthetic organisms contain, however, also extra-membraneous antenna systems which make use of different chromophores. These are the phycobilisomes of cyanobacteria and red algae which contain open chain tetrapyrroles as pigments, the so-called phycobilins which are covalently bound to proteins. The other notable exception are the so-called "chlorosomes" of the green bacteria, which are extra-membraneous antenna systems containing BChls *c*, *d*, or *e*. The photosynthetic antenna systems have been optimized by evolution during the past 2.6 billion years, after cyanobacteria and eucaryotes evolved from the archaebacteria [3]. Cyanobacteria were the first organisms capable of oxygenic photosynthesis and they evolved into the photosynthetic eukaryotes, a process which eventually led to the development of the higher plant kingdom. Light and oxygen can be extremely noxious to cells if the long-lived triplet excited states of chromophores are allowed to generate singlet oxygen ($O_2$ $\Delta_g^1$). This problem was solved during evolution by the incorporation of carotenoids which are able to efficiently quench both B(Chl) triplet states as well as singlet oxygen by thermal deactivation. The association of carotenoids with Chls is also beneficial for light-harvesting since carotenoids absorb well between 450 and 550 nm, in the so-called Chl absorption gap (c.f. spectra in Fig. 3). A third role of carotenoids is probably structural: due to their extended and rigid conformation they help in the assembly of chlorophyll-protein complexes (CP)

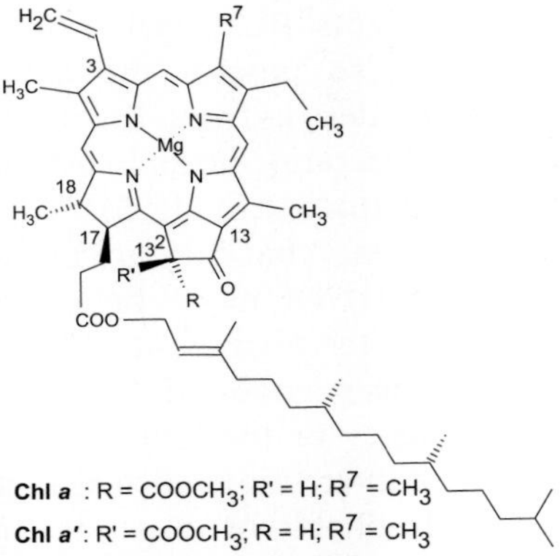

**Chl *a*** : R = $COOCH_3$; R' = H; $R^7 = CH_3$

**Chl *a'*** : R' = $COOCH_3$; R = H; $R^7 = CH_3$

**Chl b** : R = $COOCH_3$; R' = H; $R^7$ = CHO

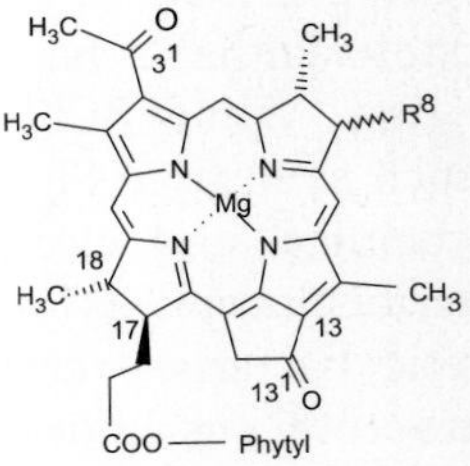

**BChl *a*** : $R^8 = CH_2CH_3$

**BChl *b*** : $R^8$ = =CH-$CH_3$

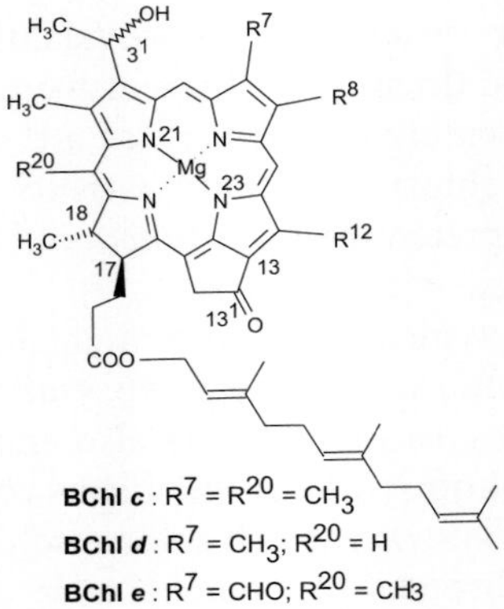

**BChl *c*** : $R^7 = R^{20} = CH_3$

**BChl *d*** : $R^7 = CH_3$; $R^{20}$ = H

**BChl *e*** : $R^7$ = CHO; $R^{20} = CH_3$

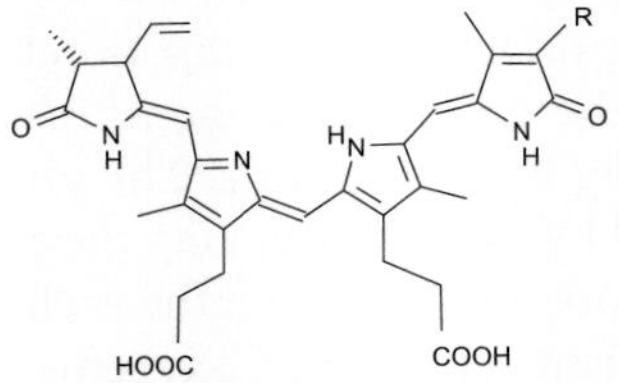

Phycocyanobilin: R = $CH_2CH_3$

Phytochromobilin: R = CH=$CH_2$

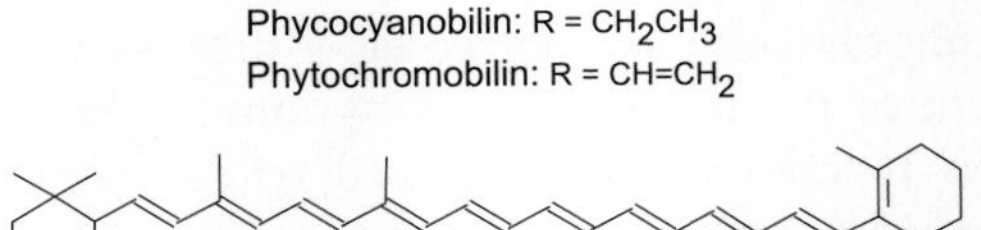

β-Carotene

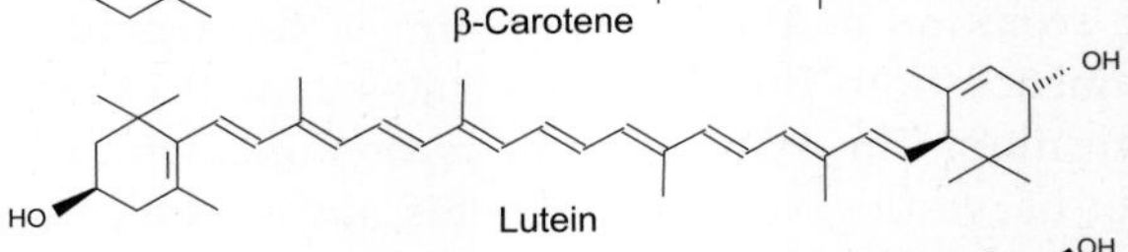

Lutein

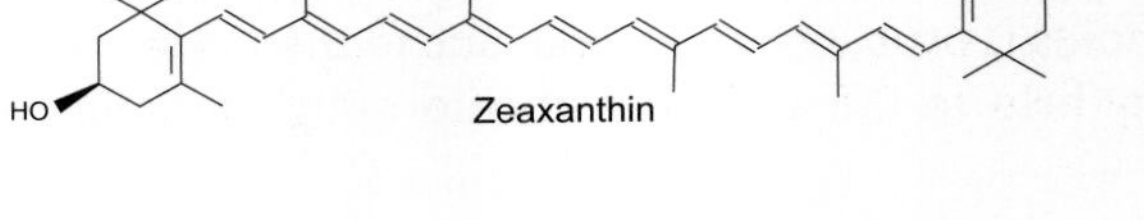

Zeaxanthin

◀ **Fig. 2** Natural chromophores involved in light-harvesting antennae. Phytol is the fatty alcohol exclusively esterifying the Chls in higher plants and other oxygen-evolving organisms while farnesol is the most abundant for the BChls. Also encountered in BChls are stearol, cetol, phytol, geranyl-geraniol, and other fatty alcohols. The BChls *c*, *d*, and *e*, of the so-called "green bacteria" which are actually Chls according to their electronic structure, occur usually as homolog mixtures with different side chains in the 7, 8, and 12 positions. The $R^8$ substituent of BChls can be either methyl, ethyl, propyl or isobutyl while the $R^{12}$ substituent can be methyl or ethyl. These BChls also contain an additional stereocentre in the $C3^1$ position and usually appear in most green bacteria as a mixture of epimers

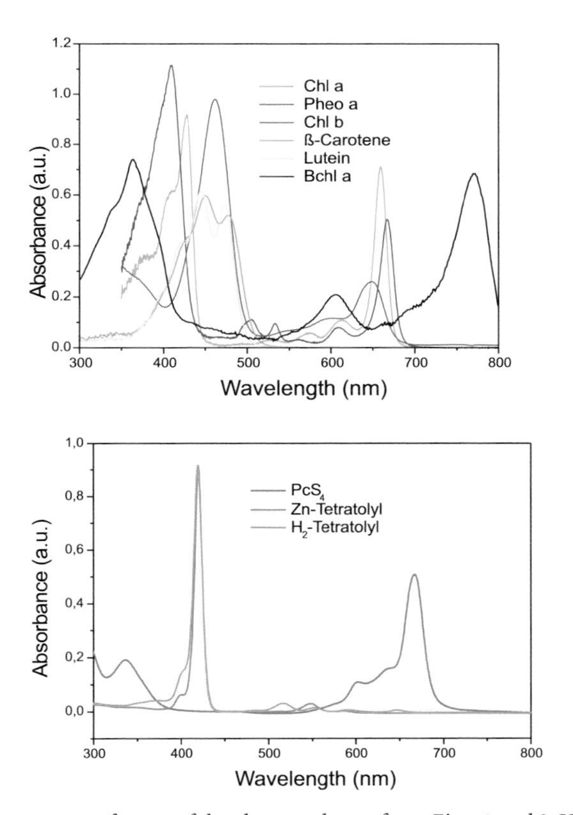


**Fig. 3** Absorption spectra of some of the chromophores from Figs. 1 and 2. Upper part: some natural chromophores. Pheo *a* stands for pheophytin *a*, the free base of Chl a after replacement of the magnesium ion by two protons; Lower part: some synthetic chromophores: dark green trace – nickel tetrasulfonated phthalocyanine (PcS4) dissolved in a water DMSO mixture (note the shoulders at 640 and 600 nm due to dimers and H-aggregates, respectively); magenta trace – *meso*-tetratolyl-porphyrin; cyan trace – zinc tetratolyl-porphyrin. Note the sharp 420 nm (Soret) bands of the porphyrins and their very low visible absorptions (the Q bands) in comparison to Chls and phthalocyanines.

conferring rigidity and mechanical stability. Carotenoidless mutants often assemble more labile and photochemically unstable antenna systems.

Light-harvesting is the primary event in photosynthesis where specialized chromophores, typically organized as pigment-protein complexes, absorb parts of the solar radiation and become excited into their singlet excited states. Excitation energy is then rapidly transferred among such chromophores on a time scale from hundreds of femtoseconds to tens of picoseconds in a partially directed random walk process and is eventually trapped within the so-called reaction centres, where photoinduced charge separation occurs. After several successive electron transfer steps, the hole and the electron become separated on opposite sides of the photosynthetic membrane. This electrochemical potential is used to pump protons across the membrane, which ultimately drive the synthesis of ATP. In oxygenic photosynthesis concomitantly the reductant NADPH is produced and these two compounds serve the organisms as fuel and redox equivalents, respectively, allowing them to perform endergonic biochemical transformations.

Apart from cellulose, Chls are among the most abundant organic compounds in the biosphere and are being continuously synthesized, degraded and recycled. Together with carotenoids the photosynthetic pigments account for about 10% of the total biomass. A simple calculation shows that if all the Chls and carotenoids produced during one year by the South American Continent were shipped by 100 m-long oil tankers each presumed to carry 1000 tons of pigments, then one would need a convoy whose length would span the Atlantic ocean from the southern tip of Argentina to London (15 000 km). This does not even take into account the marine algae and cyanobacteria which, according to remote sensing of chlorophyll fluorescence, can furnish over 10 mg Chl $a/m^3$ of sea water down to a depth of up to 100 m.

## 2 Natural Chlorin-Protein Complexes

Due to their molecular architecture, chlorins in general, but especially the naturally occurring (B)Chls *c*, *d*, and *e* can easily function as building blocks for supramolecular interactions. A hierarchy of several non-covalent interactions are used to arrange these building blocks into defined architectures. The strongest such non-covalent bonding is metal ligation and metallo-chlorins, as well as other metallated tetrapyrroles possess a very rich coordination chemistry [4].

Hydrogen bonding is the next strongest supramolecular interaction in the hierarchy. In cases where multiple hydrogen bonds come into play in a cooperative manner in a supramolecular complex, very tight and directional binding can be effected. All (B)Chls carry in the fifth ring the 13-carbonyl group which can act as an acceptor group for hydrogen bonding and most

B(Chls), except for BChl *c*, *d*, and *e*, also possess a methoxycarbonyl group in the $13^2$ position whose carbonyl group is often involved in hydrogen bonding. In the ethyl chlorophyllide a dihydrate crystal structure [5, 6] one water molecule coordinates the central magnesium atom while a second structural water molecule is doubly hydrogen bonded between the first water and the $13^2$-methoxycarbonyl group. In BChl *a* or in Chl *b* additional carbonyl groups in the $3^1$ or $7^1$ positions, respectively, can help in positioning these chlorins in supramolecular complexes by engaging in hydrogen bonding.

The extended conjugated macrocycles of chlorins and porphyrins are also ideal for forming $\pi$–$\pi$ interactions, another element in the hierarchy of the supramolecular interactions. The resulting strong dipole-dipole interactions are responsible for excitonic coupling between groups of chromophores (see below) which for example plays an important role for engineering the "special pairs" of (B)Chls which are at the core of the photosynthetic reaction centres. Due to their special optical and redox properties these special pairs function as electron donors. A simple but useful description for $\pi$–$\pi$ interactions has been given by Hunter and Sanders [7].

Finally, another important interaction which can account for a high stability of chlorin-containing supramolecular complexes within photosynthetic membranes is the hydrophobic interaction. Long chain fatty alcohols esterify the 17-propionic acid residue of all (B)Chls. These large residues are highly flexible and can thus adapt to occupy voids within hydrophobic pockets of protein matrices, or may help to solvate (B)Chls in non-polar solvents.

An often ignored but highly important aspect for the self-organization of chlorins and related macrocyclic compounds is the cooperativity of the above-mentioned supramolecular interactions [8]. Positive cooperativity leads to thermodynamic stabilization of the supramolecular assemblies beyond the mere sum of the individual non-covalent bonding contributions.

## 2.1 Self-Assembly: a Program Encoded within the Structure of the Tectons

When a molecule has functional groups which allow it to interact with partner molecules which may be of the same kind or different, *and* when the physical conditions which include temperature, medium polarity, absence of inhibitors, etc. are such that non-covalent bonds can be formed, self-assembly occurs. The algorithm which governs the process dictating the architecture and thus ultimately the function of the final nanostructure is encoded within the molecular structure of the "bricks" or tectons [1][1]. Viruses or the correct positioning of nucleotides within nuclei acid strands are examples where Nature uses with perfection self-assembly mechanisms. Misplaced components usually have weaker binding constants such that under equilibrium

[1] tecton is derived from the Greek τεκτων meaning "builder".

conditions they may be expelled from the ordered structures. Thus a repair mechanism operates which ensures that in the end a correctly performed assembly process leads to the pre-programmed nanostructure. For this efficient self-assembly to occur there must exist a very fine balance between the entropy and enthalpy terms which is dictated by the reaction conditions. With respect to the monomeric chromophores self-assembly is usually accompanied by an entropy loss. However, this loss is over-compensated typically by the desolvation of the tectons and the entropy gain of the solvent. Often the entropy term is the controlling factor in the thermodynamics of formation of the most stable nanostructure(s).

## 2.2
## Supramolecular Chemistry of Chlorins by Metal Ligation

The central magnesium atom within chlorins provides an anchoring point via metal ligation. Within Chl-protein complexes (CP), histidine residues are by far the most common ligands to the central Mg of B(Chls) but other amino acids like for example tyrosines, nitrogen atoms from glutamines or asparagines or even sulphur atoms from methionines may also take up that role. The second most frequent ligand is water. The respective Chls are typically bound within the protein matrix by additional weak interactions, like for example by hydrogen bonding to the magnesium-bound water molecule, hydrogen bonding to the 13-carbonyl group at ring V, or for Chl *b* in the 7-position.

One aspect which has been neglected so far in the biophysical/biochemical community is the importance of the diastereotopic arrangement in which the Chls are ligated within proteins [9–11]. Due to the presence of one or more chiral carbon atoms in (B)Chls and the non-planarity of the macrocycle, there exist two diastereotopic configurations when the magnesium atom is ligated, thus becoming five-coordinated: one configuration has the ligand above the tetrapyrrolic plane and the other below the macrocycle. The metal centre thus becomes an additional stereocentre and the two diastereoisomers must have different chemical and electronic properties, such as different absorption or emission wavelengths, radiative lifetimes, circular dichroism, NMR spectra, chromatographic retention times, etc. Figure 4 shows the formulae of these diastereoisomers. Provided that the time scale of the observation method employed is shorter than the average ligation lifetime, the diastereomers appear to be different.

In order to comply with current IUPAC nomenclature rules for metallated tetrapyrroles (as e.g. in hemes or Vitamin $B_{12}$ derivatives), the configuration having the fifth metal ligand below the tetrapyrrolic macrocycle, which is numbered in a clockwise fashion is denoted as $\alpha$, while the $\beta$ configuration has the fifth ligand above the macrocycle. This nomenclature complies also with the one proposed by Sharpless for the direction of attack of an asym-

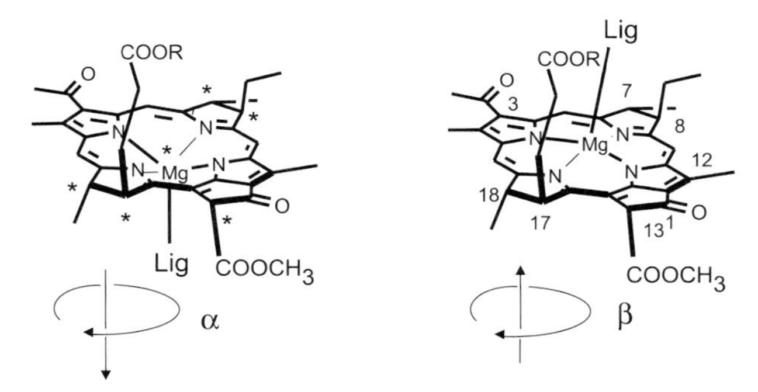


**Fig. 4** BChl *a* with chiral atoms indicated by asterisks. At right the IUPAC numbering is given. Note that in the $\alpha$ configuration the ligand to the metal is on the opposite side of the macrocycle to the 17 propionic acid residue, while in the $\beta$ configuration they are on the same side. As a mnemonic rule, the $\beta$-configuration can be derived by a left-hand rule where the thumb points to the ligand when the fingers are pointing in a clockwise (or nomenclature-wise) fashion

metric epoxidation reagent to an olefin plane [12, 13]. Furthermore, as with sugars or steroids, the $\alpha$-substituent is below the molecular plane and the $\beta$-substituent is above this plane.

When the metal ligand is a water molecule, or a group which binds only weakly (activation energy within kT), or if the ligand is present in the solution in relatively large concentrations, rapid interconversion may occur. When imidazole coordinates Chl *a* the interconversion is slow on the NMR time scale and a splitting of signals is observed [14]. Within a protein matrix, the lifetime of a particular diastereomeric configuration can be considered infinite, since a (B)Chl molecule cannot be de-ligated, rotated along the plane, and religated by the same ligand from the other side of the macrocycle. Figure 5 shows two Chl *a* molecules with histidine ligands within the Photosystem I (PS I) core complex whose structure has been solved by X-ray analysis to 2.5 Å resolution [15].

Sequence comparison showed that the diastereotopic nature of the binding sites of (B)Chls within protein complexes have been strictly conserved during evolution in the so far known antenna complexes [9]. Thus among related, but even phyllogenetically quite distant organisms an $\alpha$-bonded (B)Chl is never turned into a $\beta$-ligated one or vice versa.

Both from statistics [9] and from semi-empirical calculations [10, 11] it follows that the $\alpha$-coordination is by about 1 kcal/mol more stable than the $\beta$-coordination. In PS I out of the 96 Chl *a* molecules only 14 are $\beta$. Most remarkably, almost all are part of the inner circle of the core antenna system, immediately surrounding the reaction centre. This might suggest that there exists an energy funnel from the outer antenna Chls which are all $\alpha$, to the $\beta$ ones. Figure 6 shows this location of the $\beta$-Chls in the PS I core antenna.

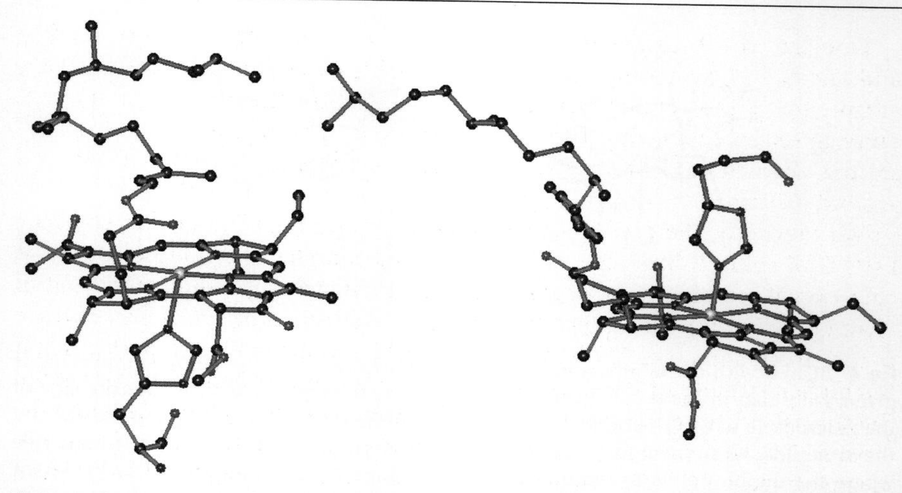

**Fig. 5** *Left*: Example of a "normal" $\alpha$-chl *a*; *Right*: a $\beta$-Chl *a* as encountered in the core of PS I [9, 15]

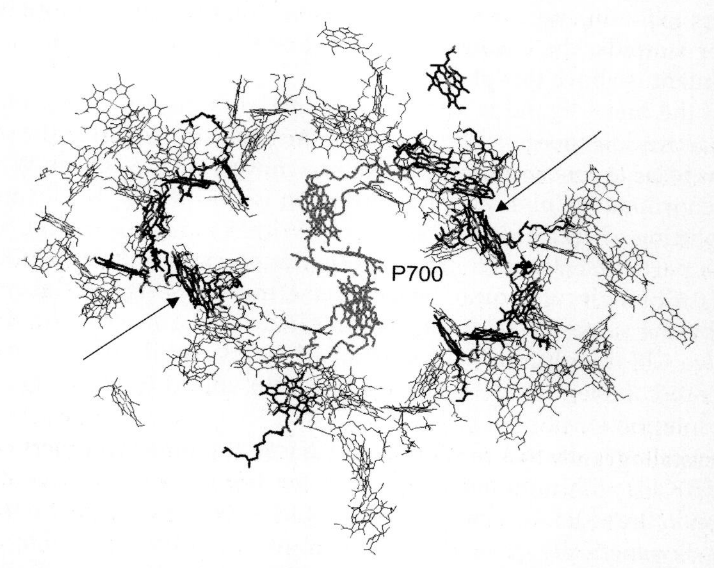


**Fig. 6** Location of the $\beta$ chlorophylls (shown with *thicker lines*) around the electron transfer chain chlorophylls (shown with *thick cyan bonds*) of the reaction centre. All the other chlorophylls shown with *thinner lines* are $\alpha$-coordinated. Adapted with permission from Elsevier ([9])

In PS I there are two $\beta$,$\beta$-dimers (indicated by arrows in Fig. 6) positioned symmetrically around the special pair P700 absorbing around 700 nm. It has been speculated that these dimers form the so-called "red Chls" which act as energy-trapping sites in the antenna, about 1.8 nm away from P700 [16].

One feature which makes these $\beta$,$\beta$-dimers "special" is that both the ligand and the 17-propionic acid residue, which carries the long phytol chain, are on the $\alpha$-side of the macrocycle. This enables the $\beta$-sides to interact quite closely, allowing in principle efficient excitonic coupling to create a red-shifted absorption in comparison to the monomeric Chls (which absorb around 670 nm).

Very recently, the crystal structure of the major light-harvesting complex LHC II of spinach has been solved [17]. This complex is the most abundant chlorophyll protein on Earth and occurs as a trimeric supramolecular unit in vivo. Within a monomer there are eight Chl *a* molecules of which only two are $\beta$-coordinated and six Chl *b* molecules of which only one is $\beta$ [18, 19]. Again the higher stability of the $\alpha$-ligation is thus confirmed. The functional role of the special $\beta$-Chls *a* has been assigned from mutagenesis studies as being the most red-shifted Chls [20, 21]. Furthermore, decreasing the Chl *a*/*b* ratio in reconstituted LHC II has structural and functional consequences [22]. Thus, the diastereotopic magnesium ligation is another structural element for engineering function in supramolecular chlorin complexes. At present, however, conclusive experimental evidence for the hypothesis that $\beta$-ligation gives rise to the red-shifted Chls is still lacking.

## 2.3 Supramolecular Chemistry of Chlorins by Self-Assembly of Protomers within Bacterial Light-Harvesting Systems

One way in which purple bacteria have managed to economize the biosynthetic and genetic effort required for building large antenna systems was to assemble three BChl *a* and one carotenoid with two short $\alpha$-helical polypeptides and then to self-assemble these so-called protomers into much larger supramolecular structures [23].

In the light-harvesting complex 2 (LH2) of *Rhodopseudomonas acidophila*, the first integral membrane-bound light-harvesting complex characterized by X-ray crystallography to a resolution of 2 Å [24, 25], nine protomers combine to form a circular structure, ideally suited for efficient exciton delocalization. Two rings of BChl *a* molecules are positioned at different heights in the photosynthetic membrane: a ring of nine loosely coupled BChl *a* molecules absorbing light at 800 nm (thus termed the B800 ring[2]) is closer to the cytoplasmic side. A second ring of eighteen BChl *a* is located closer to the periplasmic side. The nine $\beta$-$\beta$ special dimers (vide supra) formed are strongly excitonically coupled and yield a red-shifted absorption at 850 nm (the B850 ring). The carotenoids are oriented almost perpendicularly to the ring plane and are in close contact with BChls from both rings, thus increasing the electronic coupling among the BChls. At the same time the carotenoids are optimally

[2] B stands for "bulk" (B)Chl

positioned for quenching the BChl triplet states. The protomer self-assembly leads to the circular arrangement where the nine-fold symmetry is actually not functionally stringent.

Thus, in the related LH2 complex of the purple bacterium *Rhodospirillum molischianum*, an eight-fold symmetry is found with nearly identical optical properties and kinetics of energy transfer [26]. The BChls of the B800 ring show Förster energy transfer [27] across the 2.1 nm distance (Mg–Mg distances) with hopping times of $\sim$ 700 fs [28, 29]. From the B800 ring energy is transferred "down-hill" to the B850 ring within about 1 ps. Although the Mg–Mg distance within a protomer between a B800 and B850 BChl is somewhat smaller (1.8 nm), due to a rather unfavourable Förster orientation factor, the inter-ring excitation energy transfer (EET) is actually slower than the intra-ring EET. Owing to the strongly coupled B850 BChls with partially overlapping orbitals (9.1 Å distance within a protomer and 8.9 Å between protomers), delocalized excitonic states determine the optical and dynamic excited states properties. The energy migrations occur via hopping of small excitons, being delocalized over 2–4 BChls, depending on temperature etc. The exciton dynamics are controlled by both Anderson localization and spatial inhomogeneities [29]. Very recently a generalized Förster theory has been devised in order to account for the very rapid excitation energy transfer times in multichromophoric systems which are in a confined medium [30].

The crystal structure of the combined LH1/RC complex of the related purple bacterium *Rhodopseudomonas palustris* has been solved at 4.8 Å resolution [31]. This amazing arrangement proves the validity of the phrase that "many wheels make light work" [32] in bacterial photosynthesis. The crystal structure proved correct previous molecular modelling studies [33] assuming that the RC is snugly engulfed by the LH1 complex. Fifteen protomers are assembled in a slightly ovoid structure which shows an unexpected small opening within a ring of thirty tightly coupled BChls. This ring is positioned at the same height in the photosynthetic membrane as the smaller B850 rings of the LH2 complex. Thus EET can occur from one of several peripheral surrounding LH2 complexes to the central LH1 (on a $\sim$ 3 ps timescale) and from there, finally, to the special pair within the RC on a $\sim$ 25 ps timescale [28, 34]. As the $\alpha$- and $\beta$-polypeptides within the LH2 and LH1 protomers have a high degree of homology, one may speculate that the LH1 annulus is made out of similar $\beta$-$\beta$ "special dimers" (vide supra) as the LH2. Presently, the structural resolution of the LH1-RC complex is too low to allow any detailed stereochemical features of the BChls to be assigned. Under very low light illumination conditions, purple bacteria such as *Rhodopseudomonas acidophila* express an additional peripheral smaller LH3 complex, which has also been characterized by X-ray crystallography [35], allowing a detailed understanding of how energy is sequentially funnelled to the RC. The LH3 has a nine-fold symmetry similar to the LH2 and a BChl *a* ring absorbing at 800 nm. Unlike the B850 ring of LH2, the absorption of LH3 is blue-shifted to 820 nm (B820).

This shift may be due to a tilting of the 3-acetyl group out of conjugation with the BChl plane. Single-molecule fluorescence studies do indeed confirm that it is not the delocalization energy which is altered in the B820 ring as compared to the B850 ring in LH2, but rather the site energy of the monomeric BChls [36]. This is but one example where the protein matrix, by a slight geometrical alteration, drastically modifies the optical properties of CPs. There exists ample evidence illustrating the principle that the protein matrix finely tunes the absorption and emission properties of bound (B)Chls, on top of the diastereotopic Mg ligation. From both experimental and theoretical studies on porphyrins it is known that not only the absorption and fluorescence spectra but also the excited state lifetimes can be substantially altered by the surrounding protein, e.g. by ruffling the tetrapyrrolic macrocycle [37].

## 3
## Natural Self-Assembling Chlorins: the Chlorosomal Bacteriochlorophylls

Green photosynthetic bacteria have evolved a special organelle for light-harvesting under extremely low light illumination conditions, the so-called "chlorosome" (c.f. Fig. 7). This efficient light-harvesting unit enables these organisms to live at depths over 50 m under the water surface where light is extremely scarce. The light-harvesting complexes of purple bacteria or of higher plants are all pigment-protein complexes. In contrast the chlorosomes BChls *c*, *d* and *e* (see Fig. 1 for structure formulae), which are present as

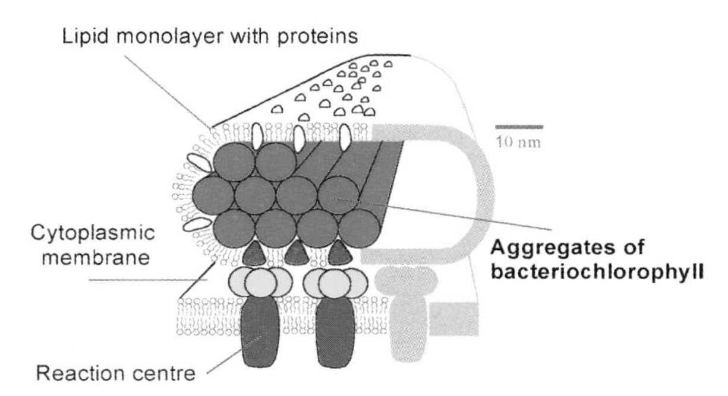


**Fig. 7** Schematic view of a model of a chlorosome from *Chlorobium tepidum* sitting on top of the FMO-complex (*yellow*) which attaches the chlorosomes to the cytoplasmic membrane containing the reaction centres (*blue*). *Green*: Rod elements (ca. 10 nm diameter, ca. 400–600 nm length) containing supramolecular aggregates of BChl *c* (c.f. Figs. 8 and 9). The chlorosome is surrounded by a lipid monolayer. This model follows the early electron microscopy picture of Staehelin et al. [59, 60]. See text for the discussion of an alternative model for the long range arrangement of the supramolecular BChl aggregates. Figure adapted from [2] with permission

a mixture of homologs, self-assemble to yield supramolecular structures only controlled and determined by pigment–pigment interactions [38, 39]. A few small proteins are only present in the envelope of the chlorosomes but do not take part in the chromophore organization or orientation. Consequently this very simple architectural principle for building the supramolecular organization can be reproduced or mimicked by completely synthetic molecules programmed for self assembly (see next section).

Although initially debated very controversially, it is now generally accepted that BChl self-assembly without protein interaction is the structural principle governing the chlorosome architecture. Such self-assembly occurs in the micelle-like hydrophobic environment in the interior of the chlorosomes which are surrounded by a lipid monolayer. Several groups, including the present authors, have provided biochemical [40, 41] and detailed spectroscopic evidence supporting this principle [42–54]. The spectroscopic evidence includes Resonance Raman [43, 44], circular dichroism [45], FT-IR [46], and especially solid state NMR data [47–53]. More recently molecular biology has provided the proof that proteins play at best only a marginal role for the chromophore organization in the hydrophobic interior of chlorosomes [54]. Since the finding that deletion of nine out of the ten small chlorosomal polypeptides, all located in the chlorosomal envelope, did not prevent native chromophore organization and development of functional chlorosomes, the self-assembly theory has gained general acceptance. Several structural models have been proposed for the self-assembly of BChls in the chlorosomes and the earlier ones have been reviewed [38, 42, 55]. Common for all the proposed models is a cooperative interaction between the central magnesium atom, i.e. a metal coordination by the $3^1$-hydroxy group of one chlorin with hydrogen bonding of the activated O – H group to the 13-keto group of a third chlorin. Additionally $\pi$–$\pi$ interactions between the chlorin macrocycles and also favourable electrostatic interactions are possible (c.f. Fig. 8). This general interaction picture is now well accepted. However, within this basic interaction unit consisting of three adjacent chlorins many different short and long-range arrangements of supramolecular chlorin architectures are in principle possible. Thus, the details of the supramolecular arrangement in chlorosomes are still debated in the absence of definite crystallographic proof which may be difficult to obtain. A parallel stacking of the macrocycles with the monomer as the building block (c.f. Fig. 8b), as initially proposed by Holzwarth and Schaffner [56], appears to be the only one that is consistent with the data from solid-state MAS $C^{13}$-NMR on both intact chlorosomes and artificial aggregates of BChl *c* [47, 51–53]. Interestingly, this arrangement is in principle very similar to the one in the water aggregates of Chl *a* [48] and in the crystal structure of ethyl chlorophyllide *a* dihydrate [5, 6], except that external water molecules now take over the role of the intramolecular hydroxy group present only in BChls *c*, *d*, and *e*. This stacking arrangement differs markedly from the antiparallel models advocated by Nozawa and Wang [57]. Very recently, Pšenčik et al. have performed small angle X-ray scattering (SAXS) spectra and transmission electron microscopy (TEM) on natural chlorosomes and have found a sharp reflex at ~ 2 nm. [58] This has led the authors to propose a novel lamellar model where the 2 nm spacing represents the Mg–Mg distance between adjacent stacks of BChls which is depicted in Figs. 8d and 8e. In that work also an an-

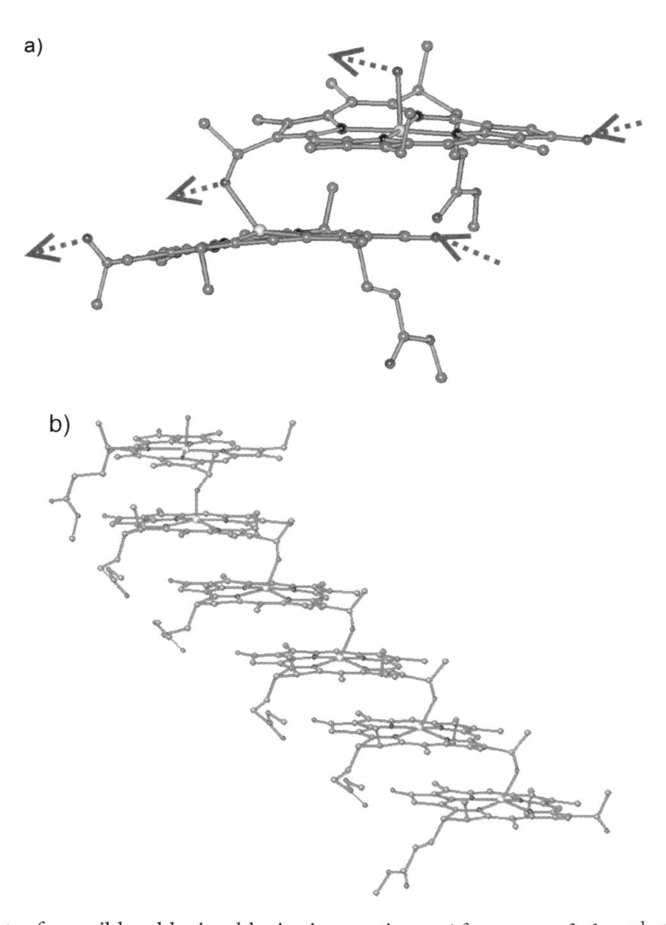


**Fig. 8** **a** Set of possible chlorin-chlorin interactions. After one of the $3^1$-OH groups ligates a magnesium atom it can also function as a hydrogen bond donor group (*arrows*) while the $13^1$-C = O groups may function as a hydrogen bonding acceptor group. **b** Stack of parallel oriented BChl *c* with the long fatty alcohol chain replaced by a methyl group. **c** Cooperative multiple hydrogen bonds between two such parallel oriented stacks. **d** Antiparallel oriented closed dimers which are concomitantly hydrogen bonded. **e** Lamellar Model of stacked BChl's as described in [58]. Thicker lines indicate BChl molecules from four dimers which are closer to the viewer. In the perpendicular direction the dimers are proposed to be bonded as shown in **d**

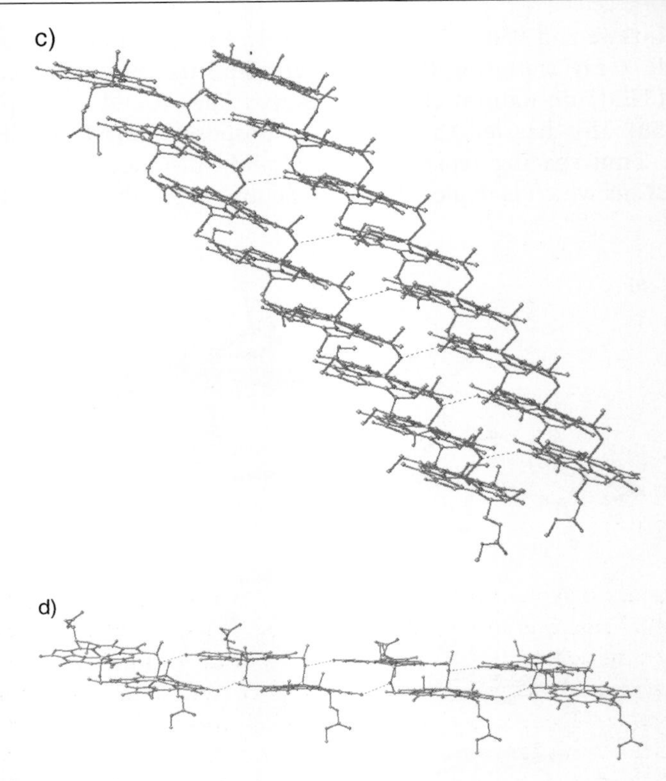


**Fig. 8** (*continued*)

tiparallel stacking was proposed which seems to be in contrast to the results from solid-state NMR data. Thus, at present there exists a controversy mainly on the long-range order and arrangement of chlorins in the chlorosomes. While the parallel stacking model supported by solid-state NMR yields a nice agreement with the original proposal of rod elements as the long-range arrangement in the chlorosomes [59, 60], the recent work of Pšenčik et al. [58] favours a lamellar structure rather than the tubular arrangement assumed so far on the basis of the early electron microscopy pictures of isolated chlorosomes. It should be noted that the methods employed by Pšenčik et al. are not sensitive to the short range arrangement but rather to the long-range order. Thus, these experiments in fact do not yield any conclusive information against or in favour of the parallel stacking arrangement. In contrast, the solid-state NMR experiments do not by themselves yield any information on the long-range order. Rather, this additional information has been provided by molecular modelling based on the information from the early electron microscopy data. The refined rod model resulting from these studies is shown in Fig. 9. While we believe that at present all the available information on the

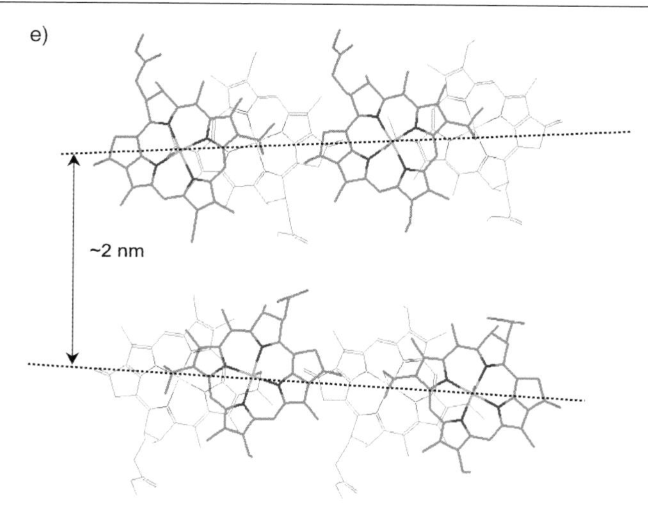


**Fig. 8** (*continued*)

short-range arrangement of the chlorins in chlorosomes is strongly in favour of the parallel stacking model, the long-range order is still an open question and a matter of debate as the data of Pšenčik et al. also falls short of providing conclusive evidence in that regard. Furthermore, there exists substantial spectroscopic evidence in favour of the parallel stacking arrangement and the rod structure in chlorosomes (see below for a discussion of the optical properties). The rod structures can be built easily from parallel stacks (Fig. 8b) by a network of hydrogen bonds for each adjacent chlorin pair in the two stacks. Additional stacks can be added in the same manner. Due to the geometric requirements for the formation of the hydrogen bonding network each stack is rotated by a small angle around the long axis relative to the adjacent

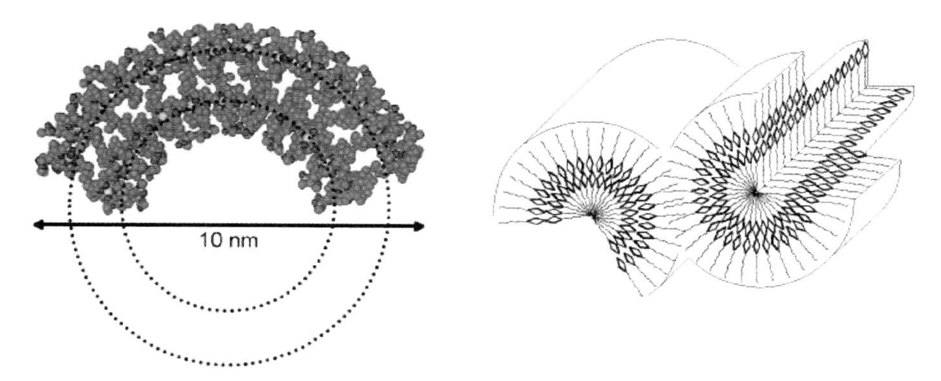


**Fig. 9** Refined rod model on the basis of solid state NMR data and molecular modelling. Reprinted with permission from ACS ([51])

one (Fig. 8c). Thus finally a rod structure with a diameter of 5–10 nm can be produced (Fig. 9). The diameter depends on the rotation angle, which in turn is a function of the side groups attached to the chlorin ring [56]. A large diameter can form double-layer tubes while the smaller diameter rods (5 nm diameter) can not. In contrast, the anti-parallel stack structures (Figs. 8d, 8e) are based on a different type of basic building block, that is the closed dimer unit. The possibility of the green bacterial BChls to indeed form rod structures has been proven recently by Würthner and his coworkers using atomic force microscopy [61]. A Zn-bacteriochlorin *d* with a modified ester side-chain, which prevents higher aggregation to multi-rod structures, has been shown to form single rods with a diameter of about 5.8 nm diameter and a length of several hundred nm [61].

## 4 Synthetic Self-Assembling Porphyrins and Chlorins

Over the recent years, in particular the groups of Tamiaki in Japan and Balaban in Germany, after their studies with Holzwarth and Schaffner, have taken a biomimetic approach for synthesizing artificial chromophores which self-assemble in vitro in a similar manner as the chlorosomal BChls. These efforts have been rewarding from two points of view. Firstly, it is now possible to program a BChl-like self-assembly to fully synthetic chromophores which may well have applications for building artificial light-harvesting [39, 62–64] and other nanostructured materials. Secondly, because in some cases these artificial self-assemblies could be crystallized, the architectural details of the natural system might become unravelled more easily. These systems will be reviewed in the following section.

### 4.1 Supramolecular Chemistry by Hydrogen Bonding and $\pi$–$\pi$ Interactions

When the central metal atom is absent, the tetrapyrrolic core can engage in hydrogen bonding with two of its nitrogen atoms. This interesting binding mode was first observed in the crystal structure of a porphyrin having two long alkyl chains (undecyl) and two 4-hydroxymethylenephenyl groups [65]. Thus a hydrogen bonding network is formed where each porphyrin molecule interacts with four different neighbouring molecules (Fig. 10). The hydroxymethylene groups are the hydrogen bonding donors while the two iminic pyrrolic nitrogens function as hydrogen bond acceptors. One hydroxymethylene group is located above the plane of the central porphyrin, while the other one binds from the opposite side of the plane.

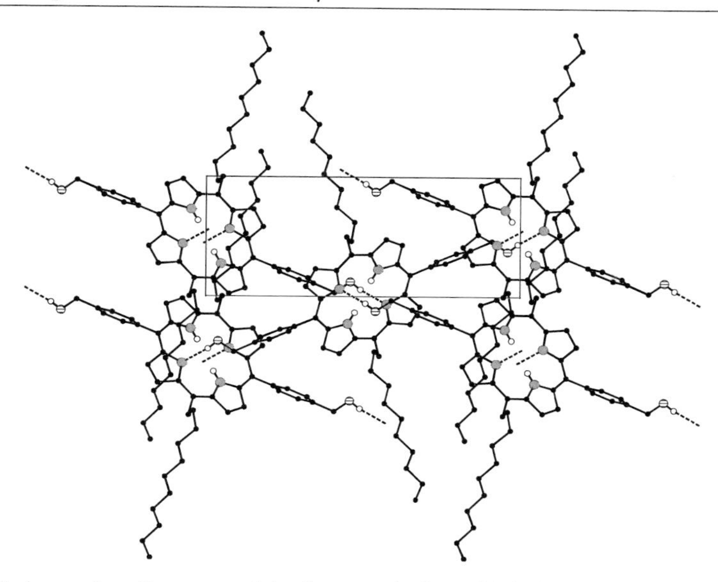

**Fig. 10** Hydrogen bonding network in the crystal of 5,15-bis(4-hydroxymethylenephenyl)-10,20-diundecyl-21*H*,23*H* porphine. The view is along the *a* crystallographic axis. Reprinted with permission Wiley-VCH ([65])

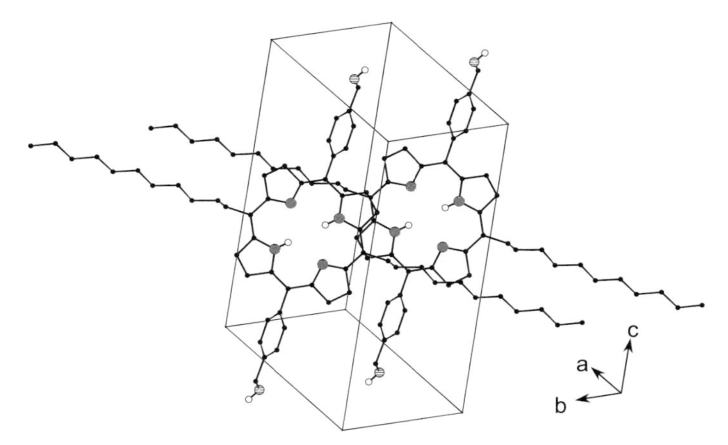


**Fig. 11** $\pi$–$\pi$ Interactions in the crystal of 5,15-bis(4-hydroxymethylenephenyl)-10,20-diundecyl-21*H*,23*H* porphine. The view is perpendicular to the porphyrin planes and hydrogen atoms bonded to carbon have been omitted for clarity. Reprinted with permission from Wiley-VCH ([65])

At the same time $\pi$–$\pi$ interactions are formed as shown in Fig. 11. The plane to plane distance is about 3.5 Å. Note the offset with which two neighbouring porphyrins overlap. This arrangement simultaneously allows for interdigitation of the long and well-ordered undecyl groups. This parallel

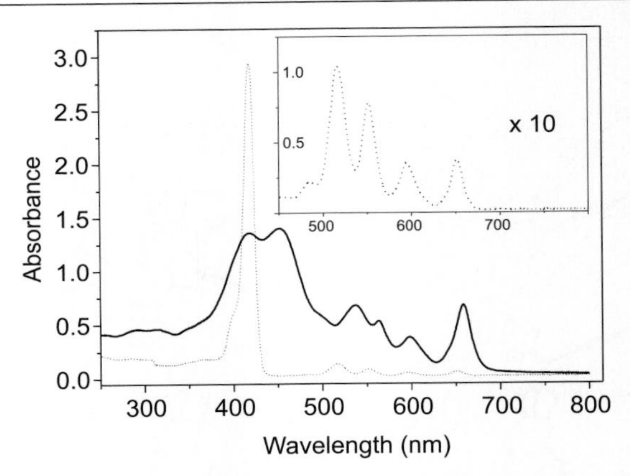


**Fig. 12** Absorption spectra of the crystal of the self-assembled porphyrin shown in Figs. 7 and 8 (*full trace*) and after dissolution in dichloromethane (*dotted trace*). Note that in the crystal the Soret band is split and that intense absorptions are present over almost the entire visible range of the spectrum, a fact beneficial for light-harvesting. Adapted with permission from Wiley-VCH ([65])

but offsetted orientation of the porphyrin rings relative to each other determines the excitonic coupling. In the crystal the Soret bands are considerably broadened and split (1960 $cm^{-1}$) (Fig. 12). The exciton coupling also leads to a redistribution of transition moment to the low-lying electronic state, which is beneficial for light-harvesting since in monomeric porphyrins the low energy transition(s) have only a quite low transition moment, in contrast to (B)Chls.

Hydrogen bonding to the inner pyrrolic nitrogen atoms similar to that shown in Fig. 10 and extended $\pi$–$\pi$ interactions have been observed also in crystal structures of 5,15-*meso*-disubstituted-diarylporphyrins by Sanders and co-workers [66, 67]. It is noteworthy that in solutions of even weakly polar solvents (such as dichloromethane, chloroform) this association is not observable at room temperature, the solvated species being monomeric. Similarly, in ethyl chlorophyllide a, upon dissolution of the crystals the hydrogen bond network is disrupted, and the absorption spectrum becomes blue shifted, typical of monomeric Chl *a*.

## 4.2 Self-Assembly of Synthetic Chlorophylls Using Hydrogen Bonding, Metal Coordination and $\pi$–$\pi$ Interactions as Light-Harvesting Antenna Models of Photosynthetic Green Bacteria

Water-locked complexes of bis(chlorophyll)s **1** have been proposed to be a model for the special pair of B(Chl)s in the photosynthetic reaction centres [68, 69]. These structures (c.f. Fig. 13) were determined by $^{1}$H NMR

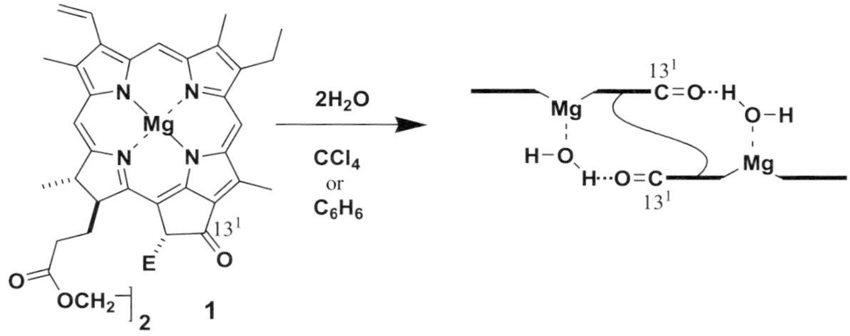

**Fig. 13** Water-locked complexes of bis(chlorophyll)s **1** (E = H or $COOCH_3$)

spectroscopy. In the supramolecule a water molecule is coordinated to the central magnesium atom which in turn is hydrogen-bonded to the 13-keto carbonyl group. Additionally, $\pi$–$\pi$ interactions between the two chlorin moieties occur. However, these structures were not confirmed to be present in natural reaction centres.

As mentioned before, the $Mg\cdots O-H\cdots O=C$ bonding and $\pi$–$\pi$ interactions similar to those in the above systems were reported in ethyl chlorophyllide a dihydrate within the crystal structures [5, 6] of a Chl-*a* oligomer where the phytyl residue had been replaced by an ethyl group [70]. From the UV-Vis and infrared analyses these particular intermolecular interactions were proposed to give a characteristic spectroscopic fingerprint.

In this section the in-vitro self-assembly models are briefly described, based upon semi-synthetic chromophores.

In order to mimic the natural systems and to test the relevant factor governing the self-organization, the self-assembly of synthetic model chromophores has been studied extensively in non-polar organic solvents and in micro-heterogeneous media, like for example aqueous micellar solutions. Chlorosomes give characteristically broad and red-shifted $Q_y$ absorption bands which stem from the longest wavelength absorption band of (B)Chls. BChl *c* is monomeric and shows a sharp $Q_y$ absorption maximum around 650 nm in polar solvents such as THF or methanol which can coordinate the magnesium atom (c.f. Fig. 14). In contrast, within the chlorosome and upon self-assembly this maximum is strongly red-shifted to 740–760 nm [71, 72]. This large red-shift is a good indicator for the presence of chlorosomal self-assemblies. Therefore, the red-shifted aggregate $Q_y$ peak position ($\lambda_{max}$) and the difference value ($\Delta$) between monomeric and oligomeric forms were analyzed from the visible spectra of various synthetic chromophores.

The chlorosomal BChls-*c*, *d* and *e* are characterized by the presence of the $3^1$-hydroxy group and the absence of the $13^2$-methoxycarbonyl group as compared to Chl *a*. Metal complexes of $3^1$-hydroxy-$13^1$-oxo-chlorins lacking the $13^2$-$COOCH_3$ group were synthesized as model compounds [73].

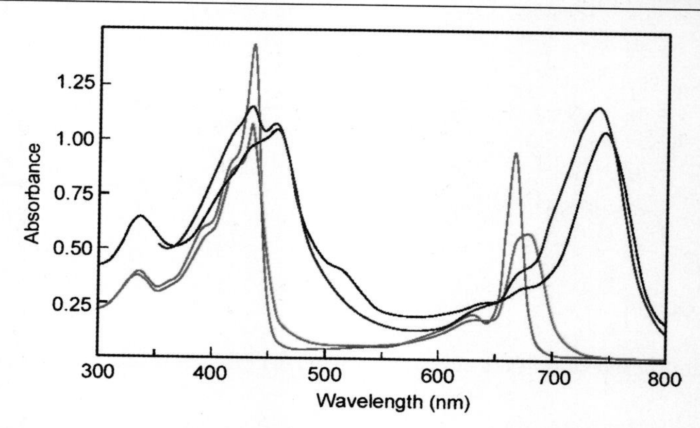


**Fig. 14** Comparison of the absorption spectra of $3^1R$-[Et,Et]-BChl *c* in various solvents and in the natural chlorosomes. *Green trace*—solution in dichloromethane; *red trace*—same solution after addition of suprastoichiometric amounts of methanol; *blue trace*—dichloromethane solution diluted into a large excess of *n*-hexane; *black trace*—natural chlorosomes isolated from Chlorobium tepidum

Zinc as a central metal is favoured due to an increased chemical stability in comparison with magnesium and cadmium chlorins which are very easily demetallated in the presence of traces of acids.

The interaction of the $3^1$-hydroxy and the 13-carbonyl groups on the one hand, and the central metal atom of a third chlorin on the other hand, plays the key role for the chlorosomal-type self-assembly [74]. In natural chlorosomal BChls, these three connecting groups are aligned along the $Q_y$ axis (N21–N23 in Fig. 1) which makes the interactions particularly favourable. The first models were synthesized with this aggregation principle in mind. One of these models, zinc chlorin **2** (Fig. 15) was synthesized by modifying Chl-*a* such that the three interaction groups were designed to be located also along the $Q_y$ axis [73, 75]. In THF, **2** is monomeric and shows a narrow $Q_y$ absorption band with a maximum at 646 nm. This monomeric solution was diluted with a 99-fold excess of *n*-hexane. A 740-nm absorbing species, ascribed to large oligomers with optical properties very similar to the chlorosomal self-assemblies is formed under these conditions. The red-shift ($\Delta$) upon self-assembly was 1970 $cm^{-1}$ (see Table 1). Vibrational analysis including FT-IR and Resonance Raman (RR) spectra clearly indicated that oligomeric **2** was formed by $Zn\cdots O-H\cdots O=C$ interactions. These aggregates show strong bands at $\nu_{max} = 1655$ in the IR and ca. 1650 $cm^{-1}$ in resonance Raman [44, 75] which is characteristic of the chlorosomal aggregation.

Several modifications of the chlorin side chain pattern were synthetically tested in order to gain more insight into the factors governing efficient self-organization and yielding stable aggregates. In one approach, the hydroxy group was moved from the $3^1$- to the $8^1$-position to yield **3** [76]. The max-

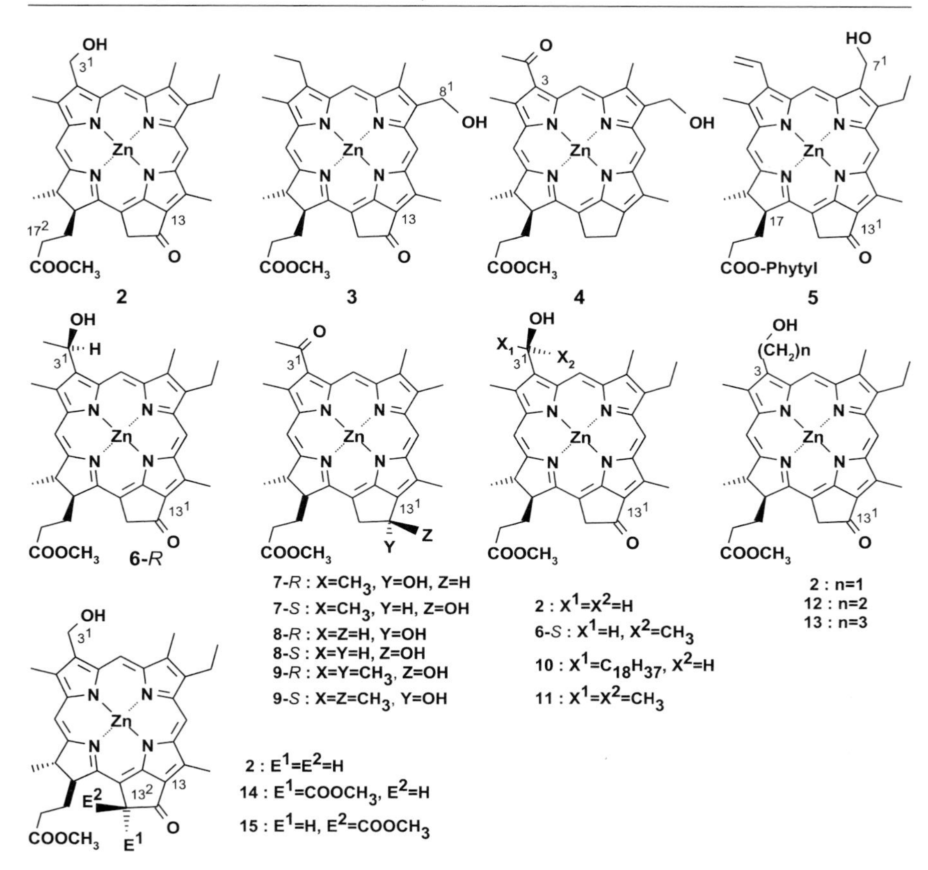


**Fig. 15** Molecular structures of self-assembling semi-synthetic zinc hydroxy-oxo-chlorins **2–15** which are mimics of BChl *c*

**Table 1** $Q_y$ absorption peaks ($\lambda_{max}$/nm) of zinc hydroxymethyl-oxo-chlorins and their red-shift values ($\Delta$/cm$^{-1}$) by self-assembly [= $\{1/\lambda_{max}\,(\text{monomer}) - 1/\lambda_{max}\,(\text{aggregates})\} \times 10^7$]

| Compound | $\lambda_{max}$ | | |
|---|---|---|---|
| | Monomer[a] | Oligomer[b] | $\Delta$ |
| **2** ($3^1$-OH/13-C=O/$17^2$-$COOCH_3$) | 646 | 740 | 1970 |
| **3** ($8^1$-OH/13-C=O/$17^2$-$COOCH_3$) | 642 | ~ 655 | 310 |
| **4** ($8^1$-OH/3-C=O/$17^2$-$COOCH_3$) | 636 | 636 | 0 |
| **5** ($7^1$-OH/13-C=O/$17^2$-COO-phytyl) | 651 | 712[c] | 1320 |

[a] In THF;
[b] In 1% (v/v) THF/hexane;
[c] In 1% (v/v) dichloromethane/hexane

ima in the $Q_y$ peaks of **3** are 642 in THF and ca. 655 nm in 1% (v/v) THF and hexane. This much smaller red-shift (310 $cm^{-1}$) indicated that **3** did not form favourable self-assemblies in the non-polar medium. Displacement of the keto carbonyl to the 3- from 13-position of **3** completely suppressed the self-organization and gave only monomeric 4 even in 1% (v/v) THF and hexane with a $\lambda_{max}$ of 636 nm. Zinc $7^1$-hydroxy-$13^1$-oxo-chlorin **5** was synthesized by modifying Chl *b* and it shows a monomeric $Q_y$ absorption peak at 651 nm in THF and an oligomeric peak at 712 nm upon self-assembly in 1% (v/v) dichloromethane/hexane [77]. The red-shift value ($\Delta = 1320\ cm^{-1}$) is smaller than that for the C3-analog **2** (1940 $cm^{-1}$) but larger than for the C8-analog **3** (310 $cm^{-1}$). Interestingly, the $3^1$-*R* and *S* diastereomers of **6** give slightly different $\Delta$-values, indicating that the self-assembly process must be diastereoselective [78]. From these results it can be concluded that a linear arrangement of the $O-H$, Zn and $C=O$ groups is also a necessary condition for efficient and stable chlorosomal-type self-assembly of the semi-synthetic model chromophores.

To further investigate the requirements for the relative arrangement of the metal, hydroxy and carbonyl groups, an exchange of the $3^1$-OH with $13-C=O$ groups was examined [79–81]. When the central metal atom is missing in that chlorin (i.e. as in the corresponding chlorin free bases), no self-assembly occurs. The same observation is made when the keto group is protected or missing [79]. In contrast zinc $3^1$(*R*)-hydroxy-$13^1$-oxo-chlorin **6**-*R* self-assembled in 1% (v/v) dichloromethane and hexane to give a red-shifted 703-nm absorbing peak as compared to the monomeric 648-nm absorption peak in dichloromethane ($\Delta = 1210\ cm^{-1}$, see Table 2) [78]. The inverse-type zinc $13^1$(S)-hydroxy-$3^1$-oxo-chlorin **7**-S also self-assembled upon dilution of the dichloromethane solution with 99-fold hexane to induce a large red-shift from 643 to 708 nm ($\Delta = 1430\ cm^{-1}$) [81]. Again the diastere-

**Table 2** $Q_y$ absorption peaks ($\lambda_{max}$/nm) of zinc hydroxy-oxo-chlorins and their red-shift values ($\Delta/cm^{-1}$) by self-assembly [= $\{1/\lambda_{max}\text{(monomer)} - 1/\lambda_{max}\text{(aggregates)}\} \times 10^7$]

| | $\lambda_{max}$ | | |
|---|---|---|---|
| Compound | Monomer[a] | Oligomer[b] | $\Delta$ |
| **6** ($3^1$-OH(*R*)/13-C=O) | 648 | 703 | 1210 |
| **7** (3-$COCH_3$/$13^1$-OH(*S*)) | 643 | 708 | 1430 |
| **8** (3-CHO/$13^1$-OH(*S*)) | 653[c] | 678 | |
| | | 709 | 1210 |
| **9** (3-$COCH_3$/$13^1$-OH,$CH_3$(*S*)) | 643 | 643 | 0 |

[a] In dichloromethane;
[b] In 1% (v/v) dichloromethane/hexane;
[c] In dichloromethane/methanol/hexane (1 : 5 : 94)

oselectivity of the self-assembly process was confirmed as the $3^1$-demethyl form **8** of **7** self-assembled with a red-shift from 653 nm to 678 or 709 nm, depending on the stereochemistry at the $13^1$-position which was determined from NOE experiments [79]. Thus, the linear arrangement of the three interacting groups in the model compounds appears to be a prerequisite for their self-assembly as is the case for the chlorosomal BChls. It is noteworthy that addition of a methyl group to **7** at the $13^1$-position fully suppressed the self-organization and thus **9** was monomeric even in 0.1% (v/v) dichloromethane and hexane ($\lambda_{max}$ = 643 nm) [81].

The above-mentioned results provided a strong indication that any small steric hindrance or strain around the three main interacting groups prevent good interactions and may severely perturb the self-assembly process. Consequently, these steric factors were examined in more detail [82–85]. The influence of (de)methylation at the $3^1$-position on the self-assembly could be inferred from the fact that the self-assemblies of **6** were known to show a 703-nm $Q_y$ absorption peak (vide supra) [77]. Demethylation of **6** at the $3^1$-position for example induced a pronounced red-shift in the oligomeric $Q_y$ peak of **2** [$\lambda_{max}$ = 646 in THF shifted to 740 nm in 1% (v/v) THF/*n*-hexane corresponding to $\Delta = 1970\ cm^{-1}$]. As steric hindrance around the $3^1$-OH group is decreased by $3^1$-demethylation, the intermolecular interaction strengthened to result in a larger, more stable and more tightly packed oligomer (see Fig. 16). However, substitution of the heptadecyl group at the $3^2$-position of **6** blue-shifted the $Q_y$ oligomeric peak of **10** at 681 nm yielding a $\Delta$ value of only $750\ cm^{-1}$ (see Table 3) [83]. Thus, a long alkyl chain as for example the $C_{18}H_{37}$ group around the interactive $3^1$-OH group interferes

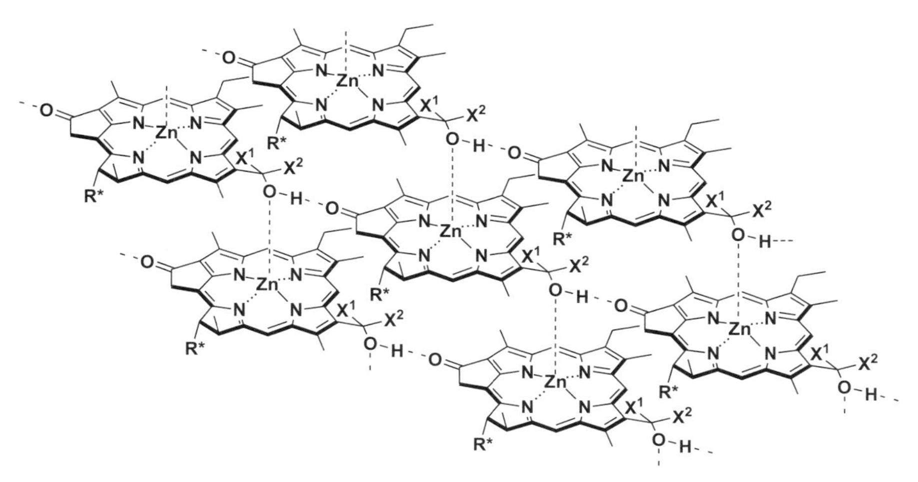


**Fig. 16** Proposed binding model for the semi-synthetic BChl *c* mimics ($R^* = CH_2CH_2COOR$)

**Table 3** $Q_y$ absorption peaks ($\lambda_{max}$/nm) of zinc $3^1$-hydroxy-$13^1$-oxo-chlorins and their red-shift values ($\Delta$/cm$^{-1}$) by self-assembly [= $\{1/\lambda_{max}$ (monomer) – $1/\lambda_{max}$ (aggregates)$\} \times 10^7$]

| Compound | $\lambda_{max}$ | | |
|---|---|---|---|
| | Monomer$^a$ | Oligomer$^b$ | $\Delta$ |
| **2** (3-$CH_2OH$) | 646$^c$ | 740$^d$ | 1970 |
| **6** (3-$CH(CH_3)OH$(*R*)) | 648 | 703 | 1210 |
| **10** (3-$CH(C_{18}H_{37})OH$(*R*)) | 648 | 681 | 750 |
| **11** (3-$C(CH_3)_2OH$) | 648 | 704 | 1230 |

$^a$ In dichloromethane;
$^b$ In 1% (v/v) dichloromethane/hexane;
$^c$ In THF;
$^d$ In 1% (v/v) THF/hexane

with the self-assembly process and results in a less red-shifted spectrum. In contrast, the addition of a second methyl group at the $3^1$-position of **6** did not affect the $Q_y$ maximum of the aggregate; $\Delta$ in **11** was 1230 cm$^{-1}$ (648 nm for the monomer and 704 nm in the aggregates) which is almost the same $\Delta$ value as in **6** (1210 cm$^{-1}$) [81]. The additional $3^1$-methyl group (thus changing from secondary to tertiary alcohol) did not further affect the supramolecular structures formed upon self-assembly, as judged by the absorption spectra of the aggregates.

In the natural chlorosomal chlorophylls the hydroxy group is connected with the chlorin $\pi$-system through only one carbon atom at the 3-position. Its reactivity is of the benzylic type. The effect of an aliphatic spacer between this hydroxy group and the chlorin chromophore was examined as well [84]. The parent compound **2** with a hydroxy-methylene group self-assembled in 1% (v/v) THF/*n*-hexane to give almost exclusively an aggregate $Q_y$ band at around 700 nm (vide supra). Insertion of a second methylene group within the $3-3^1$ bond of **2** partially suppressed the self-assembly of **12** possessing 3-$CH_2CH_2OH$ (see Table 4). In 1% (v/v) THF/hexane of **12**, 20% monomeric species still remained beside 80% oligomeric species ($\lambda_{max}$ = 701 nm). One more methylene group insertion to **12** diminished further the self-assembly in **13** which carries a 3-$CH_2CH_2CH_2OH$ substituent. Exclusively monomeric **13** was observed in 1% (v/v) THF and hexane and its aggregation $Q_y$ absorption band was observed at 702 nm in an even less non-polar medium like 0.1% (v/v) THF and hexane. Thus, the increase of the spacer length between the hydroxy group and the C3-position from **2** to **12** to **13** gradually decreased the self-assembly tendency.

Finally, the influence of steric factors around the $13^1$-oxo group has been investigated as well [85]. As described above, **2** is monomeric in THF ($\lambda_{max}$ =

**Table 4** $Q_y$ absorption peaks ($\lambda_{max}$/nm) of zinc $3^n$-hydroxy-$13^1$-oxo-chlorins and their red-shift values ($\Delta$/$cm^{-1}$) by self-assembly [= {1/$\lambda_{max}$ (monomer) – 1/$\lambda_{max}$ (aggregates)} $\times 10^7$]

| | $\lambda_{max}$ | | |
|---|---|---|---|
| Compound | Monomer[a] | Oligomer[b] | $\Delta$ |
| **2** (3-$CH_2OH$) | 646 | 740 | 1970 |
| **12** (3-$CH_2CH_2OH$) | 644 | 701 | 1260 |
| **13** (3-$CH_2CH_2CH_2OH$) | 644 | 644 | 0 |
| | | 702[c] | 1280 |

[a] In THF;
[b] In 1% (v/v) THF/hexane;
[c] In 0.1% (v/v) THF/hexane

646 nm) and self-assembles in an aqueous 0.02% (v/v ) Triton X-100 solution to give a 741-nm absorption $Q_y$ peak yielding $\Delta = 1430\ cm^{-1}$ (see Table 5). Insertion of a methoxy-carbonyl group at the $13^2$-position of **2** blue-shifted the $Q_y$ peak due to increased steric hindrance around the interacting 13-C = O group which yields unfavourable chlorin-chlorin interactions. An oligomeric $Q_y$ peak of **14** was found at 703 nm ($\Delta = 1230\ cm^{-1}$), while that of the $13^2$-epimer **15** had a more red-shifted $Q_y$ peak at 719 nm ($\Delta = 1530\ cm^{-1}$). Again, diastereotopic control of the self-assembly process might be invoked to explained the difference in the steric effects of the $13^2$-*R*/*S*-$COOCH_3$ group in a supramolecular assembly due to the five-coordination ability of the Mg through axial ligation to the $3^1$-OH group.

**Table 5** $Q_y$ absorption peaks ($\lambda_{max}$/nm) of zinc 3-hydroxymethyl-$13^1$-oxo-chlorins and their red-shift values ($\Delta$/$cm^{-1}$) by self-assembly [= {1/$\lambda_{max}$ (monomer) – 1/$\lambda_{max}$ (aggregates)} $\times 10^7$]

| | $\lambda_{max}$ | | |
|---|---|---|---|
| Compound | Monomer[a] | Oligomer[b] | $\Delta$ |
| **2** ($13^2$-$H_2$) | 646 | 741 | 1970 |
| **14** ($13^2$-$COOCH_3$,H(*R*)) | 647 | 703 | 1230 |
| **15** ($13^2$-$COOCH_3$,H(*S*)) | 648 | 719 | 1530 |

[a] In THF;
[b] In 0.02% (v/v) Triton X-100/water

### 4.3 Synthetic Self-Assembling Chlorins and Porphyrins as Mimics of the Chlorosomal Bacteriochlorophylls

While the above discussed examples of self-organization are for semi-synthetic building blocks, deriving either from Chl *a*, Chl *b* or from the natural BChls, Balaban and coworkers expanded the range of investigated compounds to fully synthetic porphyrins carrying the same recognition groups that are responsible for the self-assembly of the chlorosomal BChls [86, 87].

The synthetic concept consisted of starting from basic porphyrins, which can be easily prepared in gram quantities, and then placing various functional side groups at positions crucial for self-assembly. Using partially novel synthetic methods, several substituted porphyrins and chlorins were synthesized and studied for their self-assembly properties [86, 87]. In order to increase the solubility, which is sometimes a problem with porphyrins, two *meso* 3,5-di-*tert*-butylphenyl groups were added, a feature which also facilitates purification by chromatographic methods. A further benefit of these bulky groups is that they direct electrophilic substitutions to the remote $\beta$-pyrrolic positions, thus allowing regioselective transformations.

Tamiaki and coworkers developed novel synthetic methods which allowed transformations of octa-alkyl porphyrins, in particular of octaethylporphyrin [88]. Figure 17 shows the structure of the fully synthetic BChl mimics that were synthesized using these methods by Balaban and coworkers (**16**–**22**) [86, 87] while compounds **23** and **24** were synthesized by Tamiaki's group [88].

As for the semi-synthetic mimics discussed in the preceding section, all these compounds carry a hydroxy substituent, a carbonyl group, and a central zinc metal atom. Self-assembly occurs upon dilution with a non-polar solvent such as *n*-hexane, *n*-heptane or cyclohexane from a concentrated solution in a more polar solvent such as dry dichloromethane, chloroform or THF. A very intriguing observation was that compound **23** (R = H, $CH_3$) gave a red-shifted and broadened chlorosomal type absorption maximum typical for aggregation while compound **24** did not show signs of aggregation. It has thus been concluded initially that for self-assembly to occur a strict colinearity of the three important substituents has to be obeyed, as was found initially for the semi-synthetic chlorins [88] (see above). Interestingly, this is however not the case for it is di-*tert*-butylphenyl-substituted compounds **16**, **21** and **22** which do not obey this principle but have an angular arrangement of these three groups. All of them do self-assemble in non-polar solvents to various extents. However, for compound **22** the self-assembly is not evident directly from absorption spectra, but dynamic light scattering and circular dichroism (CD) spectra do indicate that supramolecular species are indeed formed. It seems that in this case only a very limited amount of overlap of the chromophoric systems occurs.

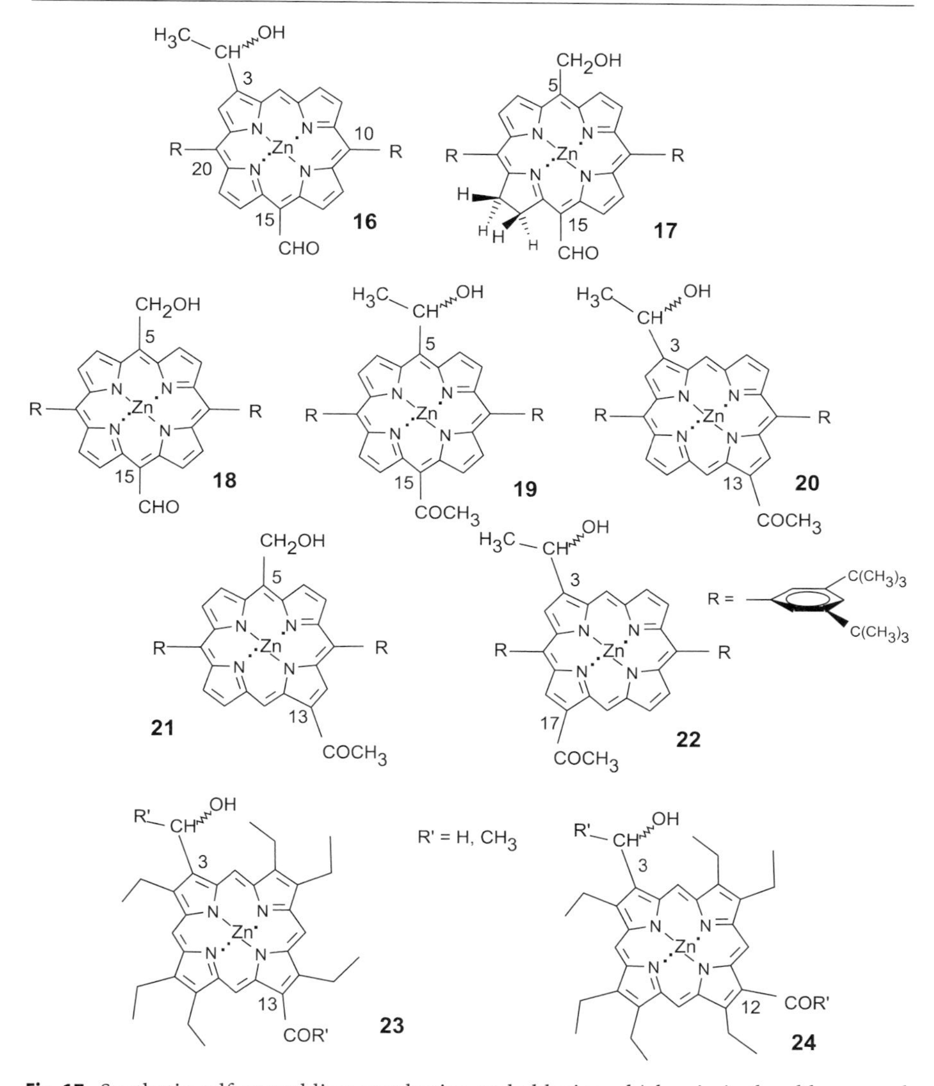


**Fig. 17** Synthetic self-assembling porphyrins and chlorins which mimic the chlorosomal BChls. Adapted with permission from Wiley-VCH ([87])

Compound **16** could be resolved into the two enantiomers. Upon self-assembly they show very intense and mirror imaged CD signals (Fig. 18). The first eluted enantiomer, which gives also a positive long wavelength Cotton effect in the aggregates, must arise from a P (plus) type helical structure according to the exciton chirality method [89]. By using two different complementary methods, the stereochemistry of the first eluting compound has been assigned recently to have a $3^1$-R stereochemistry [90].

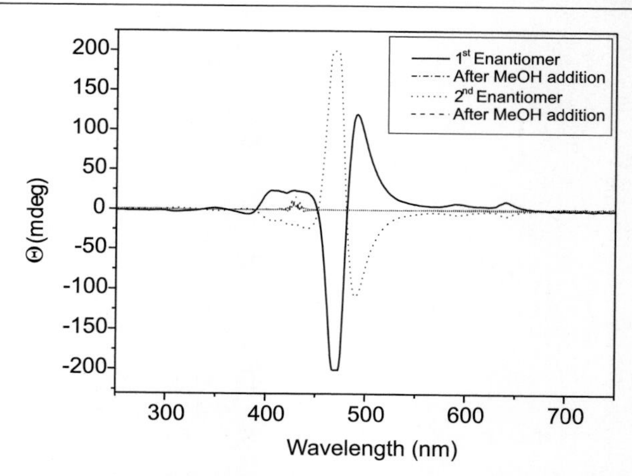


**Fig. 18** Circular dichroism spectra of the separated enantiomers of **16**. After addition of traces of methanol, complete disassembly occurs leading to the dashed traces almost superimposing the 0 line. Adapted with permission from Wiley-VCH ([87])

Chlorin **17** deserves a special mention. It was expected that by modifying the chromophore from a porphyrin to a chlorin an increase in the extinction coefficients, especially for the *Q* bands, should be observed as is typical for chlorins. Actually a drastic (almost threefold) decrease of the Soret band occurs in the chlorin in comparison to the analogous porphyrins. However, no $Q_y$ band increase was observed. This effect has been ascribed in the past as an increase of the *Q* to Soret band relative ratios of transition moments. After self-assembly, broad bands in the visible region with about the same intensity are obtained from both the chlorin and the porphyrins. As chlorins are less readily accessible (in the case of **17** a protective group strategy had to be employed which considerably increased the number of required synthetic steps) this allows us to state that for applications where large absorption coefficients are needed, the use of simple chlorins is not warranted and the cheaper porphyrins can be used instead.

Compound **19** where the interacting groups are arranged collinearly has been crystallized into two different crystal modifications and the crystal structures were solved [91]. Due to the small size of the crystals, in one case a synchrotron X-ray source was needed to obtain measurable diffraction patterns. In both crystal modifications stacks of porphyrins are encountered with the zinc atom coordinated strongly by the hydroxy oxygen group of a second porphyrin and weakly by a carbonyl of a third porphyrin as shown in Fig. 19. The Zn atom is displaced out of the tetrapyrrolic plane towards the hydroxy group.

The stacks are disordered in such a way that both upward and downward oriented stacks are present in the lattice. Further disorder was encountered due to use of the racemate of **19**. It is very surprising that the acetyl group

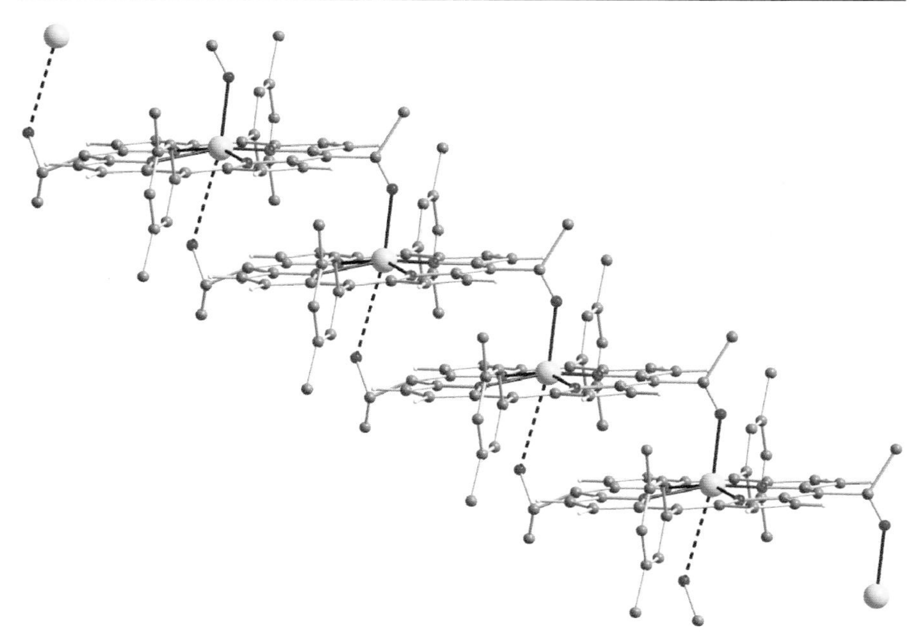

**Fig. 19** Stack of **19** as encountered in both crystal modifications shown here in an upward orientation (i.e. the short Zn – O ligation is above the porphyrin plane). Reproduced with permission from Wiley-VCH ([91])

prefers to twist out of conjugation with the macrocycle and coordinates the Zn atom weakly from the opposite side. This weak coordination (indicated by the dotted lines in Fig. 19) accounts for the fact that there is no hydrogen bonding within either crystal modification of **19**. Due to the Zn-ligation, the carbonyl group is no longer available for being a hydrogen bond acceptor.

The five-membered ring in the natural chlorosomal BChls prevents a twisting of the carbonyl group and forces a different orientation to the metal [87]. This probably hinders metal-carbonyl ligation and could render the $13^1$-carbonyl group more favourable for hydrogen bonding. In contrast, in compound **19** the acetyl group can be rotated out of the plane and become a weak ligand to the metal. However, the present structural data [91] on compounds **16–22** which include several SAXS spectra showing a pronounced 2.0 nm reflex, AFM and TEM images, all prove a striking similarity to the natural chlorosomes [91].

## 5
## Spectroscopic and Functional Properties of Chlorin Dye Self-Assemblies: Excitonic Coupling and Optical Properties of Chlorosomal Aggregates

Typically short distances between chromophores (below ca. 15 Å), depending on the relative arrangement of their transition moments, give rise to excitonic coupling. The responsible interaction is the dipole-dipole interaction of the transition moments. Excitonic coupling leads to a modification of the optical properties like absorption, CD, and fluorescence, and to a modification of the dynamic properties of the excited states (see ref. [2] for a short description on the relevance of excitonic interactions for light-harvesting systems and ref. [92] for a comprehensive treatment on excitons in photosynthetic systems). The effects can be particularly pronounced when a large number of chromophores interact, as is the case in the supramolecular arrangements present in natural chlorosomes and artificial self-assembled systems. The most obvious and immediately detectable effects are the modified absorption and CD spectra. In chlorosomes and artificial chlorosomal type aggregates of chlorins and porphyrins the optical properties are primarily determined by the excitonic interactions. The earliest effects observed were the unusually large red-shift of the long-wavelength absorption band by up to 2500 $cm^{-1}$ in the aggregates (c.f. Fig. 14) relative to the monomers. Likewise the relatively weak CD effects which are characteristic of monomeric (B)Chls are replaced by quite strong Cotton effects in the supramolecular aggregates. Thus the aggregation red-shift, the CD spectra and the linear polarization have been used early on to characterize the optical properties of chlorosomes and artificial aggregates [93–96]. However, only a few systematic theoretical studies of the optical properties of self-assembled supramolecular chlorin systems have been performed with the aim to link structure to the spectral information [97–99] and even less work in this direction was performed based on experimental structure information [100, 101]. Holzwarth and coworkers were able to show that the structure of chlorosomal aggregates based on molecular modelling [56] and solid state NMR [47, 49, 51] allows us to predict the correct optical properties like for example absorption, CD, and LD spectra of chlorosomes on the basis of exciton theory [100, 101]. The LD is polarized almost exclusively parallel to the long axis of the chlorosome [102] and this experimental result is reproduced in exciton calculations [100] which provide a strong argument in favour of tubular parallel stack arrangement within the chlorosomes [56]. Furthermore the parallel anti-($\alpha$)- and/or syn-($\beta$)-stacking of BChls seems to be the only arrangement that is consistent with the MAS solid-state $^{13}$C-NMR data [47, 49]. This arrangement has also been confirmed for aggregates of semi-synthetic chlorins [53, 103]. It has not been tested so far by detailed exciton calculations whether the recently proposed alternative structure—implying an anti-parallel arrangement of closed dimers as building blocks for the chlorosomal aggregates [58] would also yield spectral properties in agreement with experiment. It would be very surprising if largely different arrangements of chromophores, as is the case between the two structural models [56, 58], would indeed yield similar predictions for the excitonically determined optical properties.

An intriguing observation in natural chlorosome preparations has been the varying CD spectra, including even sign changes, for seemingly identical preparations of chlorosomes from the same organisms. In general the CD spectrum is a critical parameter characteristic for the arrangement and interactions of assemblies of chromophores. In this case varying CD spectra of similar chlorosome preparations thus seemed to indicate different chromophore arrangements. This long-standing problem has been solved recently by exciton calculations [100, 104]. In the cylindrical rod-like aggregates of chlorosomal structures (c.f. Fig. 9), which can be envisaged to be formed by intertwined helices in a superhelical arrangement [51, 56, 100], the CD spectrum is determined by two different contributions. Firstly, a primary CD contribution is controlled by the excitonic interactions of neighbouring chlorins. Secondly, superimposed on this basic excitonic CD is a macroscopic CD effect which derives from the superhelical arrangement. The latter contribution requires functional rod lengths which reach or exceed at least one helix turn. Thus the CD spectra, without any major influence on the other optical properties, display a marked rod length dependence. Holzwarth and coworkers have proposed that the curious observations of widely differing CD spectra for seemingly identical chromophore arrangements in chlorosomes are due to functional length variations in the excitonically coupled system of the rods caused by different growth conditions or other factors. The large aggregate sizes and strong exciton coupling allow in principle for a delocalization of the exciton states over many monomers. In practice disorder effects and dephasing of the excited states limit the delocalization lengths in most molecular aggregates to a few monomers, in particular at room temperature where dephasing can occur very rapidly. Notable exceptions are the aggregates of many cyanine dyes, the so-called J-aggregates [105], which show substantial delocalization lengths of up to 10–15 monomers [106–109]. It has thus been a surprise when time-resolved fluorescence experiments indicated the formation of superradiant states with a lower limit of 15 BChls for the delocalization lengths [110, 111]. Recent femtosecond transient absorption data actually indicate even larger delocalization lengths, that is up to at least 100 BChl monomers existing for times up to several picoseconds at room temperature (Holzwarth and coworkers, unpublished data). These are extreme delocalization lengths for molecular dye aggregates at room temperature. The data indicate very special properties of their electron phonon coupling strengths which allow the existence of such delocalized exciton states over substantial time periods. This phenomenon certainly deserves further study.

# 6
# Conclusion

The chlorosomes of green bacteria employing self-assembly from relatively simple building blocks as their organizational principle represent the most efficient natural light-harvesting system. The present authors, over many years, have been involved in studies to explore the principles that govern the natural self-assembly process and in mimicking this process with semi-synthetic and fully synthetic chromophores. Novel and optimized access routes were developed for these tectons. It is now within reach that such self-assembling chromophores can be used for designing self-assembled artificial photonic and other nano-devices. The first successful attempts have been made in this direction both for devices in solution [62–64, 73] and on solid surfaces. Novel approaches were to self-assemble BChl *c* on carbon or mica surfaces [112] or chromophores of the type **16–22** onto nanocrystalline titania [113]. Thus it is possible to not only arrange the BChls in supramolecular three-dimensional structures (vide supra) but also in highly ordered one- or two-dimensional arrays (c.f. Fig. 20) which may gain importance in solar energy conversion or charge-conducting devices. Zinc chlorins of type **2** were self-assembled also onto carbon pastes [114]. Hybrid solar cells or other photoactivated devices could in principle thus be assembled where a nanocrystalline semiconductor having a large interpenetrating network is photosensitized with a self-assembling organic antenna system. Long-term photostability and high incident photon to current conversion efficiencies are important parameters which have to be optimized. There are many reasons to believe that—based upon these studies—it will be possible in the future to combine the current emerging nanotechnologies with knowledge on how nature uses self-assembly for light-harvesting. Efficient solutions for an environmentally clean and sustainable energy source may thus be reached following research in this direction.

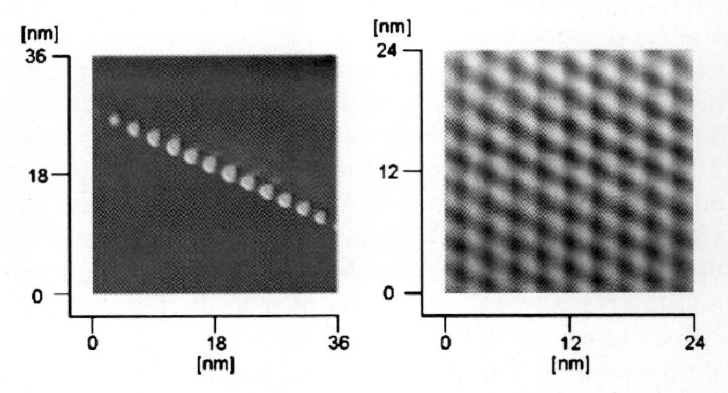


**Fig. 20** Self-assembled BChl *c* on a graphite surface [112]

**Acknowledgements** This work was partially supported by Grants-in-Aid for Scientific Research (No. 15033271) on Priority Areas (417) from the Ministry of Education, Culture, Sports, Science and Technology (MEXT) of the Japanese Government and for Scientific Research (B) (No. 15350107) from the Japan Society for the Promotion of Science (JSPS). Work at the Max-Planck-Institut in Mülheim (A.R. Holzwarth) and at Ritsumeikan University (H. Tamiaki) was in part funded by a Human Frontier Science Program Award to A.R. Holzwarth, H.J.M. de Groot (Leiden) and H. Tamiaki. In Karlsruhe, the experimental work was partially supported by the Deutsche Forschungsgemeinschaft through the Center for Functional Nanostructures at the University of Karlsruhe within former projects C3.2 and C3.6. and present project C3.5.

## References

1. Lehn J-M (1995) Supramolecular Chemistry: Principles and Perspectives. VCH, Weinheim
2. Holzwarth AR (2004) In: Barber J (ed) Molecular to Global Photosynthesis. Imperial College Press, London, pp 43–115
3. Hedges SB, Chen H, Kumar S, Wang DY-C, Thomson AS, Watanabe H (2001) BMC Evolutionary Biology 1:4
4. Kadish KM, Smith KM, Guilard R (eds) (2003) The Porphyrin Handbook. Academic Press (Elsevier), Amsterdam, vol 1–20
5. Chow H-C, Serlin R, Strouse CE (1975) J Am Chem Soc 97:7230
6. Kratky C, Dunitz JD (1975) Acta Cryst B31:1586
7. Hunter CA, Sanders JKM (1990) J Am Chem Soc 112:5525
8. Balaban TS, Leitich, J, Holzwarth AR, Schaffner K (2000) J Phys Chem B 104:1362
9. Balaban TS, Fromme P, Krauß N, Holzwarth AR, Prokhorenko VI (2002) Biochim Biophys Acta Bioenergetics 1556:197
10. Oba T, Tamiaki H (2002) Photosynth Res 74:1
11. Oba T, Tamiaki H (2002) J Photosci 9:362
12. Sharpless KB, Amberg W, Bennani YL, Crispino GA, Hartung J, Jeong K-S, Kwong H-L, Morikawa K, Wang Z-M, Xu D, Zhang X-L (1992) J Org Chem 57:2768
13. Kolb HC, Van Nieuwenhze MS, Sharpless KB (1994) Chem Rev 94:2483
14. van Gammeren AJ, Hulsbergen FB, Erkelens C, de Groot HJM (2004) J Biol Inorg Chem 9:109
15. Jordan P, Fromme P, Witt H-T, Klukas O, Saenger W, Krauss N (2001) Nature 411:909
16. Balaban TS (2003) FEBS Lett 545:97; Erratum (2003) FEBS Lett 547:235
17. Liu Z, Yan H, Wang K, Kuang T, Zhang J, Gui L, An X, Chang W (2004) Nature 428:287
18. Balaban TS (2005) Photosynth Res (in press)
19. Oba T, Tamiaki H (2005) In: van der Est A, Bruce D (eds) Photosynthesis: Fundamental Aspects to Global Perspectives. Proc XIII Photosynth Congr, Montreal, 2004. Allen Press, Lawrence, KS (in press)
20. Bassi R, Croce R, Cugini C, Santona D (1999) Proc Natl Acad Sci USA 96:10056
21. Remelli R, Varotto C, Sandonà D, Croce R, Bassi R (1999) Biophys J 47:3351
22. Kleima FJ, Hobe S, Calkoen F, Urbanus ML, Peterman EJG, van Grondelle R, Paulsen H, van Amerongen H (1999) Biochemistry 38:6587
23. Freer A, Prince S, Papiz M, Hawthornthwaite-Lawless A, McDermott G, Cogdell RJ, Isaacs NW (1996) Structure 4:449

24. McDermott G, Prince SM, Freer AA, Hawthornthwaite-Lawless AM, Papiz MZ, Cogdell RJ, Isaacs NW (1995) Nature 374:517
25. Papiz MZ, Prince SM, Howard T, Cogdell RJ, Isaacs NW (2003) J Mol Biol 326:1523
26. Koepke J, Hu X, Muenke C, Schulten K, Michel H (1996) Structure 4:581
27. Förster Th (1948) Ann Phys (Leipzig) 2:55
28. Hu XC, Ritz T, Damjanovic A, Autenrieth F, Schulten K (2002) Quart Rev Biophys 35:1
29. Abramavicius D, Valkunas L, van Grondelle R (2004) Phys Chem Chem Phys 6:3097
30. Jang S, Newton, MD, Silbey RJ (2004) Phys Rev Let 92:218301
31. Roszak AW, Howard TD, Southall J, Gardiner AT, Law CJ, Isaacs NW, Cogdell RJ (2003) Science 302:1969
32. Kühlbrandt W (1995) Nature 374:497
33. Hu X, Damjanović, A, Ritz T, Schulten K (1998) Proc Natl Acad Sci USA 95:5935
34. Cogdell RJ, Isaacs NW, Freer AA, Howard, TD, Gardiner AT, Prince SM, Papiz MZ (2003) FEBS Lett 555:35
35. McLuskey K, Prince SM, Cogdell RJ, Isaacs NW (2001) Biochemistry 40:8783
36. Oellerich S, Ketelaars M, Segura J-M, Margis G, de Rujiter W, Köhler J, Schmidt J, Aartsma TJ (2002) Single Mol 3:319
37. Drain CM, Gentemann S, Roberts JA, Nelson NY, Medforth CJ, Jia S, Simpson MC, Smith KM, Fajer J, Shelnutt JA, Holten D (1998) J Am Chem Soc 120:3781
38. Blankenship RE, Olson JM, Miller M (1995) In: Blankenship RE, Madigan MT, Bauer CE (eds) Anoxygenic photosynthetic bacteria. Kluwer Academic Publishers, Dordrecht, pp 399–435
39. Balaban TS (2004) In: Nalwa HS (ed) Encyclopedia of Nanoscience and Nanotechnology, vol 4. American Scientific Publishers, Los Angeles, pp 505–559
40. Griebenow K, Holzwarth AR, Schaffner K (1990) Z Naturforsch 45c:823
41. Griebenow K, Holzwarth AR (1990) Molecular Biology of Membrane-Bound Complexes in Phototrophic Bacteria. Plenum Press, New York, pp 375–381
42. Holzwarth AR, Griebenow K, Schaffner K (1992) J Photochem Photobiol A 65:61
43. Hildebrandt P, Griebenow K, Holzwarth AR, Schaffner K (1991) Z Naturforsch 46c:228
44. Hildebrandt P, Tamiaki H, Holzwarth AR, Schaffner K (1994) J Phys Chem 98:2192
45. Balaban TS, Holzwarth AR, Schaffner K (1995) J Molec Struct 349:183
46. Chiefari J, Griebenow K, Fages F, Griebenow N, Balaban TS, Holzwarth AR, Schaffner K (1995) J Phys Chem 99:1357
47. Balaban TS, Holzwarth AR, Schaffner K, Boender G-J, de Groot HJM (1995) Biochemistry 34:15259
48. Boender G-J, Balaban TS, Holzwarth AR, Schaffner K, Raap J, Prytulla S, Oschkinat H, de Groot HJM (1995) In: Mathis P (ed) Photosynthesis from Light to Biosphere. Kluwer Academic Publishers, Dordrecht, p 347
49. Van Rossum BJ, van Duyl BY, Steensgaard DB, Balaban TS, Holzwarth AR, Schaffner K, de Groot HJM (1998) In: Garab G (ed) Photosynthesis: Mechanisms and Effects. Kluwer Academic Publishers, Dordrecht, p 117
50. Nozawa T, Ohtomo K, Suzuki M, Nakagawa H, Shikama Y, Konami H, Wang Z-Y (1994) Photosynth Res 41:21
51. van Rossum BJ, Steensgaard DB, Mulder FM, Boender G-J, Schaffner K, Holzwarth AR, de Groot HJM (2001) Biochemistry 40:1587
52. van Rossum BJ, Schulten EAM, Raap J, Oschkinat H, de Groot HJM (2002) J Magn Reson 155:1

53. de Boer I, Matysik J, Amakawa M, Yagai S, Tamiaki H, Holzwarth AR, de Groot HJM (2003) J Am Chem Soc 125:13374
54. Frigaard N-U, Gomez MCA, Li H, Maresca JA, Bryant DA (2003) Photosynth Res 78:93
55. Schaffner K, Holzwarth AR (1997) Leopoldina 42:205
56. Holzwarth AR, Schaffner K (1994) Photosynth Res 41:225
57. Wang Z-Y, Umetsu M, Kobayashi M, Nozawa T (1999) J Am Chem Soc 121:9363
58. Pšenčik J, Ikonen TP, Laurinmäki P, Merckel MC, Butcher SJ, Serimaa RE, Tuma R (2004) Biophys J 87:1165
59. Staehelin LA, Golecki JR, Fuller RC, Drews G (1978) Arch Mikrobiol 119:269
60. Staehelin LA, Golecki JR, Drews G (1980) Biochim Biophys Acta 589:30
61. Huber V, Katterle M, Lysetzka M, Würthner F (2005) Angew Chem Int Ed 44:3147; Angew Chem 117:3208
62. Tamiaki H, Miyatake T, Tanikaga R, Holzwarth AR, Schaffner K (1996) Angew Chem Int Ed 35:772; Angew Chem 108:810
63. Miyatake T, Tamiaki H, Holzwarth AR, Schaffner K (1999) Helv Chim Acta 82:797
64. Prokhorenko VI, Holzwarth AR, Müller MG, Schaffner K, Miyatake T, Tamiaki H (2002) J Phys Chem B 106:5761
65. Balaban TS, Eichhöfer A, Lehn J-M (2000) Eur J Org Chem 4047
66. Redman JE, Feeder N, Teat SJ, Sanders KM (2001) Inorg Chem 40:2486
67. Bond AD, Feeder N, Redman JE, Teat SJ, Sanders JKM (2002) Crystal Growth Design 2:27
68. Wasielewski MR, Studier MH, Katz JJ (1976) Proc Natl Acad Sci USA 73:4282
69. Boxer SG, Closs GL (1976) J Am Chem Soc 98:5406
70. Katz JJ, Bowman MK, Michalski TJ, Worcester DL (1991) In: Scheer H (ed) Chlorophylls. CRC Press, Boca Raton, FL, pp 211–235
71. Bystrova MI, Mal'gosheva IN, Krasnovskii AA (1979) Mol Biol 13:440
72. Smith KM, Kehres LA, Fajer J (1983) J Am Chem Soc 105:1387
73. Tamiaki H, Amakawa M, Holzwarth AR, Schaffner K (2002) Photosynth Res 71:59
74. Tamiaki H (1996) Coord Chem Rev 148:183, and further references cited therein
75. Tamiaki H, Amakawa M, Shimono Y, Tanikaga R, Holzwarth AR, Schaffner K (1996) Photochem Photobiol 63:92
76. Yagai S, Miyatake T, Tamiaki H (1999) J Photochem Photobiol B: Biol 52:74
77. Oba T, Tamiaki H (1998) Photochem Photobiol 67:295
78. Balaban TS, Tamiaki H, Holzwarth AR, Schaffner K (1997) J Phys Chem B 101:3424
79. Jesorka A, Balaban TS Holzwarth AR, Schaffner K (1996) Angew Chem Int Ed Engl 35:2861; Angew Chem 108:3019
80. Tamiaki H, Miyatake T, Tanikaga R (1997) Tetrahedron Lett 38:267
81. Yagai S, Miyatake T, Tamiaki H (2002) J Org Chem 67:49
82. Yagai S, Miyatake T, Shimono Y, Tamiaki H (2001) Photochem Photobiol 73:153
83. Tamiaki H, Kitamoto H, Nishikawa A, Hibino T, Shibata R (2004) Bioorg Med Chem 12:1657
84. Yagai S, Tamiaki H (2001) J Chem Soc Perkin Trans 1:3135
85. Oba T, Tamiaki H (1999) Photosynth Res 61:23
86. Balaban TS, Bhise AD, Fischer M, Linke-Schaetzel M, Roussel C Vanthuyne N (2003) Angew Chem Int Ed 42:2140; Angew Chem 115:2190
87. Balaban TS, Linke-Schaetzel, Bhise AD, Vanthuyne N, Roussel C (2004) Eur J Org Chem 3919
88. Tamiaki H, Kimura S, Kimura T (2003) Tetrahedron 59:7423

89. Berova N, Nakanishi K (2000) In: Berova N, Nakanishi K, Woody RW (eds) Circular Dichroism. Principles and Applications, 2nd ed. Wiley-VCH, New York, pp 337–382
90. Balaban TS, Bhise AD, Bringmann G, Eichhöfer A, Fenske D, Mizoguchi T, Reichert M, Schraut M, Tamiaki H, Roussel C, Vanthuyne N (manuscript in preparation)
91. Balaban TS, Linke-Schaetzel M, Bhise AD, Vanthuyne N, Roussel C, Anson CE, Buth G, Eichhöfer A, Foster K, Garab G, Gliemann H, Goddard R, Javorfi T, Powell AK, Rösner H, Schimmel Th (2005) Chem Eur J 11:2267
92. Van Amerongen H, Valkunas L, van Grondelle R (2000) Photosynthetic Excitons. World Scientific, Singapore, pp 1–590
93. Van Amerongen H, Vasmel H, van Grondelle R (1988) Biophys J 54:65
94. Brune DC, Gerola PD, Olson JM (1990) Photosynth Res 24:253
95. van Mourik F, Griebenow K, van Haeringen B, Holzwarth AR, van Grondelle R (1990) In: Current Research in Photosynthesis II. Kluwer Academic Publishers, Dordrecht, pp 141–144
96. Ma Y-Z, Cox RP, Gillbro T, Miller M (1996) Photosynth Res 47:157
97. Alden RG, Lin SH, Blankenship RE (1992) J Luminesc 51:51
98. Somsen OJG, van Grondelle R, Van Amerongen H (1996) Biophys J 71:1934
99. Pšenčik J, Ma Y-Z, Arellano JB, Hala J, Gillbro T (2003) Biophys J 84:1161
100. Prokhorenko VI, Steensgaard DB, Holzwarth AR (2003) Biophys J 85:3173
101. Prokhorenko VI, Steensgaard DB, Holzwarth AR (2000) Biophys J 79:2105
102. Griebenow K, Holzwarth AR, van Mourik F, van Grondelle R (1991) Biochim Biophys Acta 1058:194
103. de Boer I, Matysik J, Erkelens K, Sasaki S, Miyatake T, Yagai S, Tamiaki H, Holzwarth AR, de Groot HJM (2004) J Phys Chem B 108:16556
104. Didraga C, Klugkist JA, Knoester J (2002) J Phys Chem B 106:11474
105. Mishra A, Beherea RK, Behera B, Mishra K, Behera GM (2000) Chem Rev 100:1973
106. Potma EO, Wiersma DA (1998) J Chem Phys 108:4894
107. Fidder H, Wiersma DA (1995) Physica Status Solidi B 188:285
108. Didraga C, Pugzlys A, Hania PR, von Berlepsch H, Duppen K, Knoester J (2004) J Phys Chem B 108:14976
109. Knoester J (1995) Adv Mater 7:500
110. Prokhorenko VI, Holzwarth AR, Müller MG, Schaffner K, Miyatake T, Tamiaki H (2002) J Phys Chem 106:5761
111. Prokhorenko VI, Katterle M, Holzwarth AR, Schaffner K, Miyatake T, Tamiaki H (2002) In: Femtochemistry and Femtobiology: Ultrafast Dynamics in Molecular Science. World Scientific, New Jersey, pp 782–788
112. Möltgen H, Kleinermanns K, Jesorka A, Schaffner K, Holzwarth AR (2002) Photochem Photobiol 75:619
113. Linke-Schaetzel M, Bhise AD, Gliemann H, Koch Th, Schimmel Th, Balaban TS (2004) Thin Solid Films 124–125:16
114. Kureishi Y, Shiraishi H, Tamiaki H (2001) J Electroanal Chem 496:13

Top Curr Chem (2005) 258: 39–82
DOI 10.1007/b136668

Published online: 11 August 2005

# Metallosupramolecular Dye Assemblies

Chang-Cheng You · Rainer Dobrawa · Chantu R. Saha-Möller ·
Frank Würthner (✉)

Institut für Organische Chemie, Universität Würzburg, Am Hubland, 97074 Würzburg, Germany
*wuerthner@chemie.uni-wuerzburg.de*



**Abstract** Metallosupramolecular assemblies of dyes are of increasing interest as model systems for natural light harvesting and photosynthetic complexes as well as for potential applications in solar energy conversion, photonics and molecular electronics. Considerable efforts have been made in the past years to design novel metallosupramolecular dye architectures for such applications. Often in combination with other noncovalent interactions like hydrogen-bonding and $\pi$–$\pi$-stacking, metal-ion coordination has been utilized to achieve self-organization of dye building blocks into elaborate architectures with interesting optical and photophysical properties. Diverse metallosupramolecular dye assemblies – macrocycles, polyhedra, polymers and dendrimers – have been constructed in recent years by the metal-coordination approach. Here we review the recent advances in metallosupramolecular dye assemblies with particular attention to their thermodynamic stability and photophysical functionality. Emphasis is given to metal-ion mediated self-organization of porphyrin, metalloporphyrin and perylene bisimide dyes.

**Keywords** Self-assembly · Coordination interaction · Supramolecular architecture · Light-harvesting system · Dyes/pigments

**Abbreviations**

| | |
|---|---|
| CSI-MS | cold spray ionization mass spectrometry |
| DABCO | 1,4-diazabicyclo[2,2,2]octane |
| DFWM | degenerate four-wave mixing |
| dppp | 1,3-bis(diphenylphosphano)propane |
| GPC | gel permeation chromatography |
| LH | light-harvesting |
| MLCT | metal-to-ligand charge transfer |
| NBI | naphthalene bisimide |
| PBI | perylene bisimide |
| PGSE | pulsed gradient spin echo |
| PPBI | phenoxy-substituted perylene bisimide |
| PyP | *meso*-(4-pyridyl)-porphyrin |
| SANS | small-angle neutron scattering |
| TPP | meso-tetraphenylporphyrin |
| tpy | 2,2′:6′,2″-terpyridine |
| VPO | vapor pressure osmometry |

# 1 Introduction

## 1.1 General Aspects

Supramolecular chemistry deals with different aspects of molecular architecture, organization, self-assembly and recognition. Self-assembly is generally defined as "the spontaneous assembly of molecules into structured and noncovalently bound aggregates" [1]. Accordingly, metallosupramolecular dye assemblies, the topic of this chapter, can be considered as highly organized metal-containing dye architectures created through coordination of dye ligands to metal ions. Metallosupramolecular dye assemblies have been known for a long time to exist in nature. For instance, the natural LH complexes of photosynthetic organisms comprise numerous chlorophyll and carotenoid dyes that are bound to proteins by noncovalent interactions such as hydrogen-bonding, $\pi$–$\pi$-stacking and N- and O- ligand coordination to the chlorophyll $Mg^{2+}$ ion [2, 3]. In such pigment-protein complexes, most chlorophylls serve as light-harvesting antennae, which collect the solar light and transport electronic excitation towards the reaction centers [2]. The discovery of pivotal functions of dye assemblies in the photosynthetic machinery has paved the way for the development of diverse artificial systems based on functional dyes mimicking the basic light-harvesting processes. On the other hand, the driving force for the continuing intense research activities on dye assembly is the excellent opportunity to develop novel photonic/electronic materials and devices for application in modern technology. Indeed, aggregates of metal-free and metal-containing dyes have been employed as photoconductors in electrophotography [4], highly sensitive biosensors [5], field effect transistors [6, 7], organic light-emitting diodes [8], organic solar cells [9, 10], light-harvesting systems [11, 12], and molecular wires [13].

Dye assemblies can be created through various noncovalent weak forces, including hydrophobic, electrostatic, hydrogen-bonding, $\pi$–$\pi$-stacking as well as coordinative interactions. Among them, hydrogen-bonding and coordinative interactions are more attractive due to their features of directionality and complementarity which enable the topology of the assemblies to be predictable and controllable. The hydrogen bond directed dye assemblies are treated separately in this volume, therefore, they are not considered here. This review focuses primarily on the recent development of dye assemblies mediated by metal-ion coordination with particular attention to their thermodynamic stability. We consider here only self-assembled systems generated by metal-mediated reversible processes, and kinetically inert covalent dye-metal complexes are completely excluded.

As generally known, coordinative interactions can attain quite high strengths (with bond energies ranging from 40 to 120 kJ mol$^{-1}$ per interaction) compared with other weak interactions such as hydrogen-bonding and $\pi$–$\pi$-stacking. At the upper limit, almost the strength of covalent bonds can be achieved by coordinative interactions. This characteristic endows the resulting entities with considerable stability favoring the production of discrete or infinite supramolecular species often in a quantitative manner. Similar assembly approaches are also known to occur in a number of natural self-assembled systems where the high binding strength emerges from large complementary surfaces, i.e. multiple interacting binding sites. Due to this unique combination of directionality and binding strength, the methodology of metal coordination directed self-assembly has become one of the most widely employed techniques to construct superstructures of abundant geometries [14–17]. However, high binding strength which is related to large negative $\Delta G^0$ is quite often accompanied by high $\Delta G^{\neq}$ values which undermine the reversibility and, hence, may lead to the formation of kinetic products. As a consequence some typical advantages of self-assembled materials such as quantitative formation of the thermodynamically most stable species and error correction in the self-assembly process are not always fulfilled for metallosupramolecular materials.

For the most general approach to construct supramolecular dye assemblies through metal coordination, dye molecules must possess appropriate Lewis basic coordination sites, which are, however, usually scarce in parent dye molecules. Therefore, aromatic aza ligands, especially pyridyl, bipyridyl and terpyridyl groups, are often introduced into dye molecules by established organic synthetic routes to enable metal-aza ligand interactions for the construction of dye assemblies of diverse structures and properties. Surprisingly, few classes of dyes have been subjected so far to produce supramolecular assemblies from such functionalized dyes by addition of metal ions. In contrast, for the majority of existing metallosupramolecular dye assemblies the respective metal-ion containing dyes are applied. For this purpose porphyrins are particularly attractive, and also the most intensively studied dyes to date, primarily due to their favorable photochemical properties, ease of chemical modification and, in particular, the possibility to introduce various metal ions into the center of the porphyrin macrocycle. Thus, it is not surprising that the assemblies of free base porphyrins and metalloporphyrins occupy the major part of this review article. Another class of functional dyes, namely perylene bisimides (PBIs), have attracted much attention in the past years as versatile building blocks for supramolecular self-assembly [18]. Therefore, we emphasize here also on coordination assemblies of perylene bisimides. It is the purpose of this article to discuss the recent development of metal-mediated self-assembly of functional dyes into macrocyclic and polymeric structures, and to highlight the potential application of these materials in photonics and electronics.

## 1.2 Optical, Redox and Binding Properties of Some Representative Dyes

In this review, we restrict ourselves to metal-mediated assemblies of functional $\pi$-systems, which absorb light in the visible region (i.e. $\lambda > 400$ nm). Thus, materials whose color arises from metal ions or metal-ligand charge transfer (MLCT) are not included here. Self-assemblies derived from free base porphyrins, metalloporphyrins and perylene bisimide dyes have shown promising applications in artificial LH systems. Therefore, some fundamental properties of a few representative chromophores, namely *meso*-tetraphenylporphyrin (TPP), zinc tetraphenylporphyrin (ZnTPP), tetraphenoxy-substituted perylene bisimide (PPBI) and bay area unsubstituted perylene bisimide (PBI), are briefly summarized herein. These properties are essential for the understanding of the photophysical and photochemical behavior of assemblies derived from such dyes.

As shown in Table 1 and Fig. 1, the free base porphyrin TPP exhibits an intense absorption maximum at around 420 nm and four weak absorption maxima in the region 500 to 650 nm in chloroform. The strong absorption band (the Soret or B band) is due to the transition to the second excited state ($S_0 \rightarrow S_2$), while the weak absorption bands in the visible region (the Q bands) are owing to the transition to the first excited state ($S_0 \rightarrow S_1$). Both the Soret and the Q bands arise from $\pi$–$\pi^*$ transitions, but porphyrins are the only systems known to provide a distinct pattern of split Q-bands. Because of the small oscillator strength for the $S_0 \rightarrow S_1$ transition, the fluorescence of porphyrins from $S_1$ takes place with low quantum yields. Compared with free base TPP, metalated ZnTPP shows a slight red-shift of the Soret band, implying a more electron-rich $\pi$-conjugated system. Zinc porphyrin inter-

**Table 1** Optical properties of TPP, ZnTPP, and PBI, PPBI in $CHCl_3$[a]

| | $\lambda_{abs}$ (nm) ($\varepsilon/10^3$ $M^{-1}$ $cm^{-1}$) | $\lambda_{em}$ (nm) | $S_1$(eV)[b] | $\Phi_f$ (%) | $\tau_f$ (ns) |
|---|---|---|---|---|---|
| TPP | 418 (473), 515 (18.7), 551 (7.7), 591 (5.4), 647 (4.8) | 650, 715 | 1.91 | 9[c] | 10.9[c] |
| ZnTPP | 424 (500), 553 (18.2), 594 (4.9) | 601, 652 | 2.08 | 6[c] | 2.0[c] |
| PPBI | 585 (50.5) | 614 | 2.07 | 96 | 6.5 |
| PBI | 526 (94.9) | 531 | 2.34 | 100 | 3.4 |

[a] The data presented here were measured in our laboratory and they are in accord with those reported in the literature for TPP [19, 20] and ZnTPP [20, 21] as well as PPBI [22] and PBI [23] containing other imide substituents.
[b] Calculated from the intersection of the normalized absorption and fluorescence spectra.
[c] In toluene.

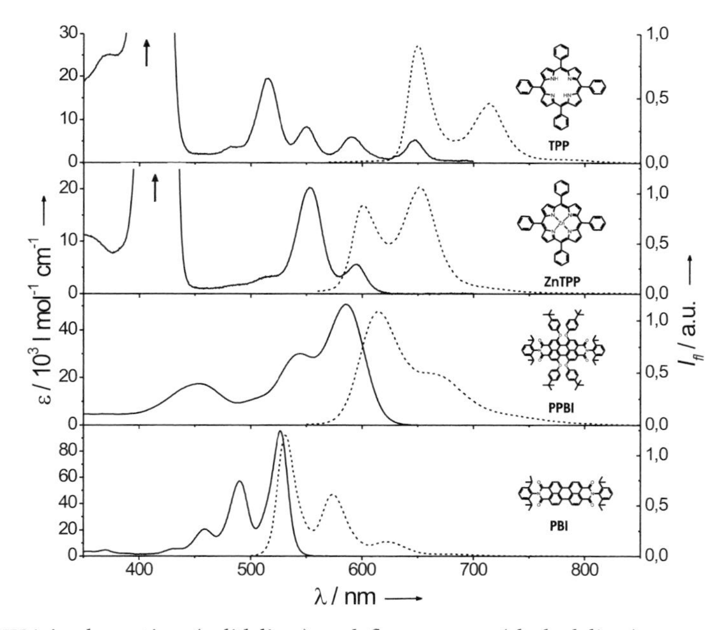


**Fig. 1** UV/vis absorption (*solid lines*) and fluorescence (*dashed lines*) spectra of TPP, ZnTPP, PPBI, and PBI in $CHCl_3$

acts readily with electron donors such as the aza ligand and the zinc(II) ion becomes penta-coordinated, which is the maximal coordination number accessible in this system due to the displacement of the zinc(II) ion out of the porphyrin plane. The coordination usually causes a red shift of both the Soret band and Q band of porphyrin, thus, this process can be monitored conveniently by UV/vis spectrometry.

In contrast to porphyrins, perylene bisimides exhibit their main intense absorption band above 500 nm due to a strongly allowed $S_0 \rightarrow S_1$ transition. Electron-donating substituents at the bay area of perylene bisimides shift the absorption maximum to a longer wavelength. Thus, at bay positions tetraphenoxy-substituted perylene bisimide PPBI exhibits an absorption maximum at 585 nm, while the unsubstituted derivative PBI shows absorption at 526 nm. Notably, the electronically excited perylene bisimides decay to the ground state almost exclusively by fluorescence emission. Thus, high fluorescence quantum yields are often observed which make perylene bisimide dyes superior to porphyrins for the fabrication of highly luminescent assemblies. It is interesting to note that ZnTPP and PPBI possess almost identical $S_0 \rightarrow S_1$ excitation energy (Table 1).

The dyes TPP, ZnTPP and perylene bisimides (PPBI, PBI) also show well-defined redox waves in cyclic voltammograms (Table 2). Both TPP and ZnTPP possess two reversible oxidation and two reversible reduction waves. As the data in Table 2 reveal, ZnTPP is somewhat easier to oxidize than TPP. Compared with TPP and ZnTPP, PBI and also the tetraphenoxy-substituted PPBI are electron-deficient chromophores, thus, they can successively accept two electrons to afford radical anionic and dianionic species, respectively, at readily accessible potentials. By contrast, the oxidation waves of PBI and PPBI appear at relatively high potentials.

In addition to the photo- and electrochemical properties of the representative dyes, we also present the binding constants of some metal-ligand interactions (Table 3) which are of relevance for the coordinative bond formation in the dye assemblies discussed in this review. The binding constants reflect the stability of the assemblies formed and are critically dependent on many factors, including ligands, metals, temperature, and solvents as well. In order to get assemblies of considerable stability, the various factors need to be adjusted properly. The binding constant of pyridine (frequently used as

**Table 2** Redox properties of TPP, ZnTPP, and PPBI, PBI in $CH_2Cl_2$[a] (in V vs. $cp_2Fe$)

| | Reduction | | Oxidation | |
|---|---|---|---|---|
| | I | II | I | II |
| TPP | −2.03 | −1.71 | +0.53 | +0.87 |
| ZnTPP | −2.21 | −1.85 | +0.42 | +0.70 |
| PPBI | −1.35 | −1.15 | +0.86 | |
| PBI | −1.24 | −1.01 | +1.29 | |

[a] For the purpose of unambiguous comparison, we have re-measured all values against the same reference electrode in $CH_2Cl_2$ with $NBu_4PF_6$ as the supporting electrolyte. The redox potentials of TPP and ZnTPP against SCE are reported in reference [24], for the redox potentials of similar PPBI and PBI derivatives see references [25, 26]

the ligand in functional dyes for metal-mediated self-assembly) and ZnTPP in chloroform is in the order of $10^3$ $M^{-1}$. In less polar cyclohexane, an increase of the binding constant by more than one order of magnitude is noted [27], which is not as pronounced as for hydrogen bonding [28]. Thus, slightly polar solvents are effective for improving the stability of assemblies but still quite high concentrations (> 1 mM) are required to effect pyridine-ZnTPP self-assembly. By introducing additional binding sites into the ligands the stability of the resultant entities can be enhanced. This may be achieved by employing either chelating ligands, such as bi- and terpyridine, or multiple monotopic functions. Indeed, an assembly simultaneously involving two pyridine-ZnTPP interactions may have a stability constant as high as $10^6$ $M^{-1}$ in $CDCl_3$, despite there being no cooperative effect present. Furthermore, the appropriate selection of the metal ion is crucial for the stability of the coordinative assemblies as the data in Table 3 reveal.

**Table 3** Selected binding constants ($K_S$) and corresponding Gibbs free energy changes ($-\Delta G^0$) of the complexation between different ligand donors and ligand acceptors at 298 K

| Ligand donor | Ligand acceptor/ metal ion | Solvent | $K_S$ ($M^{-1}$) | $-\Delta G^0$/ kJ $mol^{-1}$ | Ref. |
|---|---|---|---|---|---|
| Pyridine | ZnTPP | $CDCl_3$ | $9.2 \times 10^2$ | 16.6 | [29] |
| Pyridine | ZnTPP | Cyclohexane | $2.5 \times 10^4$ | 25.1 | [27] |
| Pyridine | (dppp)Pd or Pt | $CHCl_3$ | $\sim 10^6$ | $\sim 34$ | [18] |
| Terpyridine | $FeCl_2$ | $CHCl_3/CH_3OH$ (1 : 1) | $1.2 \times 10^7$ | 40.4 | [30] |
| Terpyridine | $Zn(OTf)_2$ | $CH_3CN$ | $> 10^8$ | > 45 | [31] |

For instance, the binding constant between pyridine and dppp chelated Pd or Pt is in the order of $10^6$ $M^{-1}$ in $CHCl_3$, which is significantly higher than the pyridine-ZnTPP interaction. When rhenium or ruthenium metal ions are employed, extremely stable and inert assemblies could be obtained as evidenced by the irreversibility of Re(II)-N and Ru(II)-N bonds under ambient conditions. Finally, it should be emphasized that only the binding constants having the same dimensions should be compared in the evaluation of assembly stability [32] or the so-called "critical self-assembly" concentration at which 50% of the self-assembled species formed. Since optical spectroscopy is typically carried out with dilute solutions ($c < 10^{-5}$ M), high binding constants ($K_S > 10^6$ $M^{-1}$) are required to study the photophysical properties of dye assemblies by these techniques.

# 2
# Macrocyclic Dye Assemblies

## 2.1
## Assemblies Mediated by External Metal Centers

One of the essential impetuses to cyclic dye architectures is the development of artificial model systems of the light-harvesting complexes in purple bacteria, where cyclic arrays of chromophores provide the fundamental structural feature for their functionality [2]. On the other hand, cavities of variable sizes and shapes are accessible upon the formation of macrocycles. The assemblies of this kind, thus, possess also potentials in substrate binding, molecular recognition, matter transportation and even supramolecular catalysis.

An efficient approach to the construction of macrocyclic dye assemblies is the coordination of dye ligands to appropriate external metal-ion mediators. Because of the directional feature of coordinative bonding, the desired product may be obtained through the spontaneous reaction between properly predesigned ligands and metal ion units. It could be envisioned that triangular, square and hexagon assemblies might be produced by the stoichiometric reaction of linear ditopic ligands and metal corner units with 60°, 90° and 120° separation, respectively [14]. However, only square dye assemblies and a few triangular dye assemblies have been reported so far, mostly owing to the abundance of 90° metal corners and, on the other hand, the poor availability of other angular building blocks.

Drain and Lehn reported the first square porphyrin arrays such as **1** and **2** by employing the concept of metal coordination directed self-assembly [33]. The equimolar combination of ditopic porphyrin ligands of 90° angle separation (5,10-PyP) and *trans*-$PdCl_2(NCPh)_2$ or, alternatively, linear ligands (5,15-PyP) and *cis*-$PtCl_2(NCPh)_2$ preferentially afforded structurally well-defined and constraint-free square tetramers **1** and **2**, respectively. However, multimers were concomitantly formed as assessed by $^1H$ NMR spectroscopy. Unfortunately, these very first porphyrin metallosquares have been characterized only by NMR spectroscopy. More recent investigations with mixtures of three different porphyrin derivatives, namely 5,15-PyP, 5,10,15-PyP and 5,10,15,20-PyP with *trans*-$PdCl_2(NCPh)_2$, afforded even more complex supramolecular tessellation of nine porphyrins (see Sect. 3.2) [34].

Based on the concept of Drain and Lehn, structurally more reliable palladium and platinum-mediated square assemblies of porphyrins such as **3** could be prepared by Stang and coworkers [35–37]. The reaction of linear or right-angle ditopic porphyrin ligands with dppp chelated Pd(II) or Pt(II) complexes at room temperature afforded the respective metallosquares in essentially quantitative yields. The square structures were unambiguously confirmed by $^1H$ and $^{31}P$ NMR, elemental analysis, and in some cases even by electrospray and FAB mass spectrometry. Because of the high bond strength

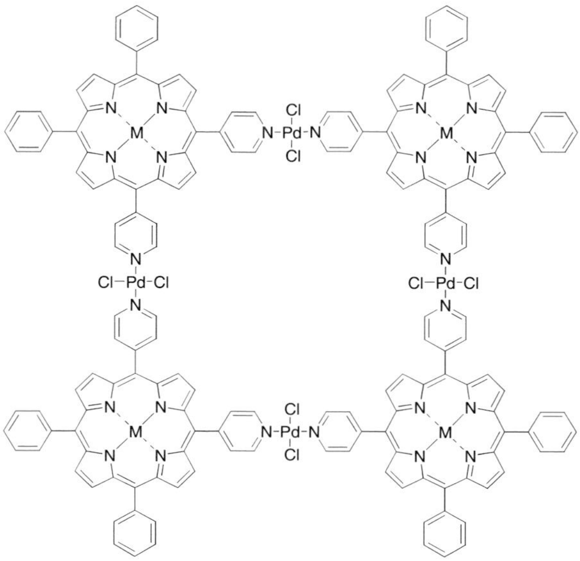

**1**
(M = 2H or Zn)

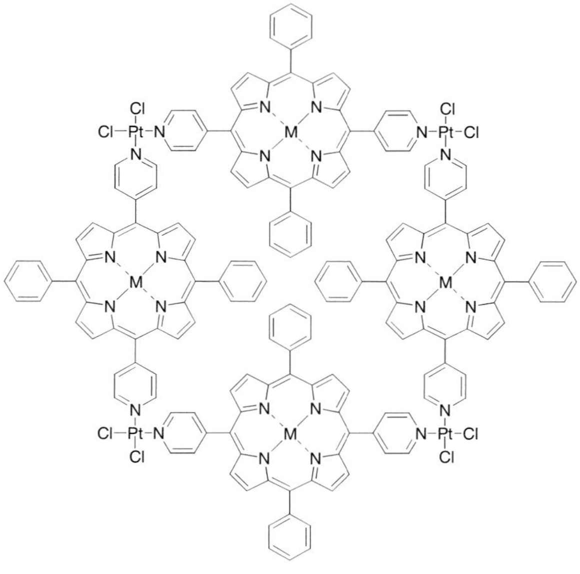

**2**
(M = 2H or Zn)

of Pd-N and Pt-N, these assemblies were found to be stable in a wide concentration range from $10^{-9}$ to $10^{-2}$ M and, therefore, suitable for spectroscopic studies under highly dilute conditions. Chirality has also been successfully introduced into the porphyrin-based molecular squares by employing metal corner units with chiral groups [37].

Since 90° metallocorners are readily available from transition metal reagents with either square planar or octahedral coordination geometries through appropriate blocking of some coordinative sites, quite a few porphyrin-based molecular squares with diverse metal corner units were constructed. Hupp and coworkers have prepared porphyrin-walled squares **4** by the reaction of 5,15-PyP with $Re(CO)_5Cl$ at elevated temperature [38]. The kinetically inert Re-N bond excludes the exchange of ligands in solution at room temperature but at elevated temperature the square is formed apparently through thermodynamic control. The tetrazinc(II)-metalated square **4b** showed strong binding affinities toward multitopic guest molecules with suitable geometries. The corresponding host-guest complexes with manganese porphyrins were used as epoxidation catalysts with enhanced catalyst lifetime and substrate selectivity [39]. Some neutral octahedral Ru(II) cornered porphyrin dimeric squares have recently been prepared by Alessio's group [40, 41]. Due to the inertness of Ru-N bond, the desired assemblies were not obtained in quantitative yields, but the assembled objects could be purified by conventional column chromatography; noteworthy, this purification method is generally not applicable to supramolecular compounds with reversible metal-ligand bonds. It was reported that trimeric and higher

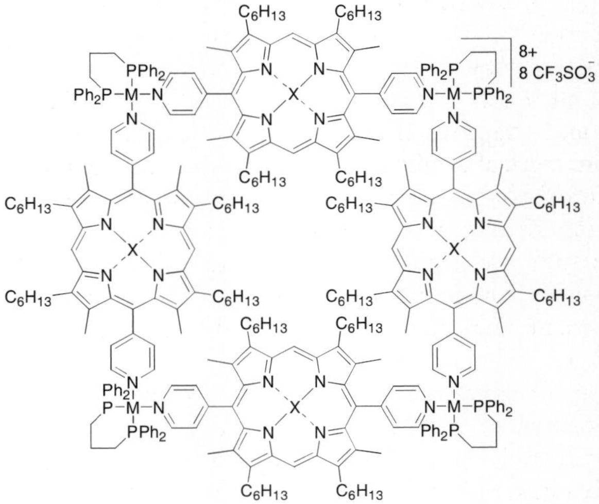

**3**

(M = Pd or Pt, X = 2H or Zn)

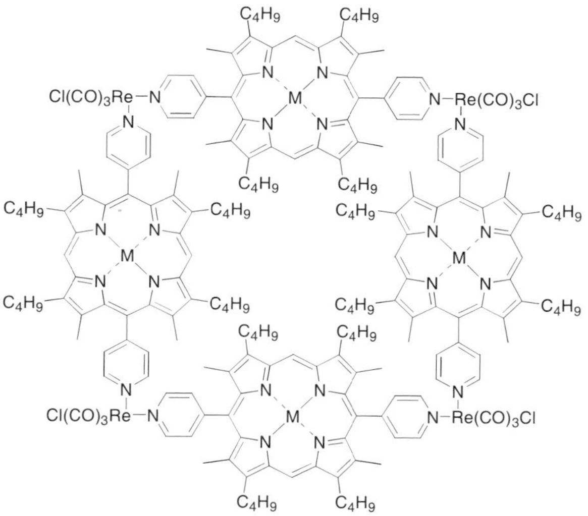

**4**
(a: M = 2H, b: M = Zn)

metallacycles were also generated in the self-assembly process of porphyrin building blocks when $RuCl_2(CO)_2$ was employed as the corner unit [42].

The combination of 90° metal corners and linear ditopic ligands usually favors the formation of square assembly, since this scaffold possesses the minimized conformational strain. However, the macrocyclic assemblies of different topologies may co-exist when further effects come into play. In this context, we have prepared a series of tetraphenoxy-substituted diazadibenzoperylene ligands [43, 44]. When these twisted dyes were used as bridging ligands, complex dynamic equilibria between triangular assemblies **5** and square assemblies **6** were observed, even though the perfectly preorganized (dppp)Pd(II) and (dppp)Pt(II) served as corner units [45]. This behavior originates from sterical strain of the twisted dyes and enthalpy and entropy factors, since the former parameter favors the generation of squares while the latter prefers the formation of triangular species. While the free diazadibenzoperylene ligands are highly fluorescent, drastic fluorescence quenching took place upon the formation of metallosupramolecular assemblies **5** and **6**. Similar phenomena were also observed for the above-mentioned porphyrin assemblies.

The undesired fluorescence quenching could be overcome in perylene bisimide-based squares **7**, which show fluorescence quantum yields of almost unity in chloroform [46, 47]. The square assemblies were quantitatively accessible by mixing an equimolar amount of perylene bispyridyl imide ligands and corresponding metal corner units in dichloromethane at room temperature. Because the pyridyl moieties are located at the imide positions of the

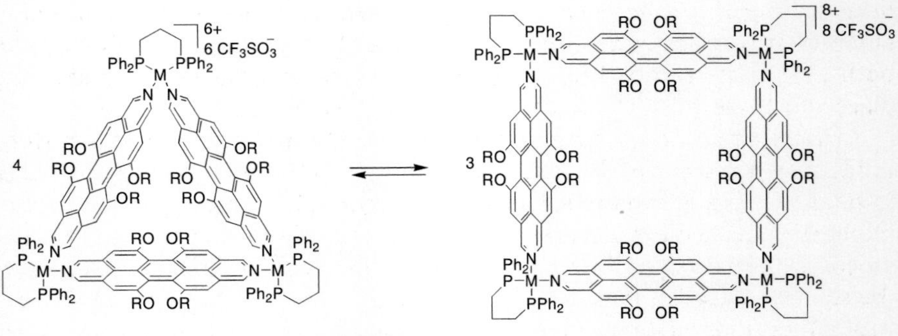

5 (R = Ph or *p*-*t*BuPh, M = Pd or Pt) 6

perylene bisimide core, where nodes exist in the HOMO and LUMO orbitals, the excellent luminescent properties of free perylene bisimide chromophores remained unchanged in the metal-assembled form. This observation implies that the binding sites of dye ligands should be decoupled (i.e. not conjugated) from the chromophore to preserve their inherent fluorescence upon binding to transition metals.

The same perylene bisimide scaffold was used to construct multichromophoric squares **8**, which contain 16 additional antenna dye units tethered to the tetrameric perylene-based square (Fig. 2) [48–50]. Such metallosupramolecular architectures are reminiscent of the cyclic dye assemblies of light harvesting bacteria, where energy transfer from outer antenna dyes to the central reaction center is essential for efficient capture of solar light. In-

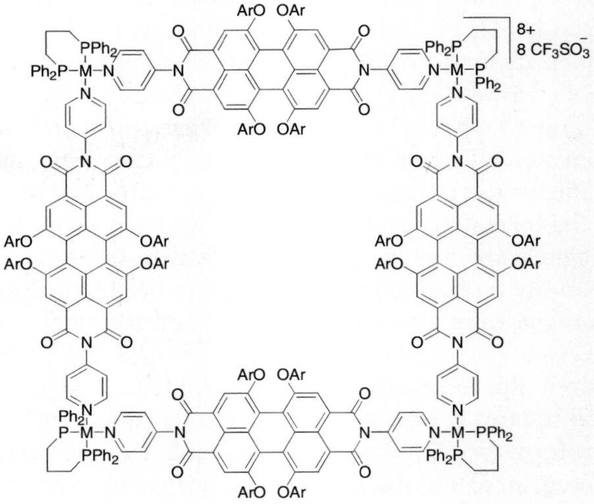

7

(Ar = *p*-*t*BuPh, M = Pd or Pt)

deed, for the assemblies **8a,b** an energy transfer was observed from the outer antenna dyes to the inner perylene bisimide dyes upon photoexcitation. In both assemblies competitive photoinduced electron-transfer processes take place which are, however, much more efficient for **8a** than for **8b** [49].

In a similar fashion, four-fold ferrocenyl-functionalized perylene bipyridyl imide dyes were assembled to square array **8c** [50]. The assemblies contain in total 20 redox active units, including 4 perylene bisimide and 16 ferrocene moieties. Due to the twisted perylene backbone, the ferrocenyl units are divided into two groups from the viewpoint of their spatial location referred to the square cavity (cf. model in Fig. 2). Indeed, the redox behavior of the ferrocenyl units depends on their spatial arrangement as established by cyclic voltammetric investigation [50].

Besides the above-mentioned porphyrin and perylene bisimide dyes, also a few dyes of other types were organized to cyclic assemblies using similar strategies. For example, diazapyrene [51], terpyridyl ruthenium complexes [52], (salen)zinc complexes [53], and 4,4′-azopyridine [54] have been used as bridging ligands to construct square assemblies with different 90° metal corners. The assemblies of 4,4′-azopyridine with Pd(II) ions are of particular interest since they undergo efficient reversible photoisomerization. The detailed investigations revealed that the square tetramer consisting of *trans*-azopyridine was transformed to the dimeric assembly of *cis*-azopyridine upon irradiation. Inversely, heating of the dimeric assembly regenerates the tetrameric square assembly [54].

1,2-Dithienylethene derivatives possess interesting applications in optical devices and switches due to their advantageous photochromic properties. Recently, a few photochromic bridging ligands derived from these dyes

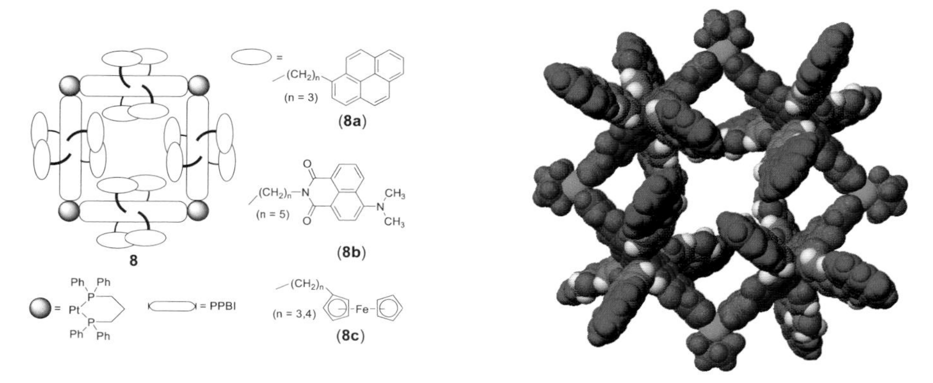


**Fig. 2** Multichromophoric molecular squares **8a–c** and molecular model of perylene-walled light-harvesting molecular square **8a**. In model the phosphane ligand is replaced by an ethylenediamine chelate ligand for simplicity. In squares **8a–c** a *p*-carbonyl-oxyphenoxy unit is used as the spacer between the perylene core and antenna dye linkage

have been subjected to supramolecular assembly with metal ions [55, 56]. Representatively, the assembly **9** was produced by the equimolar reaction of chiral 1,2-dithienylethene ligand and copper(I) triflate in deoxygenated dichloroethane [55]. Its structure was confirmed by mass spectrometry and X-ray analysis. It is noteworthy that while the photocyclization of free 1,2-dithienylethene ligands inevitably yields racemates, highly stereoselective ring-closure takes place in the metal-coordinated state.

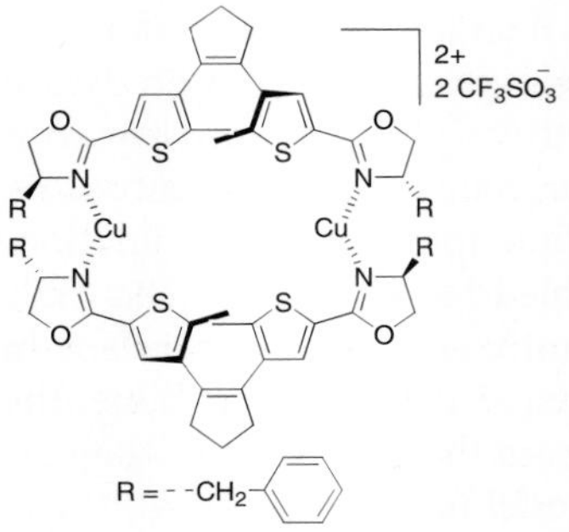

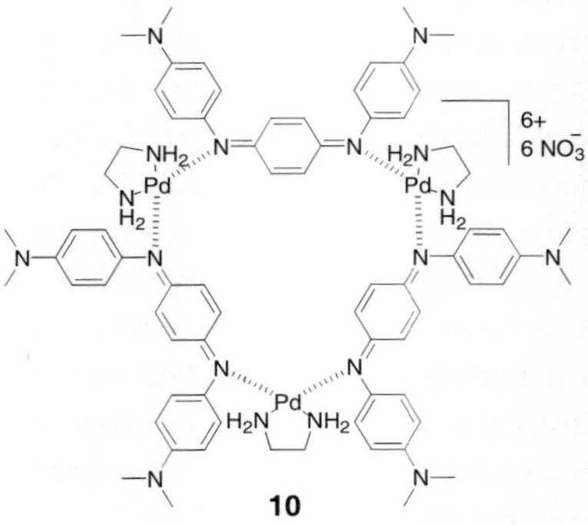

In a recent study, Hirao and coworkers have used the redox-active $\pi$-conjugated ligand *N,N′*-bis(4-dimethylaminophenyl)-1,4-benzoquinonediimine and palladium(II) complexes to build up oligomeric assemblies [57]. The conjugated trimetallic macrocycle **10** was obtained quantitatively by treating the ligand with an equimolar amount of $[Pd(NO_3)_2(en)]$ in acetonitrile. The electronic spectrum of **10** in methanol exhibited a strong, broad absorption band around 800 nm, assignable to a low-energy charge-transfer transition with significant contribution from palladium. X-ray analysis confirmed a trimetallic macrocyclic skeleton with an open three-dimensional cavity, which is reminiscent of the cone conformation of calixarenes. These redox-active supramolecular complexes are potentially important for application as catalysts and electronic materials [57].

## 2.2 Self-Cyclization of Metalloporphyrins

If the dye ligands contain their own metal centers, they are able to offer both electron lone pairs and Lewis acidic sites for the self-assembly process. Once the orientation between the ligand and the metal is appropriate, the intermolecular complementary coordination in a head-to-tail manner might lead to macrocyclic architectures. Indeed, quite a few supramolecular macrocycles were produced from metalloporphyrin derivatives through the metal-directed self-cyclization approach.

The simplest and the smallest macrocycles accessible by this strategy are the cofacially coordinated porphyrin dimers. For example, once two *N*-methylimidazolyl substituents were introduced into the facing *meso* positions of tetraethyltetramethylporphyrin, its zinc porphyrin derivative was self-organized into the dimeric assembly **11** through coordination to imidazolyl ligands [58]. In contrast to the relatively low stability of simple pyridine-zinc porphyrin interaction ($K \approx 10^3$ $M^{-1}$), dimer **11** is stable at concentration as low as $10^{-9}$ M (for a related dimer, $K_D = 3.3 \times 10^{11}$ $M^{-1}$ was determined by competitive UV/vis titration experiments [12]). Apparently, coordinative as well as $\pi$–$\pi$-stacking interactions contribute to the stability of this self-assembly. Dimeric porphyrin assemblies of a cofacial topology have also been constructed from *ortho*-pyridyl [59] or *ortho*-aniline [60] substituted zinc porphyrins and *ortho*-pyridyl substituted ruthenium porphyrins [61].

Hunter and Sarson have designed a zinc-porphyrin system, which bears a pyridyl function perpendicular to the porphyrin plane. This dye affords self-cyclized assembly **12** at $10^{-7}$ to $10^{-2}$ M concentrations in dichloromethane ($K_D \approx 10^8$ $M^{-1}$) [62]. Because this assembly has a *pseudo*-cavity with inner hydrogen-bonding sites, it can encapsulate amide guests of suitable size and shape to form host-guest complexes through the hydrogen-bonding interaction.

Through proper orientation of the pyridyl group relative to the zinc-porphyrin plane, macrocyclic assemblies of variant topologies are accessible. For example, favorable formation of the trimeric macrocycle **13** was achieved

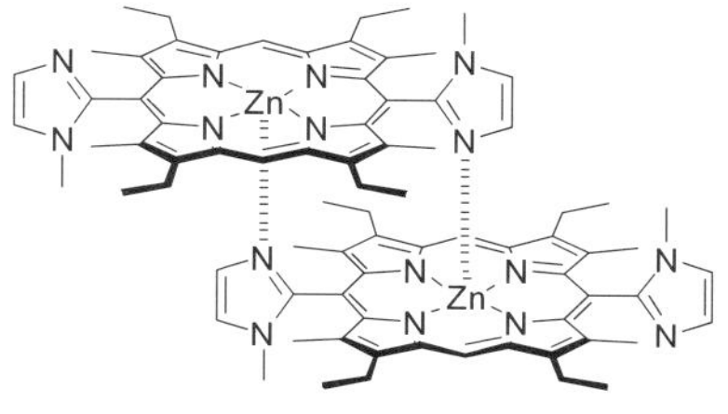

**11**

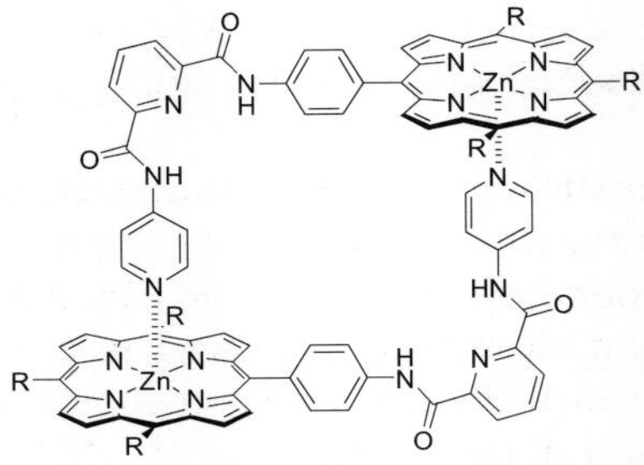

**12**
(R = 4-$C_4H_9$-Ph)

by increasing the angle between the porphyrin plane and the orientation of the pyridine ligand to 150°, while the square tetramer **14** was formed predominantly when this angle was 180° [63]. Concentration-dependent studies revealed that the assemblies **13** and **14** may exist only within a certain concentration range, while open linear oligomers/polymers grow up as preferable species at high concentrations (e.g. > 1 M).

Diverse ligand units and metal centers have been employed to achieve self-cyclized porphyrin assemblies. In the formation of trimeric macrocycle **15**, both coordinative and hydrogen-bonding forces operate cooperatively [64]. Consequently, the stability constant of the resultant assembly in dichloromethane is as high as $10^{13}$ $M^{-2}$. For comparison, the stability constant for a reference trimer system without hydrogen-bonding sites is only around $10^{8}$ $M^{-2}$. Recently, Imamura and coworkers have designed and prepared rhodium(III) pyridylporphyrin tetramer **16** [65]. While the rhodium porphyrin units in the tetramer framework are equivalent in $CDCl_3$ solution, the tetramer in the solid state deviates seriously from an ideal $C_{4h}$ symmetric

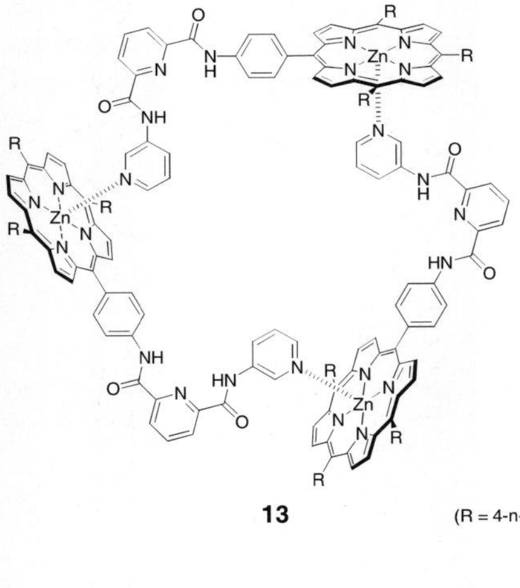

**13** (R = 4-n-$C_5H_{11}$Ph) 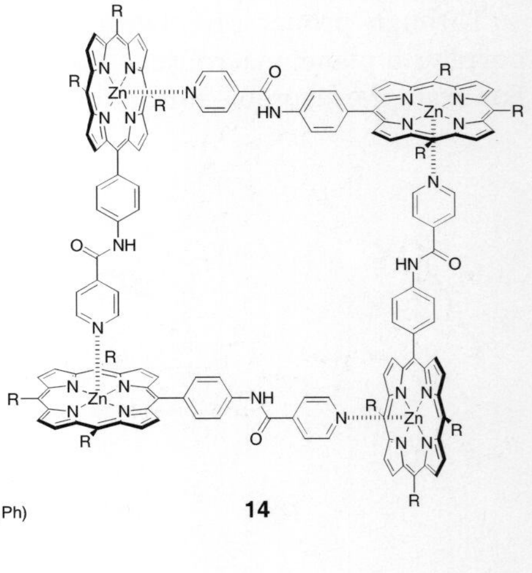 **14**

structure. By contrast, a square assembly of perfect symmetry was obtained by Osuka and coworkers from the self-cyclization of tetrameric zinc pyridyl-porphyrin [66].

The novel metal-mediated self-cyclization approach is extendable to much larger porphyrin macrocyclic architectures. Hunter's group has designed a cobalt porphyrin system, which is equipped with two different but geometrically complementary pyridine ligands. This cobalt porphyrin chromophore self-assembles to the unusually persistent complex **17** with 12 porphyrin monomers arranged in a macrocyclic array [67]. Although most of the commonly used techniques such as mass spectrometry, VPO, and $^{1}$H NMR spectroscopy are not applicable for the analysis of this system, the GPC and molecular modeling results substantiated the formation of the proposed giant macrocycle over the concentration range of 5–500 μM. When the concentration was increased further, higher molecular weight polymers began to emerge. A serious drawback of this system is that cobalt quenches effectively the porphyrin fluorescence which prevents investigations of the energy transfer properties of this synthetic model system of bacterial light-harvesting complexes. It is a pity that no spectroscopically innocent metal ion (like penta-coordinating $Zn^{2+}$, $Mg^{2+}$) is available which enables the required coordination of pyridine ligands at both axial sites (i.e. hexacoordination).

In photosynthetic bacteria, the key functional unit is composed of a bacteriochlorophyll-*a* dimer in a slipped-cofacial orientation rendered by coordination of imidazolyl residue to the central magnesium ion. Inspired by this topology, dimer **11** was synthesized by Kobuke and Miyaji already in 1994 [58]. Recently, two imidazolyl-functionalized zinc porpyrins have been connected by a *meta*-phenylene bridge to obtain the supramolecular assembly **18**, which incorporates 12 porphyrin dyes [68]. Although several oligomeric species of large molecular weight were formed initially

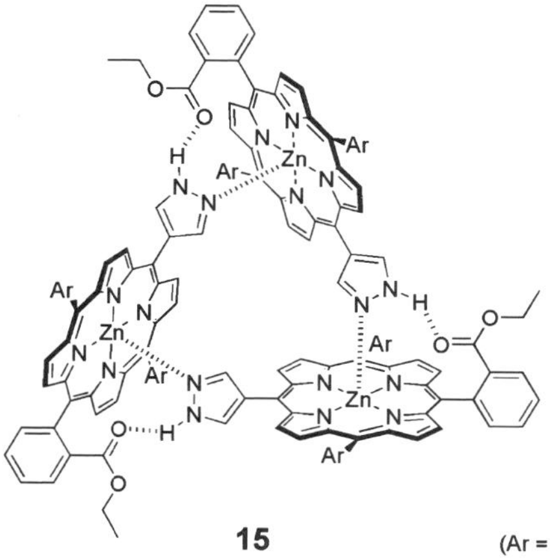

**15** (Ar = 4-$CH_3$Ph) **16**

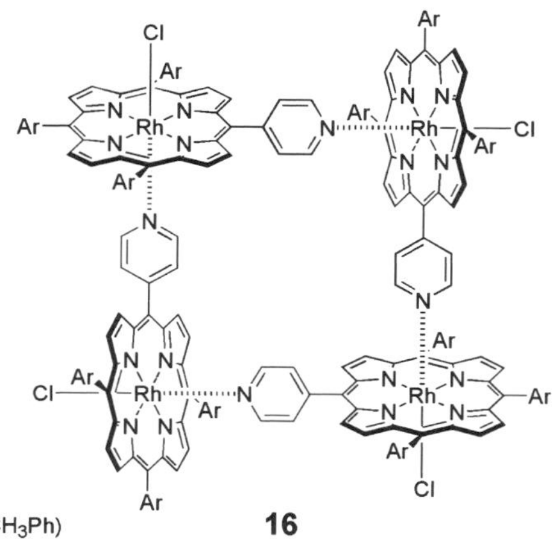

upon self-assembly, the macrocyclic hexamer **18** was evolved as the dominant supramolecule after reorganization under high-dilution conditions, as confirmed by GPC results. Small-angle X-ray scattering measurements showed that **18** possesses diameters of 42.36 and 40.26 Å according to sphere and cylinder approximations, respectively. These values agree well with the estimation of ca. 41 Å for the outer diameter of cyclic hexamer from the molecular mechanics calculation. More recently, porphyrin macrocycles composed of five and six units of *meta*-phenylene-bridged imidazolylporphyrinatozinc(II) dimers have been synthesized by self-assembly followed by ring-closing metathesis. By this means it was possible to permanently fix the macrocyclic array by covalent bond and to enable spectroscopic studies at any concentration and in diverse solvents [69]. This has been taken as an additional proof for the supramolecular macrocyclization of the so-

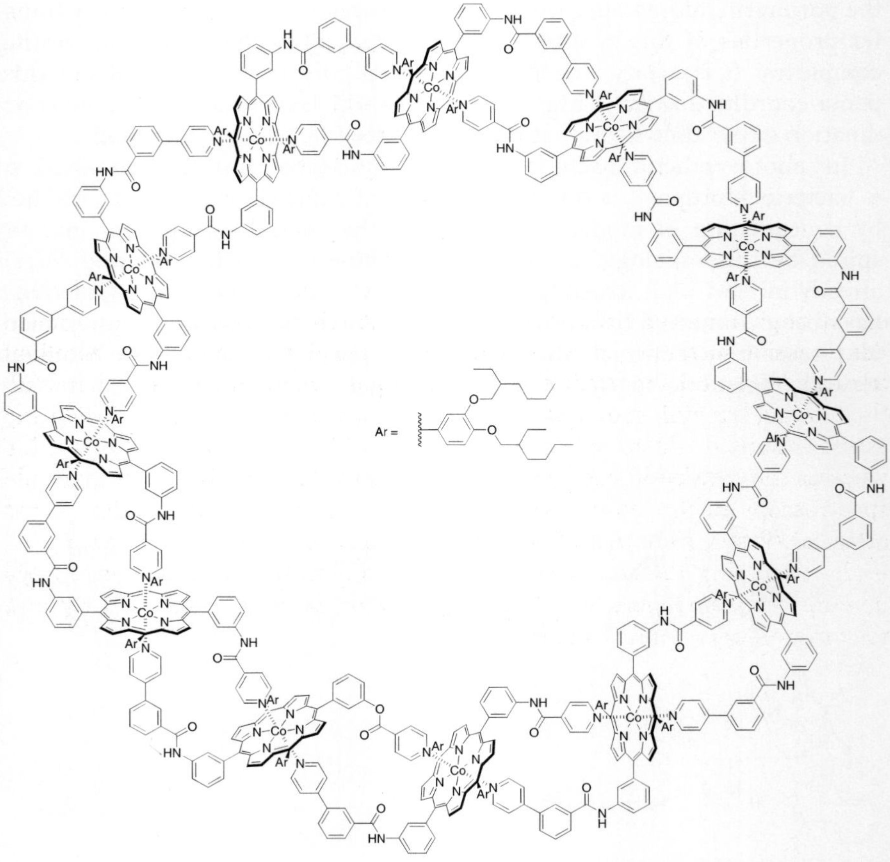

**17**

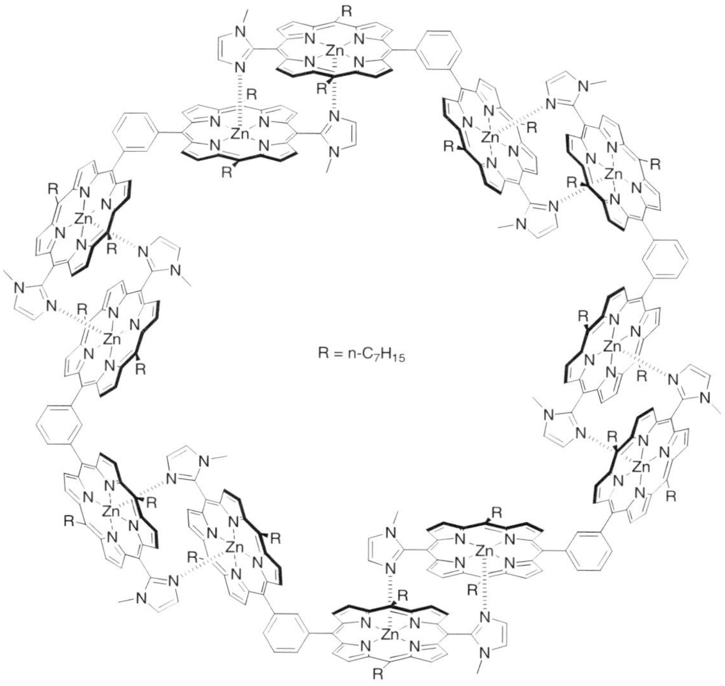

R = n-$C_7H_{15}$

**18**

called *m*-gable porphyrins to structure **18**. As the Zn ion does not quench the photoluminescence of porphyrins, these assemblies seem to be excellent model systems for the cyclic light harvesting arrays of purple bacteria. With the binding strength provided by two nitrogen-zinc interactions, binding site investigations of dilute solution of about $10^{-6}$ M should be easily possible whereas the covalently fixed system might be useful even for single molecule spectroscopy in the femtomolar regime. The successful construction of the artificial system **18** represents an encouraging step toward the mimic of naturally occurring LH complexes in photosynthetic bacteria. It convincingly demonstrates the high effectiveness of the self-assembly approach compared to conventional organic chemistry.

# 3 Metallosupramolecular Polymers

## 3.1 General Aspects

The term "coordination polymer" or "metallosupramolecular polymer" is defined in the literature quite diversely. In this article, we consider a coordination polymer as a multicomponent entity that is held together by metal-ligand coordination bonds, which are strong enough to ensure polymer character not only in the solid state but also in solution. This premise demands a proper balance of binding strength of the metal-ligand interaction to achieve coordination polymerization through reversible processes. For supramolecular polymers, the degree of polymerization $N$ depends on the binding constant $K$ and the concentration of monomers [M] [70, 71]:

$$N \sim (K[\mathrm{M}])^{1/2} \tag{1}$$

According to this relationship, a high degree of polymerization can be achieved at high monomer concentrations by involving interactions of profound binding constants or by using systems with multiple binding sites as well as chelating ligands (for a graphical representation on the relationship between the degree of polymerization and concentration see the Chapter by Ajayaghosh, George and Schenning).

## 3.2 Oligomers and Polymers based on Metal Ion-Pyridine Ligation

The first examples of porphyrin coordination polymers that have been characterized according to our definition in Sect. 3.1 were reported in 1991 by Fleischer and Shachter [72]. Upon metalation of 5-pyridyl-10,15,20-triphenylporphyrin with a $Zn^{2+}$ ion, a self-complementary building block was obtained which readily assembles to polymer **19** as confirmed by concentration dependent UV/vis spectroscopy and $^{1}$H NMR studies. The structure of the polymer **19** in the solid state was established by X-ray analysis. Recent reinvestigations of this system suggest that in solution tetrameric squares prevail, and polymer formation may take place at higher concentration ($> 1$ M) as predicted by computer simulation [12].

A structurally related system was reported by Burrell et al. with the difference that the pyridine ligand is attached to one of the pyrrole rings and bridged by an ethene group which offers *cis*/*trans* isomers [73]. While the *cis* isomer forms a defined dimer, the *trans* isomer self-assembles to a coordination polymer. Both structures were confirmed by single crystal X-ray analysis. UV irradiation (400 nm) of the coordination polymer in the *trans*

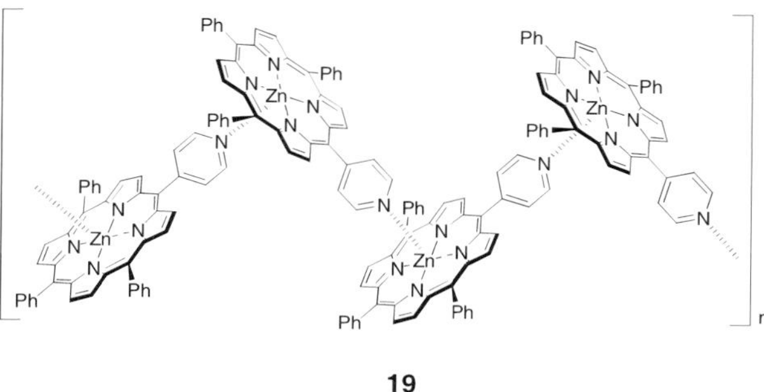


**19**

form, which shows dynamic behavior in solution, leads to the formation of dimers of *cis* isomer by photoisomerization of the double bond.

Higher binding constants are achievable using the same strategies as discussed earlier for the macrocyclization. While the macrocyclic array **17** (in Sect. 2.2) was obtained by attaching two pyridine ligands with spacers of different lengths to the porphyrin scaffold and thereby providing an angle suited for macrocyclization, Michelsen and Hunter [70] showed that the analogous polymer **20** can easily be obtained by choosing an equal length for both pyridine arms. Evidence for the polymeric structure of this assembly was provided by PGSE (pulsed gradient spin echo) NMR studies revealing a significantly smaller diffusion coefficient for polymeric assembly than the dimer reference compound. Furthermore, support for the polymeric structure was obtained by GPC measurements, since the addition of a chain terminator drastically reduced the length of the polymers to short oligomers. The GPC data also confirmed the previously mentioned relationship (see Sect. 3.1) for the present system, thus, an average chain length of $N = 95$ was obtained at a concentration of 7000 μM, while a drastic drop to $N = 12$ was observed at 55 μM.

As is the case for macrocyclic array **17**, the favorable fluorescence properties of the tetrapyrrole dye for polymer **20** are diminished by the paramagnetic cobalt ion. Thus, the very elegant supramolecular construction principle (i.e. two pyridine ligands attached to the *meso*-positions and a central metal ion with two axial coordination sites) could not provide an array with desirable functionality for artificial light harvesting. This problem could be solved again by Kobuke's approach of connecting two zinc porphyrin-imidazole receptor pairs by a covalent bond. Ogawa and Kobuke [74] have shown that the coordination polymer **21** can be obtained by self-assembly of linearly connected imidazolylporphyrinatozinc(II) dimer, while the *m*-phenylene-bridged derivate leads to the respective macrocyclic array **18** (see Sect. 2.2). These examples demonstrate convincingly that the aggregation mode (macrocyclization versus polymerization) of porphyrin derivatives through self-assembly can be controlled in a rational way by the geometry of the building blocks. For the characterization of the polymeric assembly

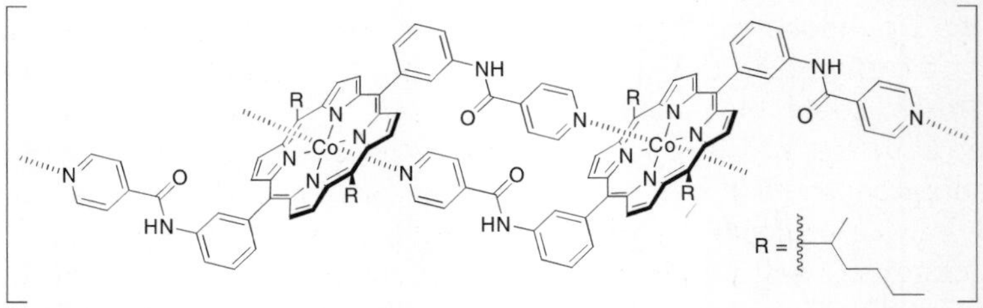

**20**

**21**, GPC proved to be the method of choice and an average length of 150 units was calculated. This linear polymer is soluble in chloroform and no fragmentation was observed when a chain-terminating compound was added indicating high stability of the polymer in this solvent. However, drastic decrease in the polymer chain length was observed when polymeric assembly **21** and a chain terminator were dissolved in a methanol/chloroform (1 : 1) mixture as assessed by GPC analysis. These results suggest that methanol acts as a competitive ligand allowing faster kinetics for the redistribution of the system. The optical properties of this coordination polymer (**21**) are indeed quite interesting since it shows all desired features of J-type aggregation as found in natural light-harvesting chlorophyll dye arrays (cp. the Chapter by Balaban et al.). In particular, the fluorescence of the porphyrin chromophore is not quenched in the polymer system. It exhibits a quantum yield of 0.053 in chloroform which is even higher than that of a comparable dimer complex (0.043 in chloroform) and monomeric ZnTPP (0.03 in chloroform). For structurally related oligomeric compounds large third-order optical nonlinearity was reported by Kobuke and coworkers [75].

Two more examples of coordination polymers have recently been introduced by this group applying $Co^{3+}$ and $Ga^{3+}$ metal ions in a porphyrin framework analogous to the dimer **11**. In both examples the porphyrin units are strongly aggregated which makes these systems potentially interesting for electronic and optoelectronic applications [13, 76].

Twyman and King [77] constructed a $(AA-BB)_n$-type coordination polymer from a rigid bis(zinc porphyrin) unit **22** and a flexible ditopic pyridine

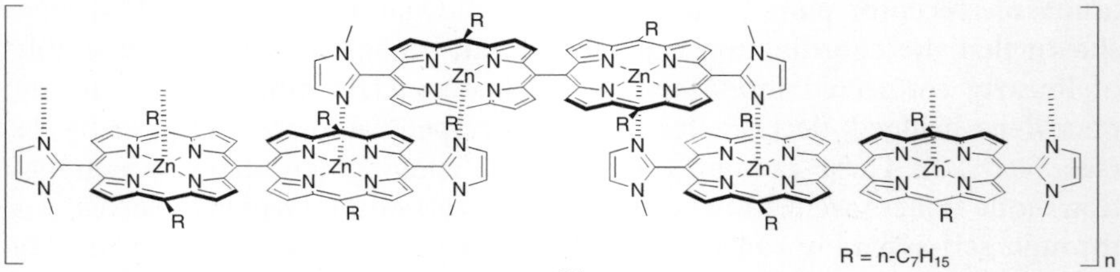

R = n-$C_7H_{15}$

**21**

ligand **23**. By choosing the combination of a rigid building block and a flexible second component, the undesired formation of cyclic assemblies, instead of polymers, could be prevented. Furthermore, the employed flexible ditopic pyridine ligand is not long enough to form a bimolecular complex. As the binding constant of pyridine to zinc porphyrin is relatively low (about $3000\,M^{-1}$ in chloroform), high concentrations are required for the formation of oligomers and polymers from these components, thus, in the case of $(\mathbf{22}\cdot\mathbf{23})_n$ polymerization could be achieved only at a concentration of $10^{-2}$ M.

A noteworthy difference between this $(AA-BB)_n$-type polymer and the self-complementary ($(AB)_n$-type) systems discussed before is that in the former case the polymer length is not only a function of the monomer concentration, but also depends crucially on the ratio of the building blocks A and B. Therefore, long chain polymers can only be obtained by employing an exact 1 : 1 stoichiometry of both components which is not always easy to realize in small-scale reactions.

A further example of a functional dye that is capable of forming coordination polymers was introduced by our group [43]. NMR studies revealed that the highly soluble tetraphenoxy-diazadibenzo-perylene ligand **24** forms complexes with a $Ag^+$ ion. Distinct signal broadening together with an increased solvent viscosity indicated the production of polymeric species at a 1 : 1 ratio. Recently, we have shown [78] that the bidentate ligand **24** coordinates with perylene-bridged diporphyrinatozinc(II) triads (**PDP**) to afford rigid zigzag oligomeric assemblies $(\mathbf{PDP}\cdot\mathbf{24})_n$ similar to $(\mathbf{22}\cdot\mathbf{23})_n$. To our knowledge, this is a rare example of a metallosupramolecular polymer in which three different dye molecules are self-assembled. However, owing to the low binding strength also in this case short chains prevail in dilute solution. An interesting feature of this system is entailed by the energetic proximity of the $S_1$ states of ZnTPP and tetraphenoxy perylene bisimide dyes (see Sect. 1.2) which afforded emission from both dyes to variable extents.

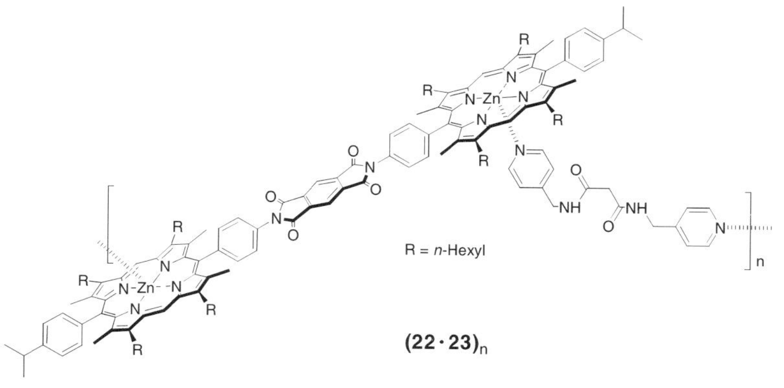

R = *n*-Hexyl

$(\mathbf{22}\cdot\mathbf{23})_n$

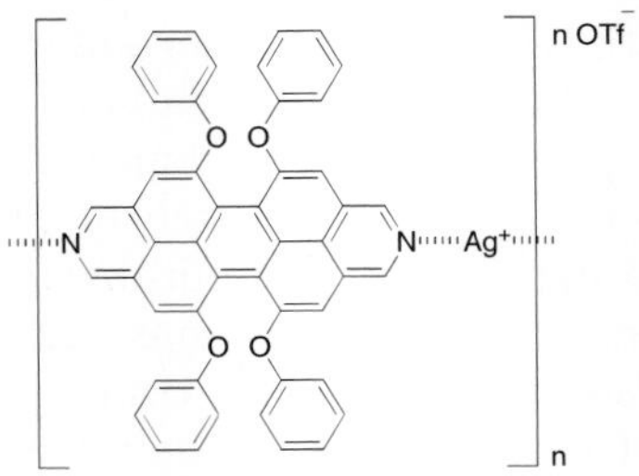

24

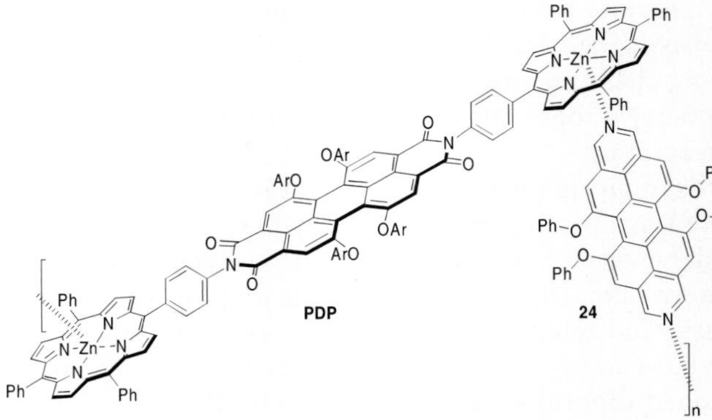

PDP

24

$(PDP\cdot24)_n$

An alternative approach to the previously discussed self-assembly of metal-containing porphyrin dyes for the construction of supramolecular polymers is the coordinative interaction of suitable external metals with aza ligands such as pyridyl groups attached at the periphery of the porphyrin ring. Drain and coworkers [34] have applied this approach to obtain mixtures of discrete linear tapes by reacting equimolar amounts of middle and end groups with geometrically suitable $[PtCl_2(NCPh)_2]$ or $[PdCl_2(NCPh)_2]$ complex. While the treatment of $Pd^{2+}$ species with a linearly coordinating dipyridyl-substituted porphyrin and endgroups containing one pyridine unit resulted in the linearly connected dimer, trimer and tetramer of type **25**, the reaction of $Pt^{2+}$ ions with the respective tetrapyridyl-substituted porphyrin and a ditopic angular endgroup yielded tape-like trimer, tetramer and pentamer of type **26**.

Three-dimensional, linear polymers **27** composed of linked cages were reported by Shinkai and coworkers [79]. Eight pyridine ligands attached to four phenyl groups of the zinc tetraphenylporphyrin are bridged by either chiral or achiral $Pd^{2+}$ complexes to form polymers with linearly aligned hollow capsules. The stoichiometry of the system was determined by UV/vis titration of the free ligand with the $Pd^{2+}$ complex which shows an inflection point of

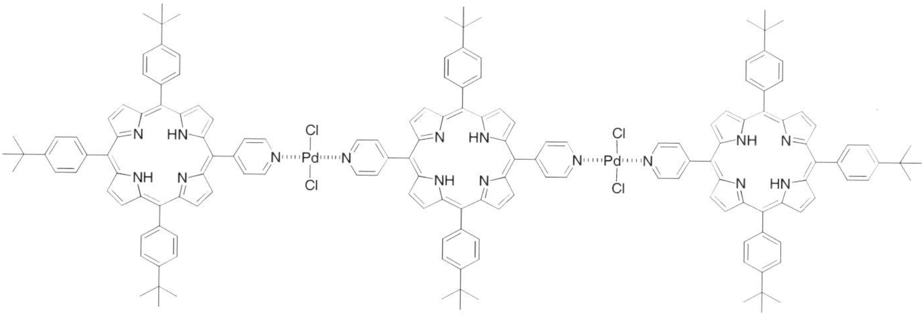

**25**

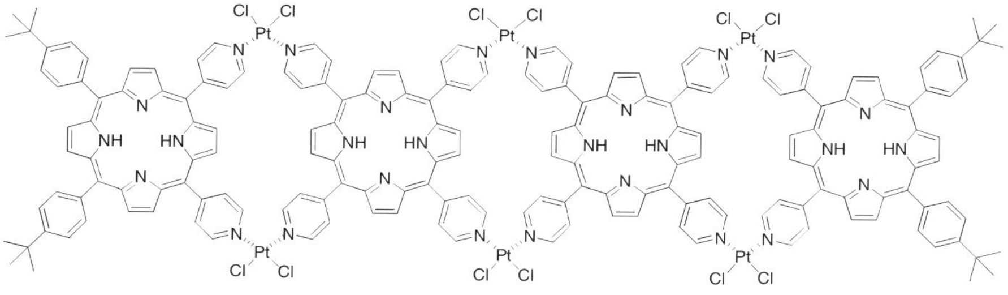

**26**

the titration curve at a metal/ligand ratio of 4 : 1 indicating the formation of a polymeric complex. Dynamic light scattering analysis and NMR studies further substantiated polymer formation and the SEM image (not shown here) revealed the formation of fibrous structures of 20–40 nm in diameter. CD spectroscopy confirmed the formation of helical polymers when a chiral $Pd^{2+}$ complex was used. As these assemblies contain free metalloporphyrin coordination sites, the open pores might be promising for catalysis.

Coordination polymers of the so-called "shish-kebab" structure (**28**) are accessible by connecting metalloporphyrin or metallophthalocyanine dyes through coordinative interaction. Hanack and coworkers prepared the supramolecular coordination polymers **28** by self-assembly of metallophthalocyanines with various ditopic nitrogen ligands such as DABCO, pyrazine and tetrazine [80]. However, for these ligands only metal ions of appropriate size and octahedral coordination, for example $Fe^{2+}$, $Co^{2+}$, $Ru^{2+}$, and $Os^{2+}$ can be used. Charge carrier mobility studies of these polymers in the solid state revealed values between $10^{-10}$ S and $3x10^{-2}$ S strongly depending on the metal center and the nature of the bridging units [80].

Anderson and coworkers [81] have extended this approach of metal-ligand mediated self-assembly of dyes to construct double-stranded, conjugated ladders **30** through cofacial bridging of metalloporphyrins. The zinc porphyrin polymer **29**, which is soluble in chloroform even in aggregated form, is transformed into the ladder complex by the addition of the ditopic ligand 4,4′-

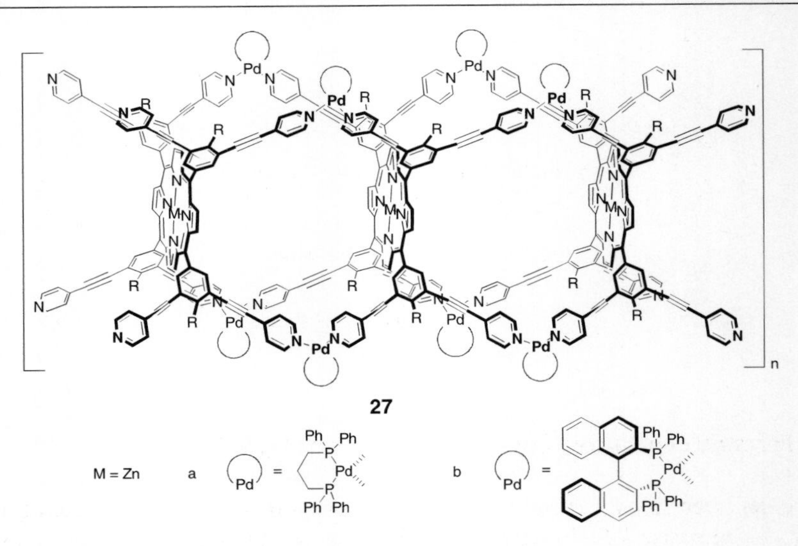


**27**

bipyridine. At high concentrations of bipyridine the duplex dissociates to the single-stranded species **31** with only one pyridine subunit of the bipyridine bound to the zinc center. The complexes **30** and **31** were characterized by UV/vis titration, NMR, GPC, and SANS (small-angle neutron scattering). Interestingly, DFWM (degenerate four-wave mixing) measurements revealed drastic enhancement of nonlinear properties of the ladder complex compared to the uncomplexed polymer. This effect was explained in terms of coordination-induced planarization of the $\pi$-conjugated polymers which increases the effective conjugation length.

M = Fe, Co, Ru, Os

**28**

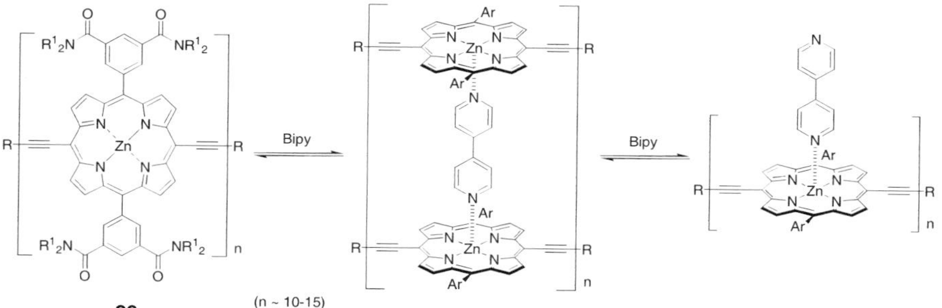

## 3.3
## Dye Polymers Constructed from Chelating Pyridine Ligands

As mentioned above, the most general approach toward high binding constants is the use of chelating ligands such as bipyridine, terpyridine or phenanthroline in combination with free metal ions. Several research groups have studied the optical and photophysical properties of linear rod-type metallosupramolecular polymers constructed from $Ru^{2+}$ and $Os^{2+}$ complexes of the afore-mentioned ligands connected through a great variety of spacers [82–84]. Although these supramolecular systems are photoactive, they are in most cases built from not photoactive ligands. As their interesting properties arise only from the metal ion or its electronic interaction with the ligand, these systems are not discussed in detail here. In addition, complexation of $Ru^{2+}$ and $Os^{2+}$ leads to kinetically inert complexes that do not fulfil our criteria of self-assembly. However, there are a few examples of supramolecular coordination polymers which are obtained from dye building blocks containing such chelating ligands upon coordination to less strongly interacting metal ions.

The 2,2′:6′,2″-terpyridine (tpy) ligand is ideally suited for the formation of coordination polymers due to its favorable structural properties. In contrast to the 2,2′-bipyridine ligand, octahedral tpy complexes can not form $\Delta$ and $\Lambda$ isomers, and as the 4′-position can be easily functionalized it allows the alignment of two residues in an exact 180° angle [82].

A red fluorescent coordination polymer (**32**) was obtained by our research group from a tpy-functionalized perylene bisimide chromophore by complexation with zinc triflate [85]. The polymer **32**, which is readily soluble in chloroform/methanol mixtures and DMF, retains the excellent fluorescence properties of the free ligand and shows reversible binding, thus, the chain length decreases upon addition of an excess amount of $Zn^{2+}$. The polymeric structure was established by $^1H$ NMR using a dimer model compound as reference as well as by DOSY NMR, UV/vis spectroscopy, fluorescence anisotropy measurements, and AFM [86]. Further superstructures were obtained by hierarchical self-organization utilizing the polyelectrolyte character of these coordination polymers [87]. By alternating adsorption of the polyanionic polystyrene sulfonate and the polycationic polymer **32**, multilayers were built-up on quartz substrates and monitored by UV/vis spectroscopy. The multilayers retained the characteristic fluorescence properties of the perylene bisimide fluorophore, but now with a significantly decreased quantum yield.

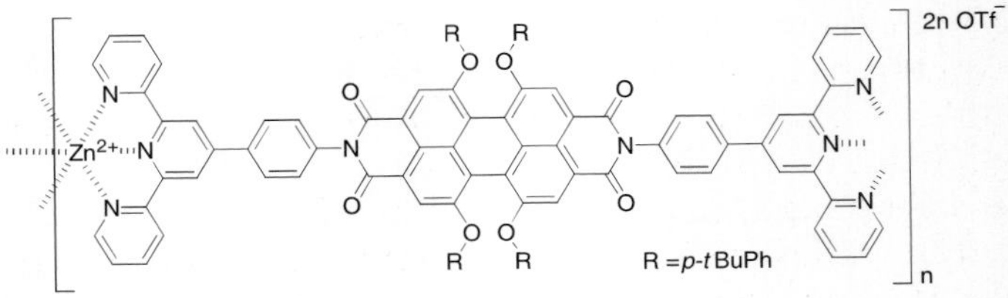

This concept was extended by Che et al. for a series of terpyridine containing chromophores [88]. The respective zinc coordination polymers obtained thereby were investigated by optical spectroscopy, NMR and viscosimetry. These coordination polymers exhibit fluorescence ranging from violet to yellow and two such polymeric compounds could be successfully incorporated into an electroluminescent device. Whereas in polymer **33** the green fluorescence originates from the fluorene unit located between the tpy ligands, the blue fluorescence in polymer **34** stems from the 4′-phenyl-substituted tpy ligand itself, so that in the latter case tpy acts as both the structural and the functional unit.

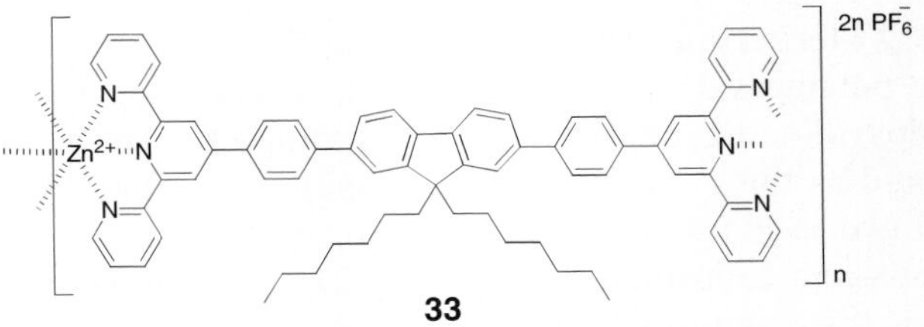

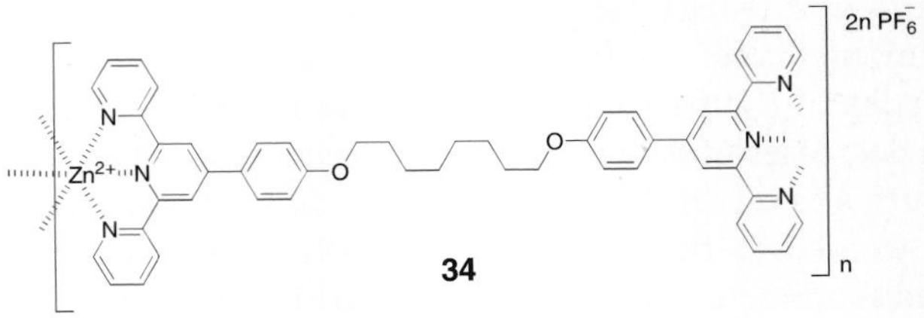

In a similar approach two tpy units attached to an oligo(phenylene vinylene) unit could be polymerized with $Fe^{2+}$ by Meijer and coworkers [30]. Formation of polymer **35** could easily be characterized by UV/vis titration with $FeCl_2$ in a chloroform/methanol mixture based on the characteristic MLCT absorption of the $Fe(tpy)_2$ unit. The binding constant was determined to be $K = 10^7\ M^{-1}$ which convincingly establishes the drastic increase of binding strength in the case of chelating ligands.

Another interesting example of metallosupramolecular dye polymers has been recently published by Groves and coworker [89]. They have constructed the polymer **36** by zinc complexation of two 9-carboxy-1,10-phenanthrolines that are connected to porphyrin. $^1H$ NMR and spectrophotometric monitoring of the complexation reaction revealed that upon addition of zinc ions the phenanthroline complex was formed in the first step prior to the metalation of porphyrin in the second step. In all experiments, an excess amount of zinc salt was used.

## 3.4
## Azo Dye Coordination Polymers

Metal complexes of azo and azomethine dyes are an important class of industrially relevant pigments. Such pigments possess more advantageous properties compared to their parent chromophore including bathochromic shift of absorption, better weather- and lightfastness and enhanced solvent resistance [90].

Coordination polymers of azo dyes were introduced by Suh and coworkers. They have employed *o,o′*-dihydroxyazobenzene and its derivatives as dye building blocks to produce coordination polymers of the basic structure **37** with various metal ions. The polymeric material obtained from the parent dihydroxy-azobenzene and iron metal showed semiconducting proper-

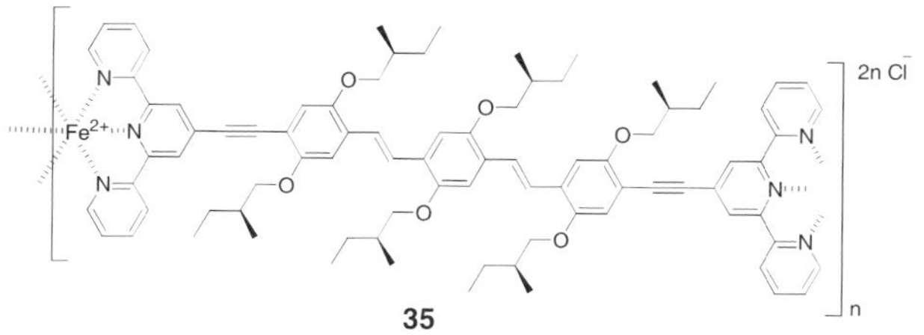

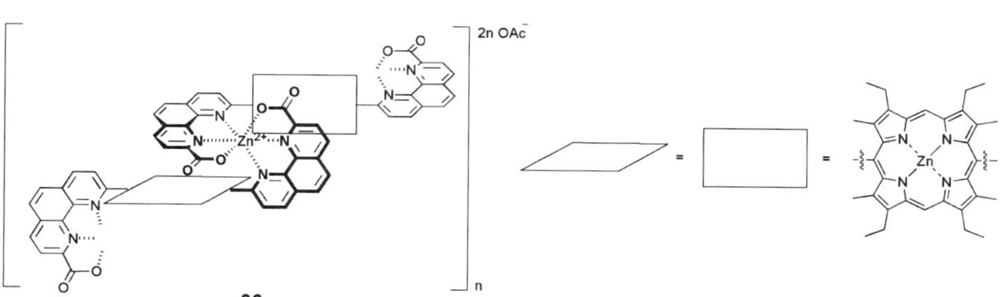

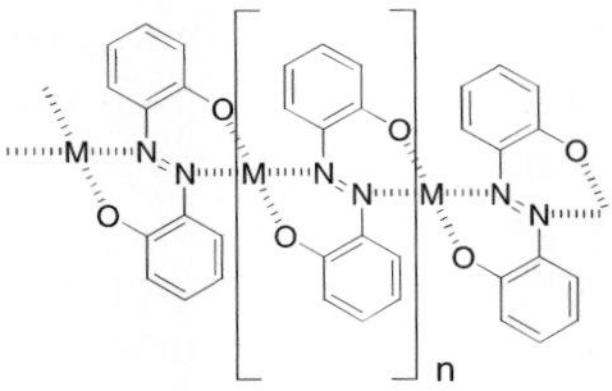

**37**

ties [91]. Substitution of the *o,o′*-dihydroxyazobenzene dye with hydrophilic or hydrophobic groups provided soluble materials for bilayer membranes, which could be stabilized by metal complexation. These metallo-crosslinked bilayer membranes were explored as model systems for metalloproteins [92]. Functionalization of the parent dye with a carboxylic acid group and subsequent complexation with $Ni^{2+}$ ions produced water-soluble polymers [93].

## 4 Miscellaneous Dye Assemblies

### 4.1 Porphyrin-based Capsules and Cages

Self-assembled capsules and cages possess three-dimensional cavities and are able to reversibly encapsulate guest molecules of appropriate size and shape. Therefore, such self-assembled entities have received much attention in supramolecular chemistry [94]. Particularly, functionalized calixarenes have been widely used for supramolecular capsules mediated by either hydrogen-bonding or coordinative interaction which is not a subject of this review article since it deals with assemblies of functional dyes. Nevertheless, during the past few years some capsules and cages derived from porphyrin dyes have been prepared through the metal-coordination strategy. Similarly, as in the case of macrocyclic porphyrin arrays, porphyrin capsules and cages can be constructed by coordination with external metals/ligands or by complementary self-assembly of metalloporphyrins.

Porphyrin-based molecular capsules **38** were obtained from two moles of respective tetra-pyridylporphyrin derivatives and four moles of *cis*-Pd(II) dppp complexes through pyridine-Pd(II) coordinative interaction [95]. The structures of **38** were confirmed by $^1$H NMR and CSI-MS. $^1$H NMR studies revealed that the capsules **38** have a highly symmetrical $D_{4h}$ structure with a large vacant cavity. The zinc porphyrin assemblies are able to accommodate large dipyridine guests such as 4,4′-trimethylenedipyridine with high affinity ($K_S = 2.6 \times 10^6$ M$^{-1}$) by a two-point simultaneous pyridine-zinc(II) interaction.

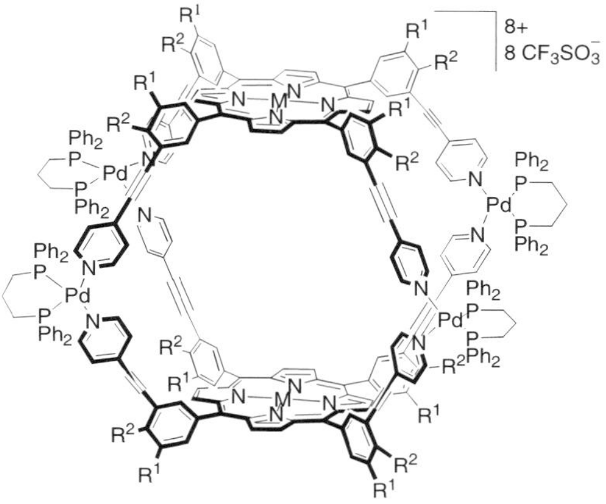

**38**

M = 2H or Zn, $R^1 = OCH_3$, $R^2 = OC_8H_{17}$

During the last decade, extensive studies were pursued by Sanders and coworkers on macrocyclic porphyrin scaffolds as templates that catalyze reactions between bound molecules [96]. Their recent work focuses on the selection and amplification properties of mixed-metal porphyrin cages through a dynamic combinatorial approach [97] by utilizing different coordinative bonds, whose strength has been evaluated by systematic investigation on the self-assembly of phosphorus and nitrogen ligands with metalloporphyrins [98]. A cage structure like **39** was created from two bisphosphine-substituted zinc(II) porphyrins as ligand donors and two rhodium(III) or ruthenium(II) porphyrins as ligand acceptors through metal-phosphorus coordination. Ditopic ligand 4,4′-bipyridine, 3,3′-dimethyl-4,4′-bipyridine or 2,7-diazapyrene serves as the template in this self-assembly process. It was

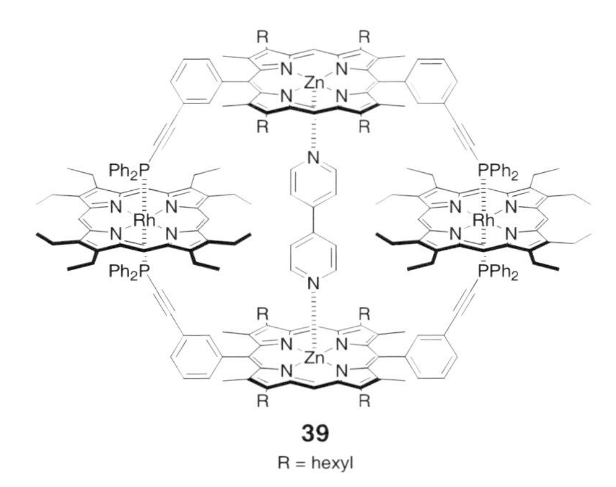


**39**

R = hexyl

found that the template plays a crucial role in the cage formation since rupture of cages takes place upon removal of the template. On the other hand, the cages with sterically demanding porphyrins can be only formed with smaller templates, while in the case of very bulky *tert*-butyl-substituted tetraphenylporphyrin, no cage formation was observed. An extension to mixed dynamic combinatorial libraries showed that only some amplification and limited selectivity were displayed by the various templates due to the formation of mixed cages. In a successive work, the research group of Sanders has convincingly accomplished the quantitative selection and amplification of diverse disulfide-linked cyclic porphyrin oligomers from a dynamic combinatorial library using bisthiol-substituted zinc(II) porphyrin units with appropriate amine donor templates [99].

Box-shaped porphyrin assemblies **40** of high thermodynamic stability were constructed by Osuka and coworkers [66] from *meso*-pyridyl-substituted *meso*-*meso*-linked zinc(II) diporphyrin in which the porphyrin moieties are arranged essentially in a perpendicular orientation. It is noteworthy that the binding sites applied in these examples are comparable to building blocks employed by Hunter and Kobuke (see assemblies **18** and **21** in Sect. 2.2 and 3.2). Only by means of their spatial organization are such diverse architectures accessible. The assemblies **40** were identified by $^{1}$H NMR, GPC and CSI-MS, and a cavity of approximately $10 \times 10 \times 8$ Å was estimated. Because of the remarkable synergistic effect in the self-assembly process, the supramolecular assembly has an association constant of at least $10^{25}\,M^{-3}$ in chloroform and remains intact even under extremely dilute conditions ($1.6 \times 10^{-8}$ M). Interestingly, the complex **40** possesses two enantiomers owing to the different *meso* substituents, and both isomers are present in equal abundance in solution (racemic mixture). This box-shaped assembly exhibits interesting photophysical properties as detailed spectrophotometric investigation revealed [100]. Johnston's group has successfully applied the self-complementary concept to synthesize a porphyrin-based capsule through dimerization of V-shaped bisporphyrin derivatives [101].

Porphyrin-based supramolecular capsules and cages are accessible not only by self-complementary building blocks but also by coordinative bonding of metalloporphyrins with appropriate external ligands. For instance, Reek et al. reported the formation of a molecular capsule based on metal-to-ligand interaction of two dizinc(II) bisporphyrins with a tetraamine ligand [102]. In this entity (structure not shown here), the dendritic tetraamine ligand is encapsulated into the self-assembled spherical superstructure. Furthermore, molecular cages were prepared by apical coordination of the terminal bases of a series of tris(2-aminoethyl)amine derivatives to a trisporphyrin derivative [103]. Baldini et al. have recently reported some calix[4]arenes equipped with two and four zinc porphyrins on the upper rim [104]. Self-assembly of calix-bisporphyrinatozinc(II) with the bidentate ligand DABCO formed

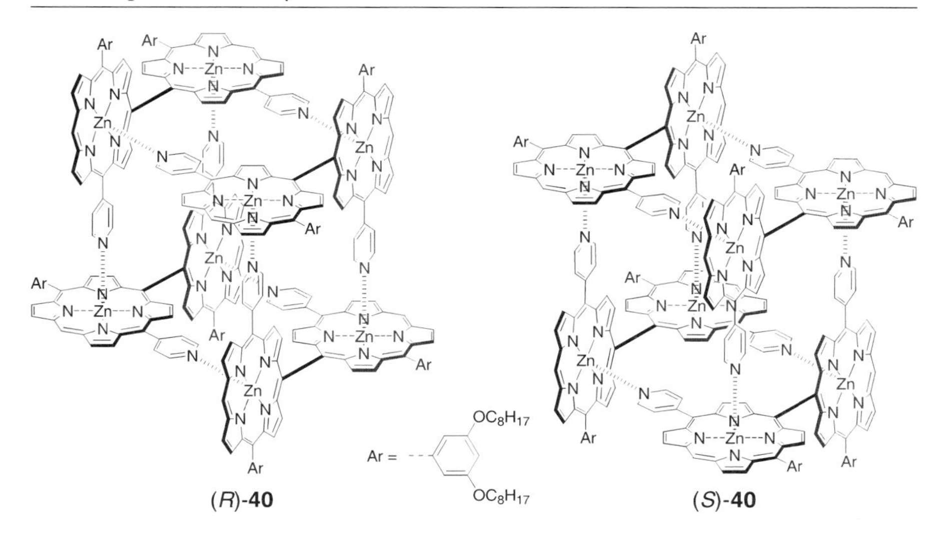


a cage with large cavity that has the potential to function as the supramolecular host [104].

## 4.2
## Other Porphyrin-based Dye Assemblies for Energy/Electron Transfer Studies

Through metal coordination, dyes can be assembled into various topologies such as catenanes [105], rotaxanes [106–108], and nanostructured materials [109–111]. Among them the electron- and energy-transfer systems are of particular interest since they may serve as the simplest models for the photosynthetic machinery of bacteria. Although quite a number of covalently joined systems have been thoroughly investigated in the past years, the noncovalently associated assemblies have drawn significant attention mostly owing to their synthetic simplicity. Compared with the covalent compounds, the assemblies sewed up through noncovalent forces are more readily accessible but they are of lower structural stability leading to fragmentation upon dilution. As outlined in the previous sections of this review, only recently dye assemblies of sufficient stability became available for photophysical studies at suitable concentrations $< 10^{-6}$ M.

Lindsey and coworkers have designed the self-assembled light-harvesting array **41** that contains six zinc-porphyrin units and one free base porphyrin [112]. Due to the cooperative binding of two cofacial zinc porphyrins through the ditipic ligand, more than 90% of the assembly kept intact at the concentration of $10^{-7}$ M. Because zinc porphyrins can serve as energy donors and the free base porphyrin as an acceptor, photoinduced energy transfer in the array could be demonstrated by using steady-state fluorescence spectroscopy. The results revealed that the energy transfer from the coordinated zinc porphyrins to the guest free base porphyrin follows a Förster through-space process, but the efficiency is only about 40%. By contrast, in array **41** the energy transfer occurs essentially quantitatively from uncoordinated zinc porphyrins to pyridyl-coordinated ZnTPPs which exhibit the lowest $S_1$ state among the porphyrin units in this system (cp. Sect. 1.2) [112].

Through coordination motif, a tetra-pyrazinyl porphyrin-based assembly with four dimeric [*meso*-tetrakis(2-carboxy-4-nonylphenyl)porphyrinato]-zinc(II) as the antenna moiety was constructed by Kuroda et al. [113]. The efficiency of energy transfer from the zinc porphyrin-pyrazine complex to the free base porphyrin has been determined to be 82%. Also for this system, the observed efficiency of energy transfer is not so high compared with those reported for covalently linked multi-porphyrin systems [114]. Hunter's group has designed a pentameric porphyrin array, which was built up by the complementary coordination of a tetra-pyridyl porphyrin and two zinc-porphyrin dimers [115]. Steady-state and time-resolved fluorescence spectroscopic studies revealed that the energy transfer efficiency of this array is around 70%.

Besides genuine metal-coordination, the cooperative effect of several weak forces has been utilized to create similar functional assemblies. Weiss and coworker have recently designed the noncovalent couple **42** by utilizing the unique recognition properties of phenanthroline-strapped zinc porphyrin towards *N*-unsubstituted imidazoles [116]. Metal coordination and hydrogen-bonding interaction simultaneously drive the formation of the assembly and

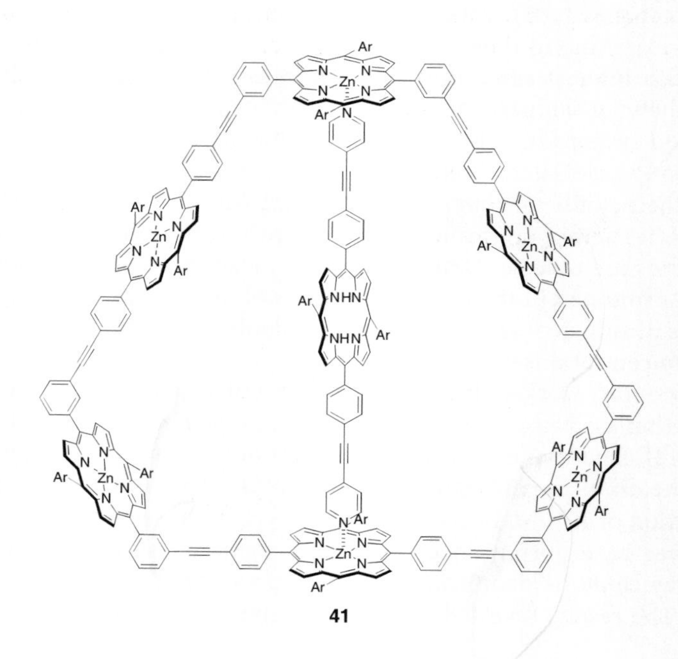

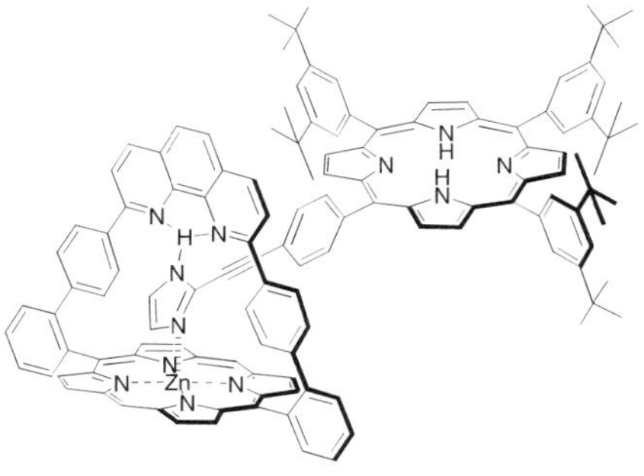

**42**

endow **42** with a high stability in $CH_2Cl_2$. Although the strength of the hydrogen bond is quite weak, it is important to establish a well-defined geometry of complex **42**. As in the case of the above-mentioned cyclic array **41**, the energy transfer from the zinc porphyrin donor to the free base porphyrin acceptor is inefficient.

Imahori et al. have reported the electron transfer assemblies **43**, in which a zinc porphyrin dimer acts as the electron donor and benzoquinone or pyromellitimide serves as the electron acceptor [117, 118]. Because of the two-sites binding of the electron acceptors with high stabilities, the assemblies **43a** and **43b** may form exclusively even at low concentrations typically used for spectroscopic studies. The electron transfer rate constants were found to be $1.6 \times 10^{10}\ s^{-1}$ for **43a** [117] and $2.1 \times 10^{10}\ s^{-1}$ for **43b** [118]. These nearly identical values suggest similar barriers for electron transfer at a comparable distance between chromophore and quencher. As the shortest through-space and through-bond pathways between the chromophore and the quencher in

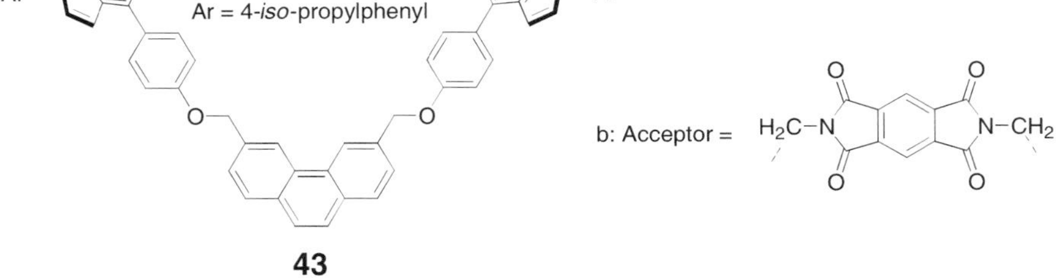

Acceptor

Ar = 4-*iso*-propylphenyl

a: Acceptor =

b: Acceptor =

**43**

**43** are similar, it is not possible to distinguish whether the electron transfer in these systems is occurring through space or through bond.

Such ambiguity can be resolved in the porphyrin-naphthalene bisimide assembly **44** designed by Hunter and Hyde [119]. As in the previous case, the bifunctional ligand bridges the two zinc centers of the porphyrin dimer with a high association constant of about $3 \times 10^8\ M^{-1}$ resulting in a close spatial approach of the free-base porphyrin chromophore and naphthalene bisimide quencher. The through-space separation of the components is less than 10 Å, whereas the through-bond distance around the ring is about 35 Å. Upon excitation, the excited state of the free base porphyrin is quenched by 70% via electron transfer to the naphthalene bisimide moieties. This process has been ascertained to operate through space [119].

Electron-transfer processes have also been investigated in porphyrin-containing [2]-rotaxanes [106, 120]. In these systems, two Zn(II)-porphyrin (ZnP) electron donors were attached as stoppers on the rod, while a macrocycle attached to a Au(III)-porphyrin ($AuP^+$) acceptor was threaded on the rod. By selective excitation of either porphyrin unit, electron transfer could be induced from ZnP to the $AuP^+$ unit that generated the same charge-transfer state irrespective of which porphyrin was excited. Additional metal ions like $Ag^+$ and $Cu^+$ were introduced into the system by coordination of phenanthrolines as shown in structure **45**. When Zn(II)-porphyrin was excited, no effect of $Ag^+$ or $Cu^+$ on the electron-transfer process was observed. However, the excitation of Au(III)-porphyrin enhanced the electron-transfer rate in the presence of $Ag^+$ as well as $Cu^+$. These results show that it is possible to tune the rate of electron transfer between noncovalently linked reactants by appropriate modification of the link.

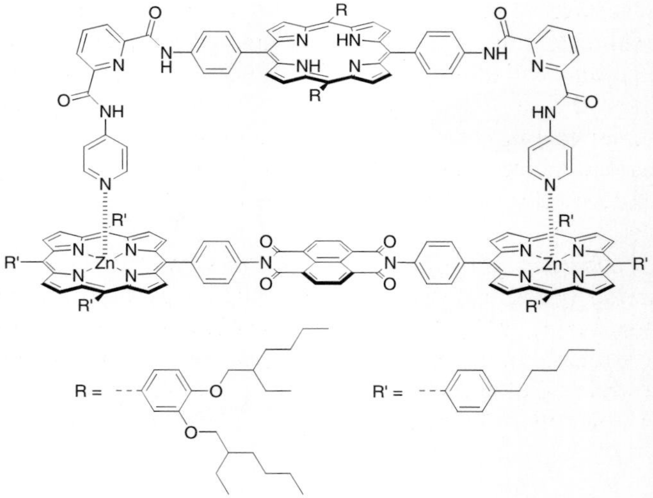

**44**

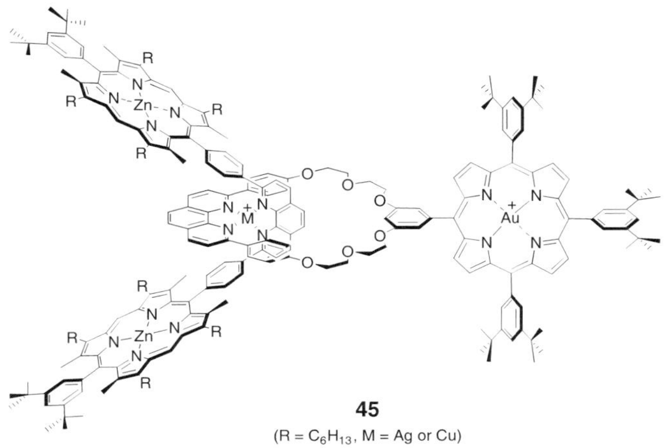

**45**
($R = C_6H_{13}$, M = Ag or Cu)

## 4.3 Metal Ion-Mediated Supramolecular Dye Dendrimers

Supramolecular dendrimers that contain photo- and/or redox-active functions have attracted a great deal of interest due to their potential applications in molecular recognition, catalysis and light harvesting [121–123]. Dye molecules are assembled into dendritic superstructures through metal-ligand mediated coordination for serving as photoactive materials.

Porphyrins have been introduced into dendritic arrays as either the core or the peripheral chromophores through the coordination motif. The group of Reinhoudt and van Veggel synthesized several metallodendrimers containing PyP on the surface by means of noncovalent synthesis [124]. The trinuclear palladium-chloride complex was reacted first with $AgBF_4$ followed by the addition of cyano palladium chloride to produce the first generation metallodendrimer, the latter was reacted further with PyP to afford dendrimer **46**. In a similar manner, a dendrimer containing 12 porphyrins on the periphery was assembled divergently. Porphyrin-containing Pd-Cl complexes were also successfully employed to construct porphyrin-cored metallodendrimers [124].

Recently, spherical porphyrin arrays were assembled through coordinative interaction between a dendrimer bearing 16 pyridyl-terminated units and zinc porphyrins [125]. Detailed investigation revealed that this pyridyl dendrimer is unable to form stable assemblies with simple zinc tetraphenylporphyrin, while a globular macromolecular assembly containing 12 terminal porphyrins is supposed to be formed by the coordination with a porphyrin trimer.

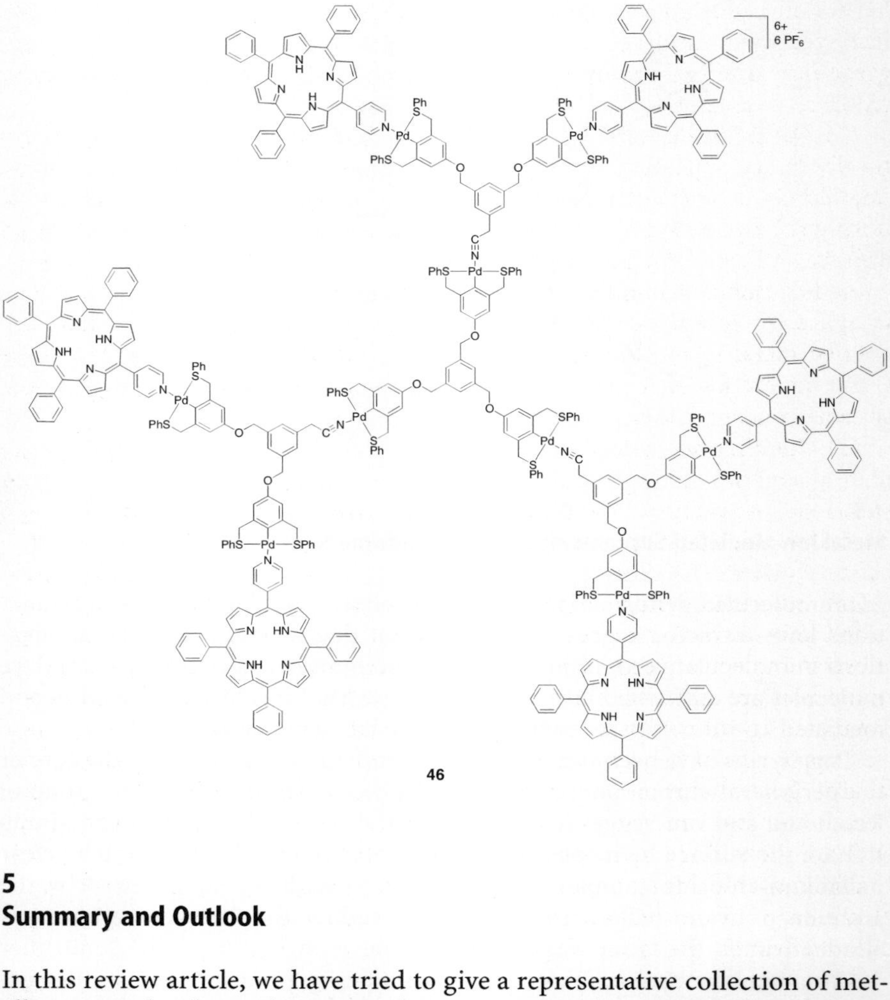

46

# 5
# Summary and Outlook

In this review article, we have tried to give a representative collection of metallosupramolecular architectures that incorporate dye molecules. Looking at these examples – and taking also into account examples given in this book for other dye assemblies like hydrogen-bonded ones – we can draw the following conclusions:

(1) Most work on metallosupramolecular dye architectures has focused on porphyrin systems. This choice was reasonable for two reasons: First, porphyrins are structurally closely related to natural chlorin dyes which give their assemblies the appeal to be "biomimetic". Second, porphyrins are tetradentate ligands which easily complex various metal ions in their center. As most metal ions are able to coordinate a fifth or even a sixth ligand, metalloporphyrins are Lewis-acidic building blocks which can be organized by appropriate ligands to establish otherwise not easily accessible supramolecu-

lar architectures. In particular, coordination of porphyrin-appended ligands like hydroxy or imidazole groups to the central metal ions leads to J-type aggregation to afford highly mobile excitons as demonstrated by the natural chlorin dye assemblies (see the Chapter by Balaban).

(2) Metal-ion directed self-assembly has enabled the design of the structurally most appealing examples of supramolecular dye architectures so far. This is in particular demonstrated by the well-defined cyclic arrays of porphyrin dyes **17** and **18** synthesized by the groups of Hunter [67] and Kobuke [68] and the box-shaped arrays **40** of Osuka [66]. These examples contain significant numbers of chromophores positioned at exact locations in space. To realize such architectures it was of major importance that the applied metal-ligand bonds are highly directional and exhibit just the right Gibbs interaction energy required to direct the reversible self-assembly at reasonable concentrations.

(3) There are only two strategies which enable the formation of supramolecular assemblies in highly dilute solution which are a necessity for most spectroscopic studies. Thus fluorescence spectroscopy is typically done at micromolar concentration and single molecule spectroscopy even at femtomolar concentration. The first approach relies on receptors with multiple noncovalent interaction sites or large complementary van der Waals surfaces as found in many biological systems but which are difficult to design in artificial substrate-receptor pairs. The second approach relies on a single highly directional bond of sufficient binding strength. Such bonding is only realized for the metal-ligand coordinative bonds. The zinc(II) ion-terpyridine interaction was introduced recently as a particularly promising unit for supramolecular dye chemistry as it combines a high binding strength with kinetic lability (which enables formation of thermodynamic structures under equilibrium) without quenching the fluorescence of appended dyes [85–87]. Thus it can be considered as a strong supramolecular "glue" similar to the extended complementary surfaces of proteins in nature. In addition, like proteins Zn ions do not affect the electronic functionality of the dye assembly.

After these positive statements on the achievements in the field of metallosupramolecular dye assemblies we may, however, now raise some critical points which will draw ones eyes towards challenges for the future:

(4) From the examples shown it became clear that most studies have been motivated so far from the structural point of view. As a typical example, light-harvesting chlorophyll dye assemblies motivated researchers to design porphyrinoid model systems. For synthetic reasons chlorin dyes were substituted by porphyrin dyes and the magnesium ion was substituted for other metals like cobalt (e.g. Hunters cyclic array **17**). As a consequence even though the metal-assembled structures look quite related to the natural counterparts, the properties are so different that photophysical studies on these assemblies have not even been initiated (because cobalt totally quenches the fluorescence of the porphyrins). This example clearly points towards our future goal

of directing our scientific attention towards the realization of supramolecular functionalities and applications of metallosupramolecular dye assemblies in sensing, solar energy conversion, electronics and photonics. Chances are quite promising if we consider the wide applications of, for example, metallophthalocyanines as pigments [90], photoconductors [4] or $n$-type semiconductors [7] or ruthenium, iridium or osmium pyridine complexes as photo- and electroluminescent materials [126, 127]. Therefore, application of metallosupramolecular strategies to these building blocks might enable exciting new functional properties on the nano- and mesoscopic scale.

(5) The field of supramolecular photochemistry has focused quite strongly on multichromophoric structures that are formed by purely covalent bonds or strong and inert coordinative bonds. In part this might be for historic reasons due to the very inspiring work by Balzani, Scandola and others already done in the late 1980s and early 1990s [128]. Other reasons are given by the fact that the investigation of covalent systems is less demanding as the once characterized compound at millimolar concentration (using the NMR technique) prevails to exist also under more dilute conditions where the optical properties are investigated. Nevertheless, highly interesting properties arising from $\pi$–$\pi$-aggregation of dyes, which are meanwhile recognized to be of utmost importance for the functional properties of natural light-harvesting dye assemblies as well as organic electronics, cannot be realized by covalent or inert coordinative bonds.

(6) If metal coordinative bonds are taken as a primary noncovalent interaction of sufficient strength the incorporation of additional weaker noncovalent interactions like hydrogen-bonding and $\pi$–$\pi$-stacking might be useful to fine-tune desired functional properties like excitonic interactions. This approach is clearly outlined by the chlorin dye assemblies found in green bacteria and is discussed in the Chapter of Balaban et al. In these dye assemblies the advanced functionality of highly mobile excitons is provided by the most beautiful arrangement of chlorin dyes as J-type aggregates in highly ordered tubules [129, 130].

Moreover, the major advantage of metallosupramolecular chemistry has not yet been systematically developed, neither in the field of dye assemblies nor in general. This major advantage is given by the fact that combinations of metal ions and ligands can be found with any desirable binding strength (thermodynamic property), exchange rate (kinetic property) as well as optical and electronic functionalities. Thus, metallosupramolecular chemistry offers the tools to synthesize large well-defined architectures by self-assembly under proper thermodynamic control in quantitative yield at elevated temperature (and if necessary in special solvents). After isolation the obtained structures may be fully inert and can be easily handled as long as the temperature remains low enough (or the particular solvent is avoided). No other single noncovalent interaction is able to provide such useful possibilities for materials science!

Based on such a multitude of new avenues we are convinced that the field of metallosupramolecular dye assemblies can evolve significantly during the coming years and enable novel functional and materials properties. Nevertheless, it is important that future research goals are defined by functional demands, whereas over the last decade research was more driven by the need to develop the methodology for gaining control of supramolecular structure.

**Acknowledgements** Our coworkers and cooperation partners are acknowledged with great appreciation for their contributions presented here. C.C. You thanks the Alexander von Humboldt Foundation for a postdoctoral fellowship (2002–2004). Generous financial support of our research work by the Deutsche Forschungsgemeinschaft (grant project: Wu 317/3-1), the Volkswagen Foundation (Priority Program "Physics, Chemistry and Biology with Single Molecules") and the Fonds der Chemischen Industrie is gratefully acknowledged.

## References

1. Whitesides GM, Mathias JP, Seto CT (1991) Science 254:1312
2. Hu X, Ritz T, Damjanovič A, Autenrieth F, Schulten K (2002) Q Rev Biophys 35:1
3. Pullerits T, Sundstöm V (1996) Acc Chem Res 29:381
4. Law KY (1993) Chem Rev 93:449
5. Jones RM, Lu L, Helgeson R, Bergstedt TS, McBranch DW, Whitten DG (2001) Proc Natl Acad Sci USA 98:14769
6. Würthner F (2001) Angew Chem Int Ed 40:1037
7. Newman CR, Frisbie CD, da Silva Filho DA, Brédas J-L, Ewbank PC, Mann KR (2004) Chem Mater 16:4436
8. Kulkarni AP, Tonzola CJ, Babel A, Jenekhe SA (2004) Chem Mater 16:4556
9. Wöhrle D, Meissner D (1991) Adv Mater 3:129
10. Coakley KM, McGehee MD (2004) Chem Mater 16:4533
11. Li X, Sinks LF, Rybtchinski B, Wasielewski MR (2004) J Am Chem Soc 126:10810
12. Satake A, Kobuke Y (2005) Tetrahedron 61:13
13. Nagata N, Kugimiya S, Fujiwara E, Kobuke Y (2003) New J Chem 27:743
14. Leininger S, Olenyuk B, Stang PJ (2000) Chem Rev 100:853
15. Swiegers GF, Malefetse TJ (2000) Chem Rev 100:3483
16. Holliday BJ, Mirkin CA (2001) Angew Chem Int Ed 40:2022
17. Würthner F, You CC, Saha-Möller CR (2004) Chem Soc Rev 33:133
18. Würthner F (2004) Chem Commun 1564
19. Bookser BC, Bruice TC (1991) J Am Chem Soc 113:4208
20. Strachan JP, Gentemann S, Seth J, Kalsbeck WA, Lindsey JS, Holten D, Bocian DF (1997) J Am Chem Soc 119:11191
21. Gogan NJ, Siddiqui ZU (1970) J Chem Soc D 284
22. Gvishi R, Reisfeld R, Burshtein Z (1993) Chem Phys Lett 213:338
23. Langhals H, Karolin J, Johansson LBA (1998) J Chem Soc Faraday Trans 94:2919
24. Bhyrappa P, Bhavana P (2001) Chem Phys Lett 349:399
25. Würthner F, Thalacker C, Diele S, Tschierske C (2001) Chem Eur J 7:2245
26. Salbeck J, Kunkely H, Langhals H, Saalfrank RW, Daub J (1989) Chimia 43:6
27. Vogel GC, Stahlbush JR (1977) Inorg Chem 16:950

28. Würthner F, Thalacker C, Sautter A, Schärtl W, Ibach W, Hollricher O (2000) Chem Eur J 6:3871
29. Rudkevich DM, Verboom W, Reinhoudt DN (1995) J Org Chem 60:6585
30. El-Ghayoury A, Schenning APHJ, Meijer EW (2002) J Polym Sci: Part A: Polym Chem, 40:4020
31. Dobrawa R, Ballester P, Saha-Möller CR, Würthner F (2005) In: Newkome GR, Manners I, Schubert US (eds) ACS symposium book on metal-containing and metallo-supramolecular polymers and materials (in press)
32. Ercolani G (2003) J Am Chem Soc 125:16097
33. Drain CM, Lehn JM (1994) Chem Commun 2313
34. Drain CM, Goldberg I, Sylvain, I, Falber A (2005) Top Curr Chem 245:55
35. Stang PJ, Fan J, Olenyuk B (1997) Chem Commun 1453
36. Schmitz M, Leininger S, Fan J, Arif AM, Stang PJ (1999) Organometallics 18:4817
37. Fan J, Whiteford JA, Olenyuk B, Levin MD, Stang PJ, Fleischer EB (1999) J Am Chem Soc 121:2741
38. Slone RV, Hupp JT (1997) Inorg Chem 36:5422
39. Merlau ML, Mejia MP, Nguyen ST, Hupp JT (2001) Angew Chem Int Ed 40:4239
40. Iengo E, Milani B, Zangrando E, Geremia S, Alessio E (2000) Angew Chem Int Ed 39:1096
41. Iengo E, Zangrando E, Minatel R, Alessio E (2002) J Am Chem Soc 124:1003
42. Iengo E, Zangrando E, Alessio E (2003) Eur J Inorg Chem 2371
43. Würthner F, Sautter A, Thalacker C (2000) Angew Chem Int Ed 39:1243
44. Würthner F, Sautter A, Schilling J (2002) J Org Chem 67:3037
45. Sautter A, Schmid DG, Jung G, Würthner F (2001) J Am Chem Soc 123:5424
46. Würthner F, Sautter A (2000) Chem Commun 445
47. Würthner F, Sautter A, Schmid D, Weber PJA (2001) Chem Eur J 7:894
48. Würthner F, Sautter A (2003) Org Biomol Chem 1:240
49. Sautter A, Kaletaş BK, Schmid DG, Dobrawa R, Zimine M, Jung G, van Stokkum IHM, De Cola L, Williams RM, Würthner F (2005) J Am Chem Soc 127:6719
50. You CC, Würthner F (2003) J Am Chem Soc 125:9716
51. Stang PJ, Cao DH, Saito S, Arif AM (1995) J Am Chem Soc 117:6273
52. Sun SS, Lees AJ (2001) Inorg Chem 40:3154
53. Splan KE, Massari AM, Morris GA, Sun SS, Reina E, Nguyen ST, Hupp JT (2003) Eur J Inorg Chem 2348
54. Sun SS, Anspach JA, Lees AJ (2002) Inorg Chem 41:1862
55. Murguly E, Norsten TB, Branda NR (2001) Angew Chem Int Ed 40:1752
56. Qin B, Yao R, Zhao X, Tian H (2003) Org Biomol Chem 1:2187
57. Moriuchi T, Miyaishi M, Hirao T (2001) Angew Chem Int Ed 40:3042
58. Kobuke Y, Miyaji H (1994) J Am Chem Soc 116:4111
59. Stibrany RT, Vasudevan J, Knapp S, Potenza JA, Emge T, Schugar HJ (1996) J Am Chem Soc 118:3980
60. Gardner M, Guerin AJ, Hunter CA, Michelsen U, Rotger C (1999) New J Chem 23:309
61. Imamura T, Funatsu K, Ye S, Morioka Y, Uosaki K, Sasaki Y (2000) J Am Chem Soc 122:9032
62. Hunter CA, Sarson LD (1994) Angew Chem Int Ed Engl 33:2313
63. Chi X, Guerin AJ, Haycock RA, Hunter CA, Sarson LD (1995) J Chem Soc, Chem Commun 2567
64. Ikeda C, Tanaka Y, Fujihara T, Ishii Y, Ushiyama T, Yamamoto K, Yoshioka N, Inoue H (2001) Inorg Chem 40:3395

65. Fukushima K, Funatsu K, Ichimura A, Sasaki Y, Suzuki M, Fujihara T, Tsuge K, Imamura T (2003) Inorg Chem 42:3187
66. Tsuda A, Nakamura T, Sakamoto S, Yamaguchi K, Osuka A (2002) Angew Chem Int Ed 41:2817
67. Haycock RA, Hunter CA, James DA, Michelsen U, Sutton LR (2000) Org Lett 2:2435
68. Takahashi R, Kobuke Y (2003) J Am Chem Soc 125:2372
69. Ikeda C, Satake A, Kobuke Y (2003) Org Lett 5:4935
70. Michelsen U, Hunter CA (2000) Angew Chem Int Ed 39:764
71. Ciferri A (2002) Macromol Rapid Commun 23:511
72. Fleischer EB, Shachter AM (1991) Inorg Chem 30:3763
73. Burrell AK, Officer DL, Reis DCW, Wild KY (1998) Angew Chem Int Ed 37:114
74. Ogawa K, Kobuke Y (2000) Angew Chem Int Ed 39:4070
75. Ogawa K, Zhang T, Yoshihara K, Kobuke Y (2002) J Am Chem Soc 124:22
76. Ikeda C, Fujiwara E, Satake A, Kobuke Y (2003) Chem Commun 616
77. Twyman LJ, King ASH (2002) Chem Commun 910
78. You CC, Würthner F (2004) Org Lett 6:2401
79. Ayabe M, Yamashita K, Sada K, Shinkai S, Ikeda A, Sakamoto S, Yamaguchi K (2003) J Org Chem 68:1059
80. Hanack M, Dürr K, Lange A, Bacina JO, Pohmer J, Witke E (1995) Synthetic Metals 71:2275
81. Screen TEO, Thorne JRG, Denning RG, Bucknall DG, Anderson HL (2002) J Am Chem Soc 124:9712
82. Sauvage JP, Collin JP, Chambron JC, Guillerez S, Coudret C, Balzani V, Barigelletti F, De Cola L, Flamigni L (1994) Chem Rev 94:993
83. De Cola L, Belser P (1998) Coord Chem Rev 177:301
84. Ziessel R, Hissler M, El-Ghayoury A, Harriman A (1998) Coord Chem Rev 178–180:1251
85. Dobrawa R, Würthner F (2002) Chem Commun 1878
86. Dobrawa R, Lysetska M, Ballester P, Grüne M, Würthner F (2005) Macromolecules 38:1315
87. Dobrawa R, Kurth DG, Würthner F (2004) Polymer Preprints 45(1):378
88. Yu SC, Kwok CC, Chan WK, Che CM (2003) Adv Mater 15:1643
89. Phillips-McNaughton K, Groves JT (2003) Org Lett 5:1829
90. Herbst W, Hunger K (1997) Industrial organic pigments: production, properties, applications. VCH, Weinheim, p 390
91. Suh J, Oh E, Kim HC (1992) Synthetic Metals 48:325
92. Suh J, Moon SJ (1998) Bioorg Med Chem Lett 8:2751
93. Lee K, Suh MP, Suh J (1997) J Polym Sci Pol Chem 35:1825
94. Conn MM, Rebek JJr (1997) Chem Rev 97:1647
95. Ikeda A, Ayabe M, Shinkai S, Sakamoto S, Yamaguchi K (2000) Org Lett 2:3707
96. Anderson S, Anderson HL, Sanders JKM (1993) Acc Chem Res 26:469
97. Stulz E, Scott SM, Bond AD, Teat SJ, Sanders JKM (2003) Chem Eur J 9:6039
98. Stulz E, Scott SM, Bond AD, Otto S, Sanders JKM (2003) Inorg Chem 42:3086
99. Kieran AL, Bond AD, Belenguer AM, Sanders JKM (2003) Chem Commun 2674
100. Hwang IW, Cho HS, Jeong DH, Kim D, Tsuda A, Nakamura T, Osuka A (2003) J Phys Chem B 107:9977
101. Johnston MR, Latter MJ, Warrener RN (2002) Org Lett 4:2165
102. Reek JNH, Crossley MJ, Schenning APH, Bosman AW, Meijer EW (1998) Chem Commun 11

103. Felluga F, Tecilla P, Hillier L, Hunter CA, Licini G, Scrimin P (2000) Chem Commun 1087
104. Baldini L, Ballester P, Casnati A, Gomila RM, Hunter CA, Sansone F, Ungaro R (2003) J Am Chem Soc 125:14181
105. Tong Y, Hamilton DG, Meillon JC, Sanders JKM (1999) Org Lett 1:1343
106. Linke M, Chambron JC, Heitz V, Sauvage JP (1999) Chem Commun 2419
107. Andersson M, Linke M, Chambron JC, Davidsson J, Heitz V, Sauvage JP, Hammarström L (2000) J Am Chem Soc 122:3526
108. Johnstone KD, Bampos N, Sanders JKM, Gunter MJ (2003) Chem Commun 1396
109. Fedorova OA, Fedorov YV, Vedernikov AI, Yescheulova OV, Gromov SP, Alfimov MV, Kuz'mina LG, Churakov AV, Howard JAK, Zaitsev SY, Sergeeva TI, Möbius D (2002) New J Chem 26:543
110. Ajayaghosh A, Arunkumar E, Daub J (2002) Angew Chem Int Ed 41:1766
111. Camerel F, Strauch P, Antonietti M, Paul CFJ (2003) Chem Eur J 9:3764
112. Ambroise A, Li J, Yu L, Lindsey JS (2000) Org Lett 2:2563
113. Kuroda Y, Sugou K, Sasaki K (2000) J Am Chem Soc 122:7833
114. Wagner RW, Lindsey JS (1994) J Am Chem Soc 116:9759
115. Haycock RA, Yartsev A, Michelsen U, Sundtröm V, Hunter CA (2000) Angew Chem Int Ed 39:3616
116. Paul D, Wytko JA, Koepf M, Weiss J (2002) Inorg Chem 41:3699
117. Imahori H, Yoshizawa E, Yamada K, Hagiwara K, Okada T, Sakata Y (1995) J Chem Soc, Chem Commun 1133
118. Yamada K, Imahori H, Yoshizawa E, Gosztola D, Wasielewski MR, Sakata Y (1999) Chem Lett 235
119. Hunter CA, Hyde RK (1996) Angew Chem Int Ed Engl 35:1936
120. Andersson M, Linke M, Chambron JC, Davidsson J, Heitz V, Hammarström L, Sauvage JP (2002) J Am Chem Soc 124:4347
121. Zeng F, Zimmerman SC (1997) Chem Rev 97:1681
122. Bosman AW, Janssen HM, Meijer EW (1999) Chem Rev 99:1665
123. Grayson SM, Fréchet JMJ (2001) Chem Rev 101:3819
124. Huck WTS, Rohrer A, Anilkumar AT, Fokkens RH, Nibbering NMM, van Veggel FCJM, Reinhoudt DN (1998) New J Chem 22:165
125. Ballester P, Gomila RM, Hunter CA, King ASH, Twyman LJ (2003) Chem Commun 38
126. Nazeeruddin MdK, Humphry-Baker R, Berner D, Rivier S, Zuppiroli L, Graetzel M (2003) J Am Chem Soc 125:8790
127. Juris A, Balzani V, Barigelletti F, Campagna S, Belser P, von Zelewsky A (1988) Coord Chem Rev 84:85
128. Balzani V, Scandola F (1991) Supramolecular photochemistry. Ellis Horwood Ltd, Chichester
129. Holzwarth AR, Schaffner K (1994) Photosynth Res 41:225
130. Huber V, Katterle M, Lysetska M, Würthner F (2005) Angew Chem Int Ed 44:3147

Top Curr Chem (2005) 258: 83–118
DOI 10.1007/b135681

Published online: 15 August 2005

# Hydrogen-Bonded Assemblies of Dyes and Extended $\pi$-Conjugated Systems

Ayyappanpillai Ajayaghosh[1] (✉) · Subi J. George[1,2] · Albertus P.H.J. Schenning[2] (✉)

[1]Photosciences and Photonics Division, Regional Research Laboratory, CSIR, 695019 Trivandrum, India
*aajayaghosh@rediffmail.com*

[2]Laboratory of Macromolecular and Organic Chemistry, Eindhoven University of Technology, P.O. Box 513, 5600 Eindhoven, The Netherlands
*a.p.h.j.schenning@tue.nl*



**Abstract** Recent developments in the area of hydrogen-bonded supramolecular assemblies of functional dyes and extended $\pi$-conjugated systems are described. Emphasis is given to the hydrogen-bonded assemblies of dyes such as porphyrins, perylene bisimides, cyanines and azo compounds. In addition, a state-of-the-art summary of the recent developments in the design and properties of hydrogen-bonded supramolecular architectures of extended $\pi$-conjugated systems, that is oligomers and polymers, is presented. Finally, the current status of the design and studies of hydrogen-bonded $\pi$-conjugated gel systems and their application is described.

**Keywords** Dyes · Chromophores · Hydrogen bonds · Self-assembly · $\pi$-Conjugated oligomers · Organogels · Phenylenevinylene

# 1 Introduction

Dyes are molecules with fascinating colors, and are traditionally used in staining a manifold of different materials [1]. Recently, dyes have been extensively used in advanced applications such as ink-jet printing, imaging and in electronics. For these uses, functional dyes are required which are able to form organized supramolecular assemblies, the properties of which can be controlled as a function of the self-assembly process. Supramolecular control over dye arrangement is important for the improved performance of existing devices and to create new dye-based materials with tunable optical and electronic properties. With these views, considerable efforts are being focused toward the modification of the structure of organic dyes to program the self-organization. These studies have generated a wealth of knowledge concerning the design of a variety of materials with intriguing properties.

Hydrogen bonds are the ideal noncovalent interactions to construct supramolecular architectures since they are highly selective and directional [2, 3]. In nature, beautiful examples of these properties are present in DNA and proteins. Hydrogen bonds are formed when a donor with an available acidic hydrogen atom interacts with an acceptor carrying available nonbonding electron lone pairs. The strength depends mainly on the solvent and the number and sequence of the hydrogen-bond donors and acceptors [4]. In order to construct a significant number of desired hydrogen-bonded assemblies, high association constants are required (Fig. 1). In many cases, however, relatively weak hydrogen-bond interactions are used, so that

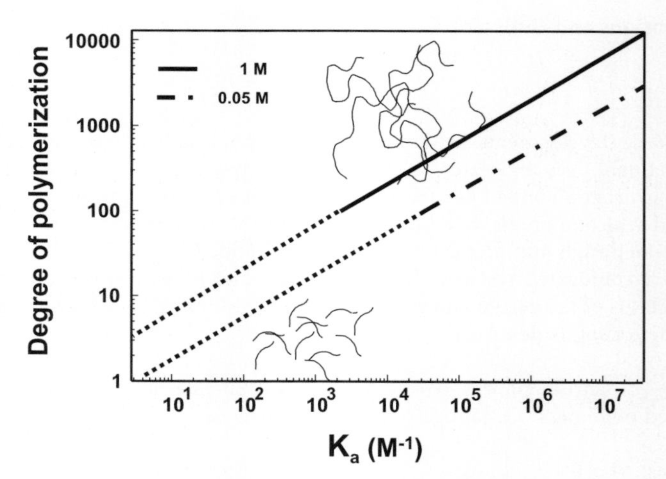


**Fig. 1** Theoretical plot of the relation between the association constant $K_a$ and the degree of polymerization as function of concentration

additional supramolecular interactions are required to obtain stable dye assemblies.

This review will focus on the most significant developments in the last 5 years until the end of 2004 in the field of hydrogen-bond-directed self-assembly of dyes and extended $\pi$-conjugated systems. We will discuss the most common dyes and chromophoric systems, such as porphyrins, perylene bisimides, cyanines and $\pi$-conjugated oligomers and polymers. Though hydrogen-bonded metal complexes have recently been reported in solution [5], which can also be regarded as dye assemblies, this subject is beyond the scope of this survey. For the recent results on self-assembled dyes by other supramolecular interactions, the reader is referred to other chapters in this issue.

## 2 Hydrogen-Bonded Porphyrin Assemblies

Porphyrins represent an important class of dyes with promising application in many areas, such as optoelectronics, chemosensors and catalysis [6, 7]. These dyes exhibit strong absorption and emission in the visible region and show electrochemical activity. The self-assembly of porphyrins is mainly inspired by the photosynthetic systems in nature and is not only important for a better understanding of natural processes, but is also valuable for applications in optoelectronic devices.

In nature, large numbers of porphyrins, that is chlorophyll dyes, are organized in light-harvesting photosynthetic systems of different shape and symmetry. In the case of purple bacteria, the photosynthetic complexes comprise ringlike structures of chlorophyll molecules and the continuously overlapping pigments inside these rings serve as light-harvesting antennae, ensuring high absorption of sunlight. Coherent transfer then funnels the absorbed energy to a reaction center, around which another ring of bacteriochlorophylls is organized, having redshifted absorption. This transfer pathway eventually leads to a charge-separated state inside the reaction center. Green bacteria contain chlorosomes that have several thousand bacteriochlorophylls arranged in rodlike aggregates held together by hydrogen bonding and coordination bonds [8, 9] (Fig. 2). Recently, the ability to characterize these complex systems was greatly enhanced by the construction of artificial systems [10], and the the implementation of powerful NMR [11] and time-resolved fluorescence techniques [12].

Porphyrins have been used as artificial receptors for the recognition of a variety of (chiral) guest molecules by hydrogen-bond interactions [13, 14]. Also energy- or electron-acceptor molecules have been complexed to porphyrin dyes in order to study photoinduced electron and energy transfer reactions. In a pioneering work, Tecilla et al. [15] designed porphyrin-

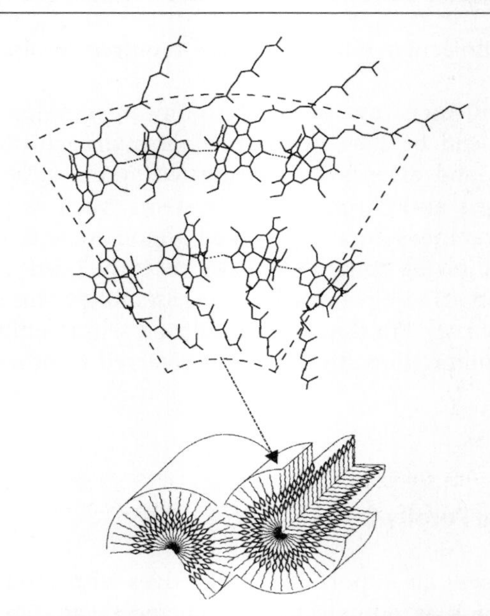

**Fig. 2** The high degree of organization, partly as a result of hydrogen bonding, in the chlorosome organization in green bacteria. (Reprinted with permission from Refs. [8, 9]. Copyright 2000/2001 American Chemical Society)

based multichromophoric hydrogen-bonded dyads and triads comprised of various fluorescent or redox-active naphthalene, ferrocene and dansyl chromophores and studied the energy and electron transfer processes within these complexes. The self-complementary interactions were based on the hexa-hydrogen-bonding complementarity between barbiturate derivatives and two 2,6-diaminopyridine units, linked through isophthalate units ($K_a = 1.1 \times 10^{-6}\ M^{-1}$). Some of the recent examples pertain to porphyrins complexed with phenoxynaphthacenquinone [16] and naphthalenebisimides [17, 18]. In these cases, the supramolecular interactions are purely based on relatively weak hydrogen bond interactions ($K_a < 300\ M^{-1}$).

Monopyrazolylporphyrins self-assemble to form dimers and tetramers (**1**) by hydrogen bonding between the pyrazole units (Fig. 3). The association constant of both processes in chloroform, $K_a = 39\ M^{-1}$ and $K_a = 9.3 \times 10^3\ M^{-3}$, respectively, is, however, low [19].

By increasing the number of the directional hydrogen-bonding interactions, Drain et al. [20] could enhance the stability of the resulting supramolecular porphyrin assemblies. The discrete squares **2** obtained by self-complementary hydrogen bonding between diacetamidopyridyl recognition groups linked to the porphyrin dye showed a $K_a$ value of $7 \times 10^9\ M^{-3}$ (Fig. 3). Such squares could also be obtained upon using two sets of complementary porphyrin building blocks [21]. Supramolecular tapes are obtained if the

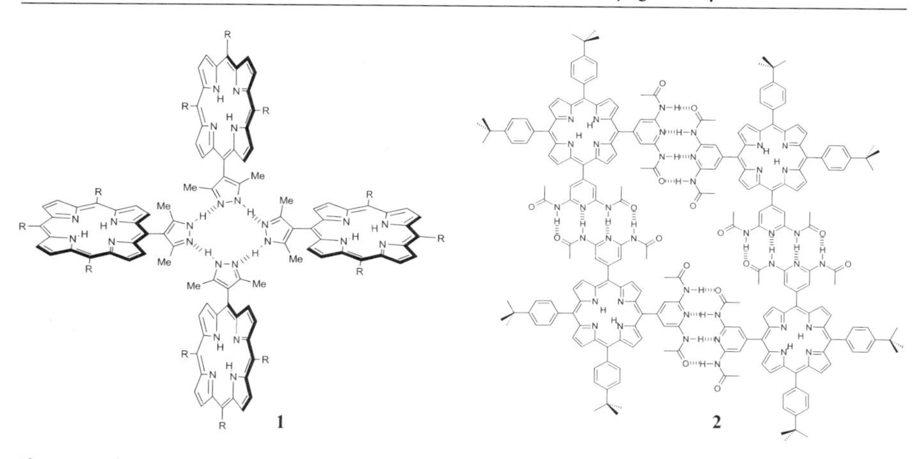

**Fig. 3** Hydrogen-bonded tetramer (**1**) formed by monopyrazolylporphyrins and the self-assembled porphyrin square (**2**) mediated by self-complementary quadruple hydrogen bonds

hydrogen-bonding units at the dye core are substituted differently to form linear arrays.

Clip molecules (**3**) have been complexed to porphyrin dyes (**4**) by hydrogen-bond interactions (Fig. 4) [22]. Four host molecules could be bound to a porphyrin core and, interestingly, the resulting supramolecular complex further self-assembles into dye arrays by additional $\pi$–$\pi$ interactions. Addition of acetone to the porphyrin ensemble results in the dissociation of the arrays.

Formation of the dimeric hydrogen-bonded assembly between carboxylic groups has been extensively used to construct supramolecular assemblies of dye molecules. However, additional supramolecular interactions are required, since the dimerization constant between carboxylic acids is low. Kuroda and coworkers applied additional metal–ligand interactions to construct, in an el-

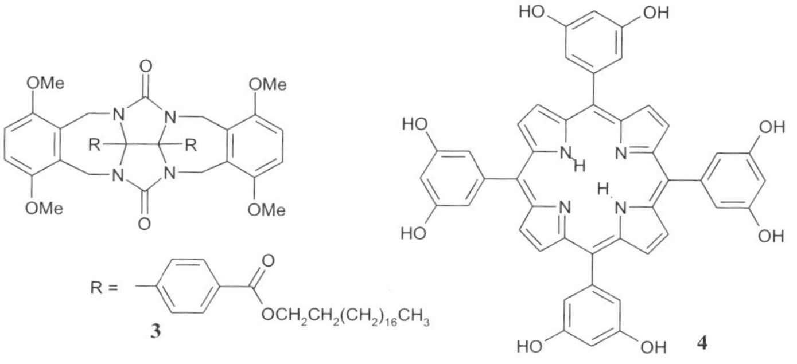

**Fig. 4** Molecular structures of porphyrin host (**3**) and clip guest molecule (**4**)

egant way, nonameric [23] and heptadecameric [24] porphyrin assemblies. Dimeric zinc porphyrins (**5**) are first obtained which bind to a free-base porphyrin focal core bearing four or eight pyrazine ligands at the periphery (**6**), by metal–ligand interactions (Fig. 5). Energy transfer takes place from the zinc porphyrins to the free-base porphyrin core in these antenna complexes.

Additional $\pi$–$\pi$ interactions with guest molecules can also be used to strengthen the hydrogen bonding of carboxylic acids. Recently, supramolecular peapods have been constructed from hydrogen-bonded zinc porphyrins to form nanotubes through fullerene-directed one-dimensional supramolecular polymerization of the zinc porphyrins bearing six carboxylic acid functionalities (**7**, Fig. 6) [25]. The untangled peapods have a high aspect ratio and are thermally stable.

Kobuke et al. [26] have used imidazole hydrogen-bond interactions to self-assemble porphyrin dyes (**8**, Fig. 7). Though the hydrogen-bond interactions were rather weak, stable aggregates could be obtained through hydrogen bonds at multiple sites, combined with hydrophobic and $\pi$–$\pi$ interactions between the porphyrin rings. Bis(imidazolyl)porphyrins were able to form supramolecular stacks in various organic solvents and efficient energy and electron transfer took place from the assemblies to a variety of external acceptors [27]. Upon metallation of the porphyrin core by zinc, the stacks could be further stabilized by additional metal–imidazole interactions [28]. In water, liposomes were observed in which the existence of hydrogen bonds is unclear [29].

The groups of Ihara and Sagawa have prepared fibrous assemblies of porphyrin **9** and pyrene **10** substituted by L-glutamic acid (Fig. 8) [30]. By amide hydrogen bonding, both chromophores self-assemble cofacially into chiral fibers several micrometers in length which can be controlled by temperature. The porphyrins form a physical gel and optical studies confirmed the

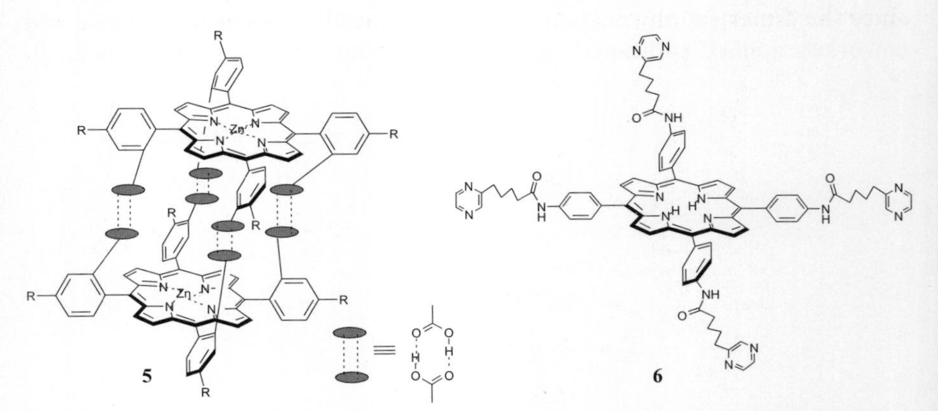


**Fig. 5** Porphyrin assemblies constructed from **5** and **6** using hydrogen bonding and metal–ligand interactions

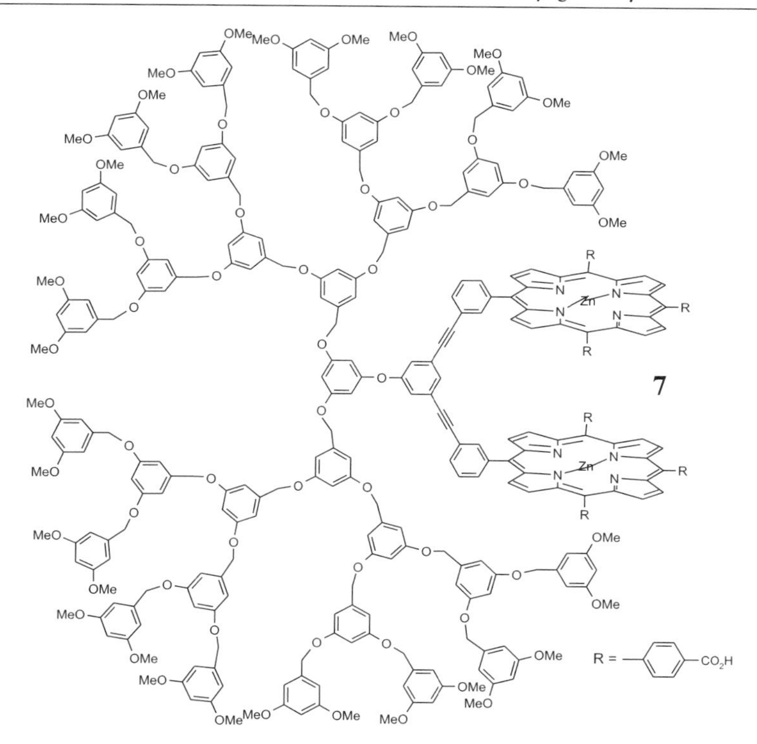


**Fig. 6** Zinc porphyrin 7 functionalized with six carboxylic acid groups

presence of highly ordered aggregates even below the critical gel-forming concentration. Although the pyrene did not produce a gel state at room temperature, it does form ordered aggregates in solution. Preliminary results indicate energy transfer within mixed assemblies in solution from pyrene excimers to porphyrin traps. Moreover, the L-glutamic acid induces helicity into

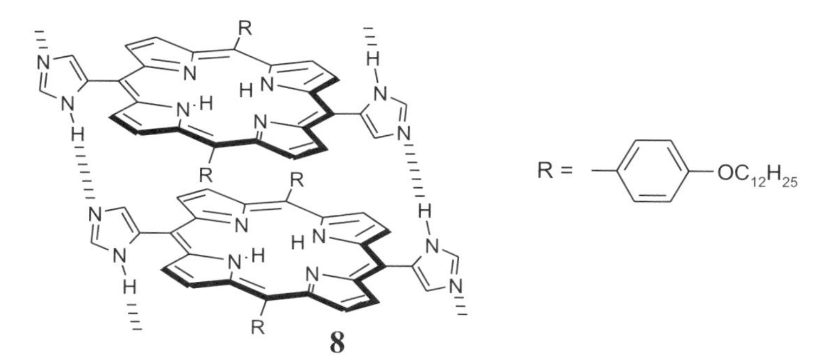


**Fig. 7** Imidazole porphyrin derivatives that self-assemble via $\pi$–$\pi$ interactions and hydrogen bonding

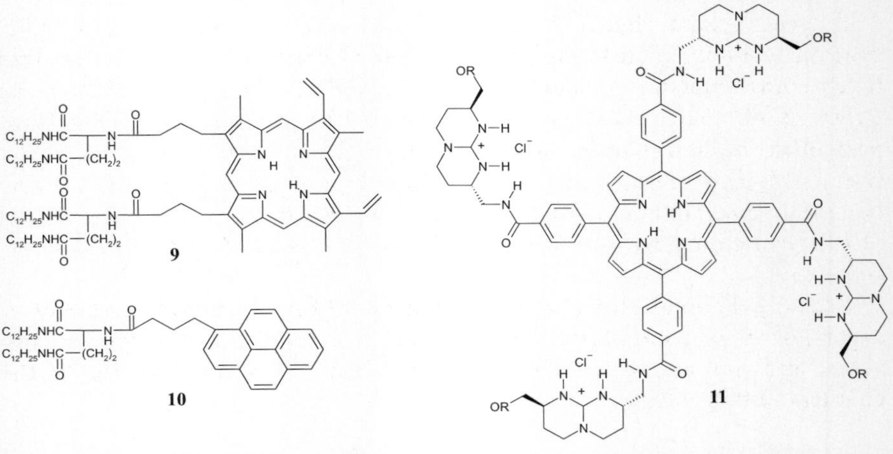

**Fig. 8** Didodecyl L-glutamic acid substituted porphyrin **9** and pyrene **10** and the porphyrin–guanidine conjugate **11**

the fibers, as can be concluded from the observed Cotton effects for both the porphyrin and pyrene systems.

Cationic porphyrins bearing one, two or four guanidines (**11**, Fig. 8) have been self-assembled into highly oriented chiral assemblies using different anions. The self-assembly occurs through hydrogen-bonding and Coulombic interactions between peripheral guanidines and small anions, which are further strengthened by $\pi$–$\pi$ interactions between the porphyrin dyes [31].

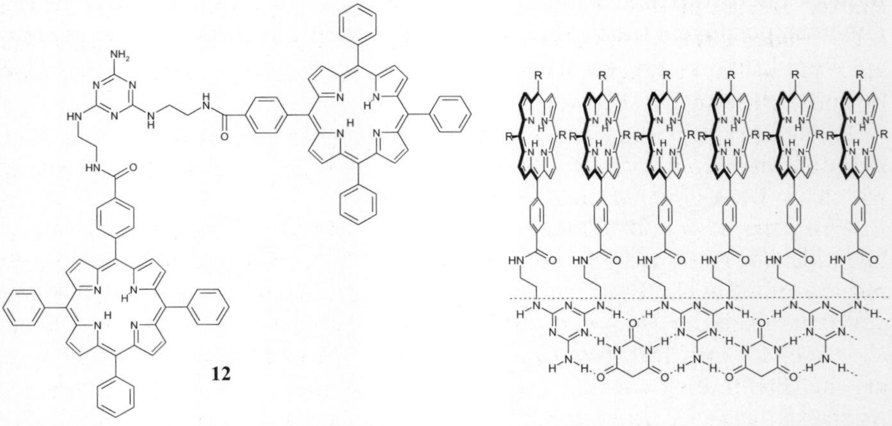

**Fig. 9** Schematic representation of the hydrogen-bonded structure between porphyrin triazine derivatives (**12**) and barbituric acid at the air–water interface

A cyclic hexaporphyrin rosette, composed of three triaminotriazine units bearing two appended tetraphenyl porphyrins or their zinc complexes and three complementary dialkylbarbituric acid derivatives, was reported by Drain et al. [32], which closely resembles the chlorophyll architectures present in the light-harvesting complexes of photosynthetic bacteria. When two porphyrins are attached to a triaminotriazine head group (**12**), stable Langmuir and Langmuir–Blodgett films can be constructed upon addition of complementary barbituric acids (Fig. 9) wich showed enhanced fluorescence [33].

Recently, Shinkai et al. have reported one-dimensional self-assembly of sugar-, urea- or amide-appended porphyrins leading to the gelation of solvents, and this is discussed in detail by Ishi-i and Shinkai in one of the chapters of this volume.

## 3
## Hydrogen-Bonded Perylene Bisimide and Related Assemblies

Perylene bisimides are widely applied as red dyes for industrial purposes owing to their favorable chemical, light and thermal stability. These dyes are also used in electronic applications as n-type semiconductors or in electrophotography. Self-assembled perylene bisimides are of great interest to tailor defined materials with improved or novel optical and electronic properties [34].

Wang et al. [35] have used hydrogen bonding, in an elegant way, to control the stacking interactions between perylene bisimides by incorporating them into the backbone of DNA. In the case of a chromophoric pentamer system surrounded by DNA hairpin structures, the perylene bisimides are stacked as a result of $\pi$–$\pi$ interaction and the hydrogen bonding between the self-complementary hairpin structures (Fig. 10). Upon addition of equimolar amounts of the complementary strand to the hybrid polymer, the perylene bisimide stacking was disrupted.

Related perylene bisimide based DNA conjugates have also been reported in which the dye increased the stablization of a hairpin triplex [36] and in which the DNA directed the dye assembly [37].

Würthner et al. [38] have explored the use of the imide units in perylene bisimides (**13**) as a hydrogen-bonding entity in self-assembled systems. In an initial study the binding constants of monotopic melamines with monotopic perylene bisimides were determined. The association constants range from $K_a = 240\ M^{-1}$ in chloroform to $K_a = 54\,000\ M^{-1}$ in methylcyclohexane, showing that the binding constant increases upon going to a more apolar solvent. As stated in the "Introduction", for hydrogen-bonded supramolecular polymers a high degree of association is required and the authors calculated that for a supramolecular polymer of ten units a concentration of 0.75 mol/L is re-

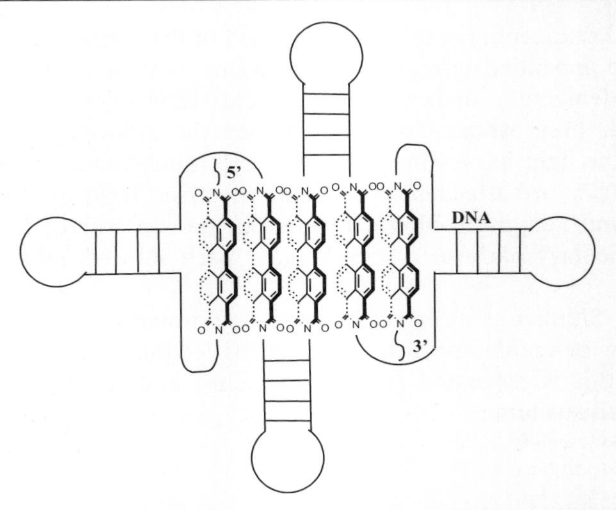


**Fig. 10** A perylene bisimide chromophoric pentamer surrounded by DNA hairpin structures that can be unfolded through binding to the complementary DNA

quired in chloroform, while only 0.003 mol/L is required in methylcyclohexane. Therefore, in an apolar solvent, hydrogen-bonded polymers are formed between ditopic perylene bisimides and dialkylmelamines (**14**) that further self-assemble by additional $\pi$–$\pi$ interactions into an intertwined network of nanostrands (Fig. 11) [39]. By using optically pure melamine derivatives chiral assemblies could be obtained as was concluded from circular dichroism (CD) experiments [40].

Recently, hydrogen-bonded perylene bisimide rows have been observed on a Ag/Si(111) surface by scanning tunneling microscopy (STM). Upon deposition using ultrahigh vacuum conditions rows were formed in which two hydrogen bonds exist between adjacent perylene bisimides. Upon addition of melamine a two-dimensional honeycomb network was formed in which $C_{60}$ clusters could be bound in the pores [41].

Li et al. [42] attached fullerene derivatives bearing a diacylaminopyridine unit (**15**) to perylene bisimides (**16**) by hydrogen-bond interactions (Fig. 12). A fluorescence experiment indicated electron transfer between the perylene bisimide and the fullerene electron acceptor. Perylene bisimides containing diacylaminopyridine units (**17**) were also synthesized and complexed to complementary perylene bisimides (**16**) yielding well-defined long fibers via hydrogen-bond and $\pi$–$\pi$ interactions (Fig. 12) [43]. Based on similar type of interactions bisurea perylene bisimides also gave rodlike structures having a uniform distribution [44].

Related napthalene bisimides have been attached to a self-complementary quadruple hydrogen-bonding unit, that is a 2-ureido-4[1*H*]-pyrimidinone (UPy), yielding hydrogen-bonded dimers (**18**, Fig. 13) [45]. Selective forma-

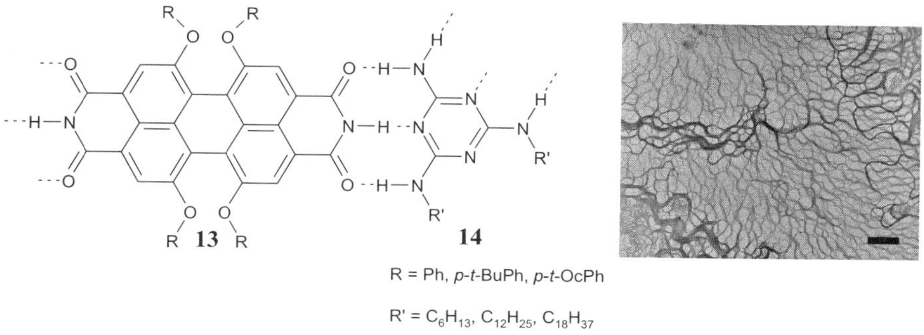

R = Ph, *p*-*t*-BuPh, *p*-*t*-OcPh

R' = $C_6H_{13}$, $C_{12}H_{25}$, $C_{18}H_{37}$

**Fig. 11** Hydrogen-bond interactions between perylene bisimide and dialkylmelamine leading to supramolecular hydrogen-bonded polymers. Transmission electron micrograph of superstructures of an evaporated methylcyclohexane solution of a perylene bisimide/dialkylmelamine mixture. (Reprinted with permission from Ref. [38]. Copyright 1999 John Wiley & Sons)

tion of a heterodimer was achieved by mixing a crown ether equipped with the same hydrogen-bonding motif. The heterodimer formation is the result of donor–acceptor and intermolecular hydrogen-bond interactions. Remarkably, this heterodimer, with a high binding constant, can be dissociated by a complementary naphthyridin derivative that competes strongly with the self-complementary UPy units.

Hydrogen bonds can also be used to control the morphology of unsymmetrically substituted naphthalene bisimides in organic field-effect transis-

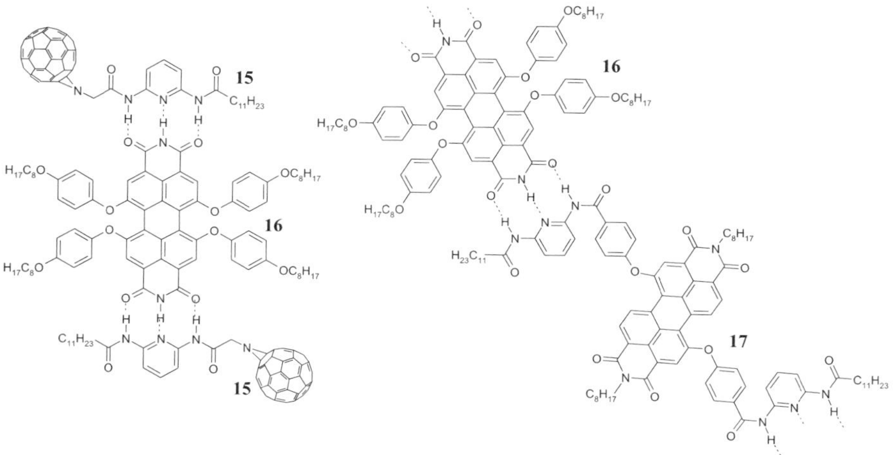

**Fig. 12** Hydrogen-bond interactions between perylene bisimide **16** and diacylaminopyridine derivatives **15** and **17**

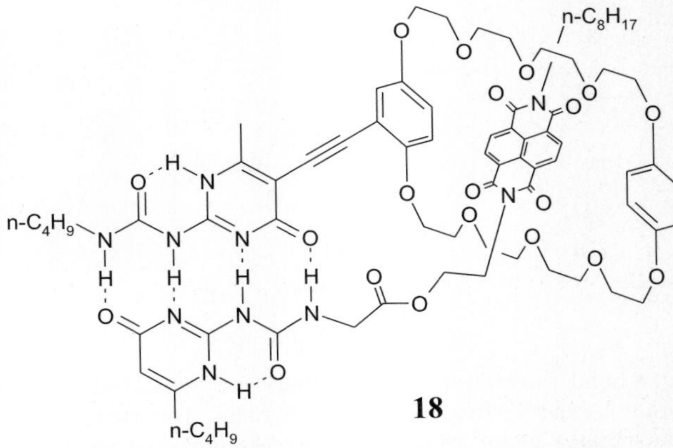

**Fig. 13** A self-assembled heterodimer **18** containing a naphtalenebisimide

tors [46]. On the basis of X-ray and IR studies it was established that dimers are formed through hydrogen bonding between the aminoimide groups, which show that hydrogen bonding can be helpful in the design of high mobility organic semiconductors

Supramolecular hydrogen-bonded systems could also be constructed using ditopic diazabenzoperylenes (**19**), which are accessible from perylene bisimides, and isophthalic acids (Fig. 14) [47]. Since the association constant for hydrogen bonding is low (for benzoic acid $K_a = 140\ M^{-1}$) [48] apolar

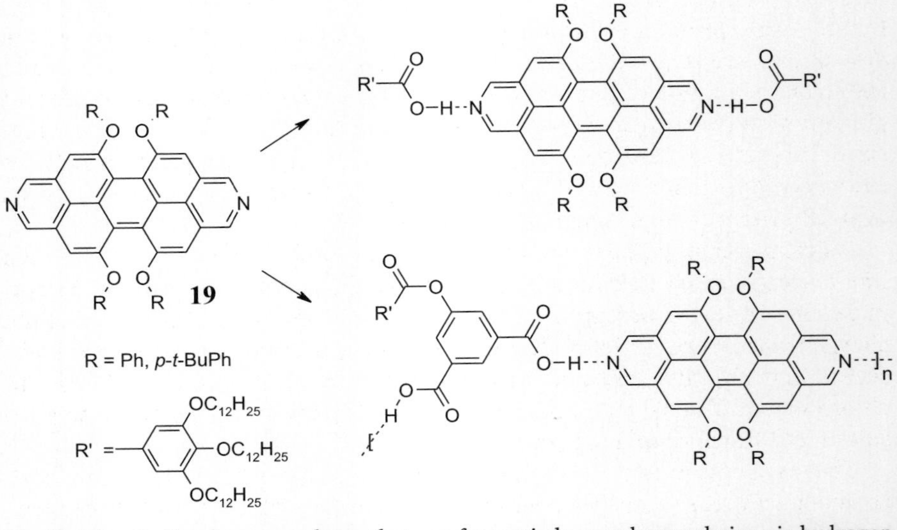

**Fig. 14** Ditopic diazabenzoperylenes that can form triads or polymer chains via hydrogen bonds

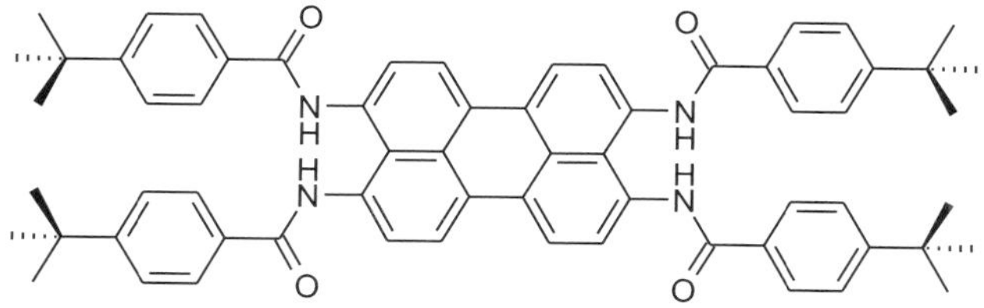

20

**Fig. 15** Perylene **20** that forms rod-shaped structures in dimethyl sulfoxide

solvents were used in order to obtain extended assemblies as a result of additional $\pi$–$\pi$ interactions. Upon addition of tridodecyloxybenzoic acids a liquid-crystalline phase could be introduced [49].

Tetra-amino-substituted perylene can be reacted with acid chlorides to yield tetra-N-substituted perylene **20** (Fig. 15), which form surprisingly rod-shaped structures in dimethyl sulfoxide (DMSO) [50]. On the basis of X-ray data of a related compound the organization is probably controlled by amide hydrogen bonds between adjacent perylenes and the solvent.

## 4 Hydrogen-Bonded Merocyanine and Phthalocyanine Assemblies

Cyanine dyes often used as sensitizers and IR absorbers belong to the class of polymethines and contain a conjugated chain of double bonds with an odd number of carbon atoms between two terminal groups. The end groups affect the absorption and fluorescence spectra and are typically part of a ring system and are then called merocyanines. These cyanine dyes have recently been applied in molecular recognition studies. Polycarboxylate, galactose and glucosamine containing carbocyanine fluorescence probes have been synthesized [51] and showed enhanced uptake of the proliferating tumor cells as evidenced by preliminary in vivo studies [52]. These dyes are of interest as near-IR probes in analyzing tissue samples.

Würthner et al. [53] explored the hydrogen-bond interactions between the imide groups in several merocyanine dyes of the general structure of **21** and melamine **14** (Fig. 16). Remarkably, despite the pronounced differences of the charge transfer properties within the dyes studied minor variations were observed in the binding constants ranging from $K_a = 120\ M^{-1}$ to $K_a = 170\ M^{-1}$ in chloroform. Colloidal assemblies were formed in methylcyclohexane as a result of additional dipolar aggregation [54].

Thermotropic liquid-crystalline merocyanine dyes have also been constructed using the same hydrogen-bonding motif. Combined triple hydrogen-bonding and $\pi$–$\pi$ interactions between a trialkoxybenzene-functionalized melamine and a merocyanine dye yielded a columnar mesophase [55].

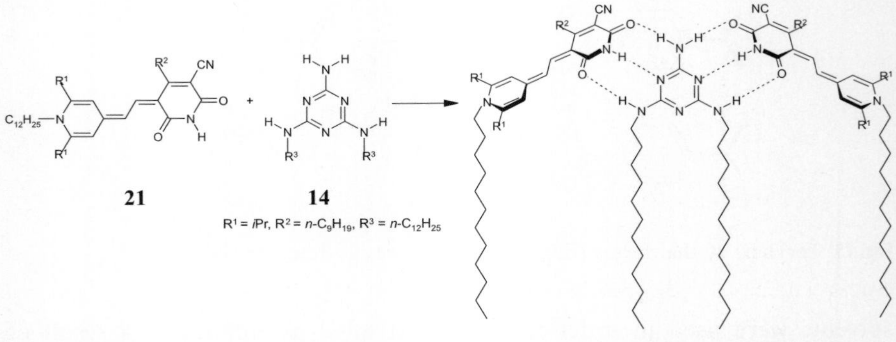

**21** **14**

$R^1 = i\text{Pr}$, $R^2 = n\text{-}C_9H_{19}$, $R^3 = n\text{-}C_{12}H_{25}$

**Fig. 16** Hydrogen-bond interactions between merocyanine dye **21** and melamine **14**

Six merocyanine dyes (**22**) could be arranged together with three complementary calixarene-based dimelamine chiral building blocks (**23**) in a chiral screw sense within a well-defined rosette supramolecular architecture (Fig. 17) [56]. A strong Cotton effect could be observed as a result of excitonic coupling between the chromophores.

A similar complementary hydrogen-bonding system has been used to construct chiral supermolecules and supercoiled self-assemblies in chloroform [57]. The barbituric acid **24** and melamine derivative **14** first form a hydrogen-bonded cyclic hexameric structure that further self-assembles into both left- and right-handed super coils (Fig. 18). When using a dimeric

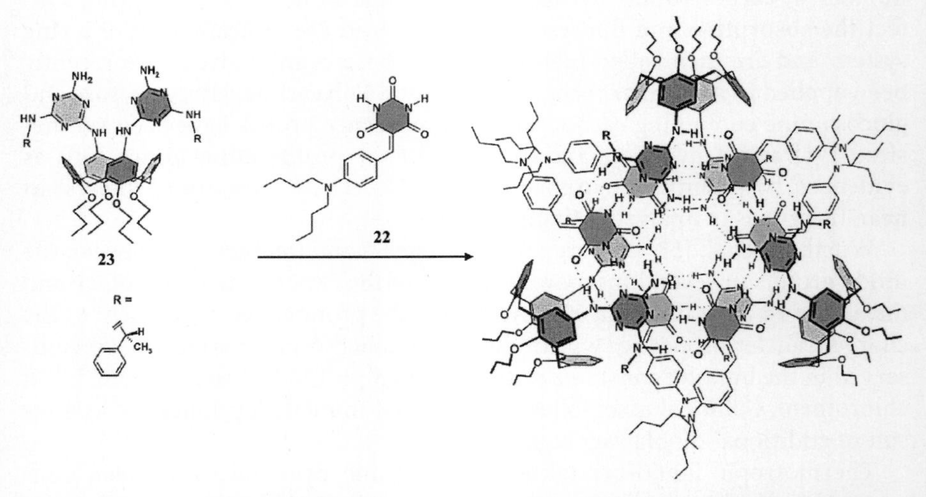


**Fig. 17** Schematic representation of the nine-component hydrogen-bonded assembly between a merocyanine dye (**22**) and a chiral melamine derivative (**23**). (Reprinted with permission from Ref. [56]. Copyright 2001 National Academy of Sciences, USA)

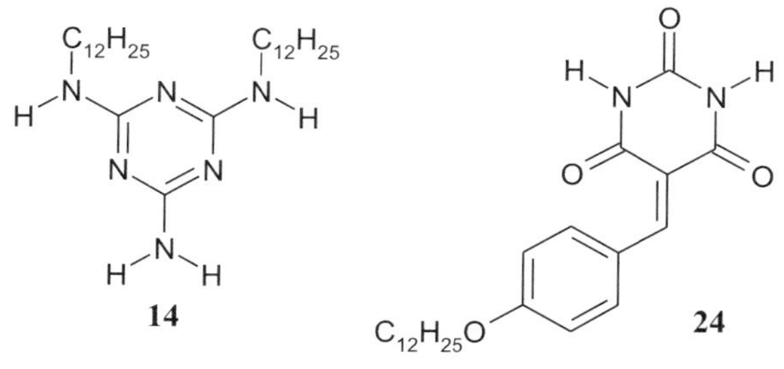

**Fig. 18** Structures of the melamine and barbituric acid derivatives **14** and **24**

equivalent of the barbituric derivative, tapes were formed on a variety of substrates using dimethylformamide as a solvent [58].

Barbituric acid derivatives (**25**) have also been used to prepare Langmuir–Blodgett layers [59] and self-assembled monolayers using hydrogen-bond interactions (Fig. 19). Remarkably supramolecular chirality was observed in Langmuir–Blodgett layers of achiral amphiphilic barbituric acid derivatives as a result of hydrogen-bond and π–π interactions between neighboring barbituric acid molecules [60]. Other work showed that mixed films of barbituric acid and triaminopyrimidine derivatives can been constructed at the air–water interface [61]. Monolayers containing receptors based on bis(2,6-diaminopyridine)amide of isophthalic acid **26** were capable of binding barbituric acid derivatives [62]. Such systems can be used as model systems for the study of interfacial recognition events and for selective and specific detection of compounds for drug screening processes.

In an interesting report by Kimura et al. [63] amphiphilic pthalocyanines were organized in fibrous assemblies via hydrogen-bonding and π–π

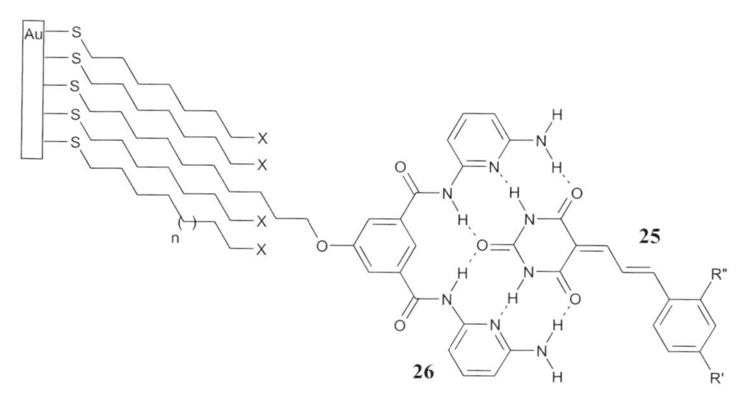


**Fig. 19** Molecular recognition between barbituric acid derivatives (**25**) and a diaminopyridine-based receptor functionalized alkanethiol (**26**) on the surface of a self-assembled thiolate monolayer on a gold film

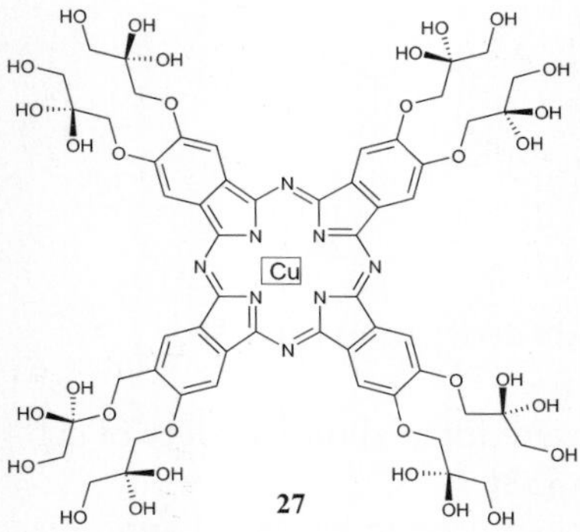

Fig. 20 Molecular structures of Kimura's chiral phthalocyanines

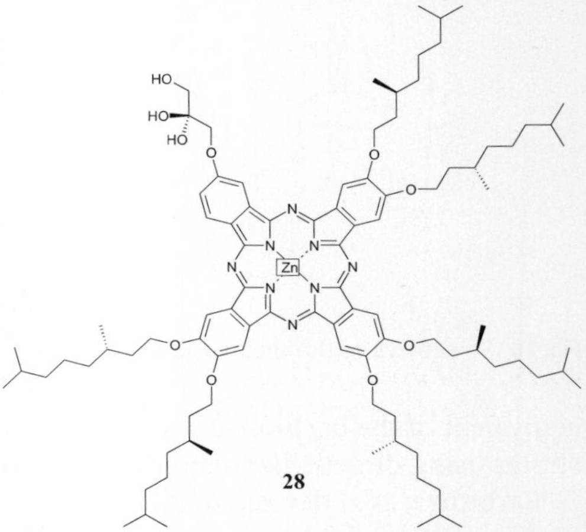

interactions. For this purpose the phthalocyanines were decorated with chiral alkane diols (**27**, Fig. 20). In a water–DMSO mixture a CD effect was observed indicative of a helical arrangement of the chromophores. X-ray diffraction and transmission electron microscopy (TEM) showed the formation of fibrous assemblies. Self-organization of hydrogen-bonded dimers of Zn(II) phthalocyanines **28** (Fig. 20) equipped with six optically active side chains and one chiral diol group results in two different supramolecular structures, and is controlled by temperature [64]. At room temperature the hydrogen bonds among the diol groups allowed the construction of a lamellar sheet. However, the cleavage of the hydrogen-bond network at 130 °C caused the structure to change to a hexagonal columnar phase, in which zinc phthalocyanine molecules are arranged in a left-handed helical manner.

## 5 Hydrogen-Bonded Azo Dyes

Azo dyes account for over 50% of all commercial dyes and have been studied more than any other class of dyes [1]. Azo dyes contain at least one azo group that can be in the *trans* or *cis* form and switching between these two isomeric form is possible with light. Remarkably, hydrogen-bond interactions have been rarely applied in tuning the properties of these dyes. Ammonium-derivatized azobenzene cyanuric acids (**29**) and glutamate-derivatized melamines (**30**) have been organized in water [65, 66] and at the air–water interface [67] by complementary hydrogen bonding (Fig. 21). In water, helical superstructures were formed and the aggregation process could

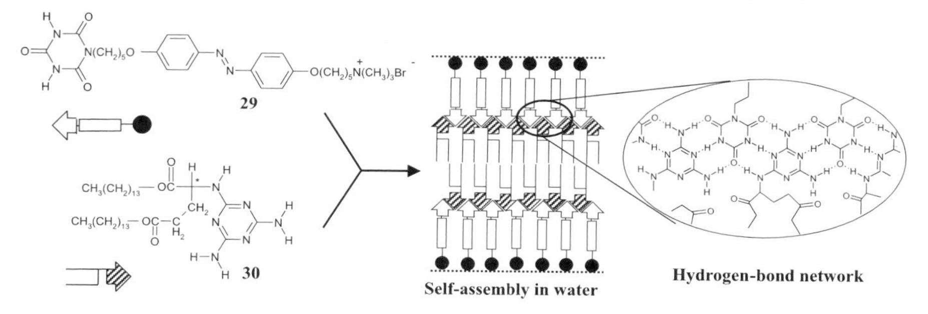


**Fig. 21** Schematic illustration of the hydrogen-bonded complexation and self-organization of azobenzene cyanuric acid **29** and a melamine **30** in water

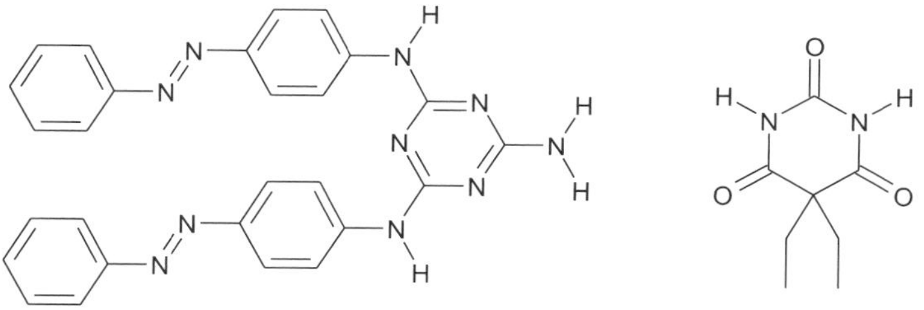

**31**

**32**

**Fig. 22** Chemical structures of melamine derivative **31** and barbiturate **32**

be controlled by photoisomerization of the azobenzene dye resulting in segregation of the two components.

Recently, Yagai and coworkers [68, 69] reported the photoswitching properties of *trans*-azobenzene incorporated melamine **31** and barbiturate **32** hydrogen-bonded assemblies (Fig. 22). Photoisomerization of azobenzene units on melamine enhances the thermodynamic stability of the rosette assembly in chloroform solution and suppress its transformation into insoluble tapelike polymers. Interestingly, the resulting stable rosette supramolecular structure could be retained in the solid state when the solvent was removed.

In another report, carboxylic acids were used to self-assemble azo dyes, which could be controlled by photoisomerization of the azobenzene dye [70]. The *trans*-azobenzene isomer **33** of this dye associates into hydrogen-bonded linear tapes, while the *cis*-azobenzene **34** yields hydrogen-bonded self-assembled tetramers that form rodlike aggregates by additional $\pi$–$\pi$ stacking interactions (Fig. 23).

Recently DNA–azo dye conjugates have been constructed by Asanuma et al. [71], yielding aggregates that could be reversibly converted into a new type of dye aggregate by hybridization with complementary DNA. Further-

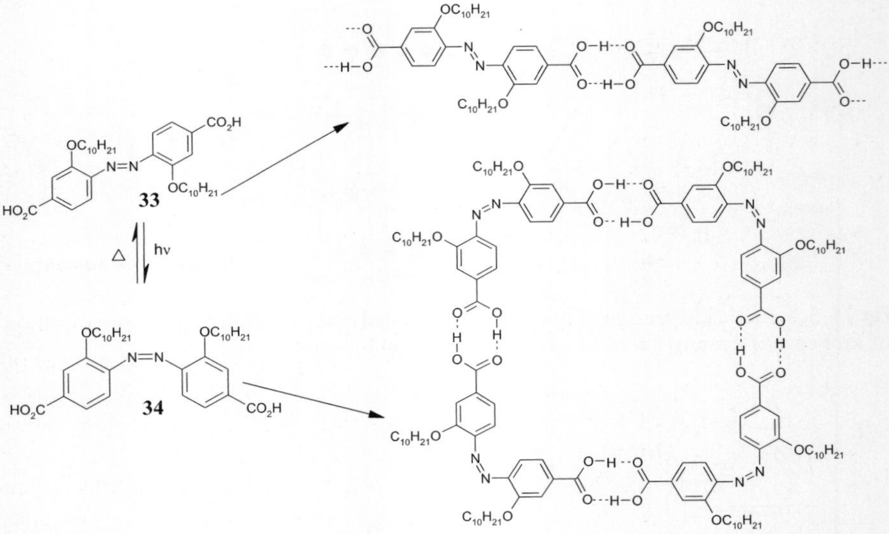

**Fig. 23** Control of organization of hydrogen-bonded azo dyes by photoirradiation

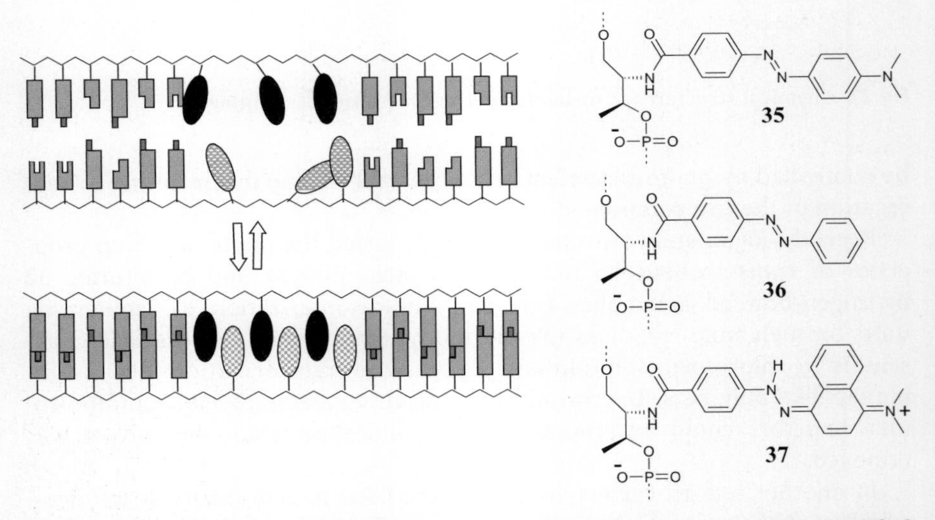


**Fig. 24** Schematic illustration of the formation of dye heteroaggregates from modified DNA–azo dye conjugates **35–37**

more heteroaggregates in which two dyes are stacked alternately could be constructed by interstrand stacking of two DNA–azo dye conjugates (**35–37**, Fig. 24) [72].

# 6
# Hydrogen-Bonded Assemblies of Extended $\pi$-Conjugated Systems

$\pi$-Conjugated molecules by virtue of their loosely bound $\pi$ electrons possess intriguing optical and electronic properties, such as strong absorption, emission, electron mobility and conductivity. These properties make them useful in the design of a plethora of advanced materials, such as LEDs, field-effect transistors, solar cells, nonlinear optics, sensors and switches. Therefore, control of the optical and electronic properties of conjugated systems is of utmost importance in the domain of advanced materials research, particularly in the area of molecular electronics and in the emerging area of supramolecular electronics. Reversible tuning of the electronic properties via the creation of nanoscopic and mesoscopic supramolecular architectures is a challenging task in the design of functional nanoscopic devices. As already illustrated, among the different noncovalent interactions, hydrogen bonding is the most crucial and powerful tool in the crafting of supramolecular assemblies. Combining the advantages of extended $\pi$-conjugated systems and the power of hydrogen-bond-directed supramolecular interactions could provide an elegant way of creating functional nanoscopic as well as macroscopic assemblies with desired properties.

Recently, Tew et al. [73] reported the synthesis and characterization of triblock rod–coil molecules consisting of diblock coil segments of polystyrene and polyisoprene, and oligo($p$-phenylenevinylene) (OPV) trimers as rod segments. On the basis of TEM and X-ray diffraction data, these molecules formed self-organized nanostructures, which in turn assemble into monolayers. The same group has also synthesized dendron rod–coil

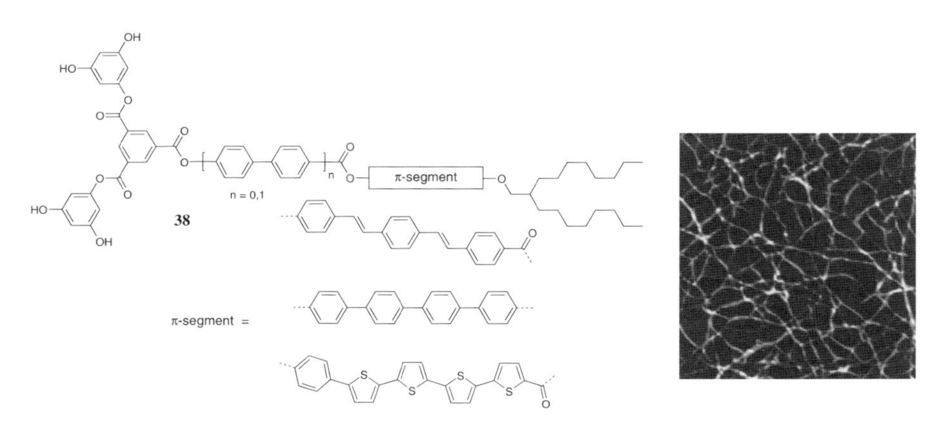


**Fig. 25** Chemical structure of the dendron rod–coil molecules and the formation of nanoribbons. (Reprinted with permission from Ref. [74]. Copyright 2004 American Chemical Society)

molecules that contain a conjugated segment of oligothiophene, OPV and oligo(phenylene) (**38**, Fig. 25). Despite the structural difference all three molecules self-assemble into high-aspect-ratio ribbonlike nanostructures on mica [74]. The authors proposed that the ribbons consist of two molecules held together by hydrogen-bond interactions in the hydroxyl-rich dendrons that further self-assemble in ribbons by $\pi$–$\pi$ stacking of the conjugated segments. Interestingly, electric field alignment of the assemblies was possible, creating arrays of self-assembled nanowires on a device substrate.

Inspired by nature, in a preliminary communication, peptide–oligothiophene conjugates, that is a silk-inspired peptide linked to an oligo(3-alkylthiophene), have been reported with the aim to create $\beta$-sheets by multi-hydrogen-bond interactions [75].

Meijer, Schenning and coworkers have made significant contributions in the understanding of the supramolecular organization of OPVs. They have designed a variety of OPV supramolecular assemblies based on the strong dimerization of the self-complementary 2-ureido-4[1*H*]-ureidopyrimidinones (**39**) or ureido-*s*-triazine (**40**), which are quadruple hydrogen-bonding units

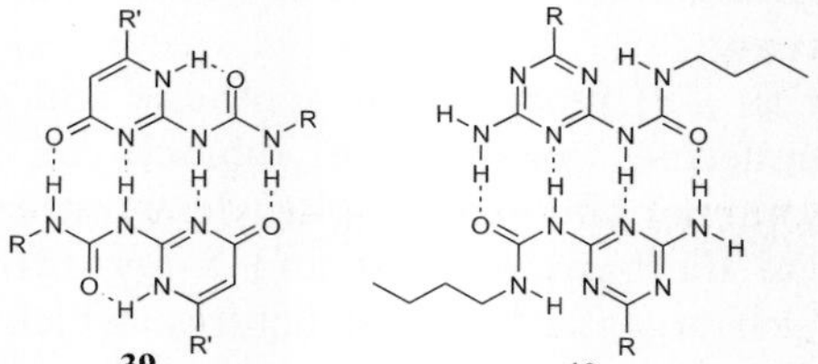

**Fig. 26** Quadruple hydrogen bonding between ureidotriazines and ureidopyrimidinones

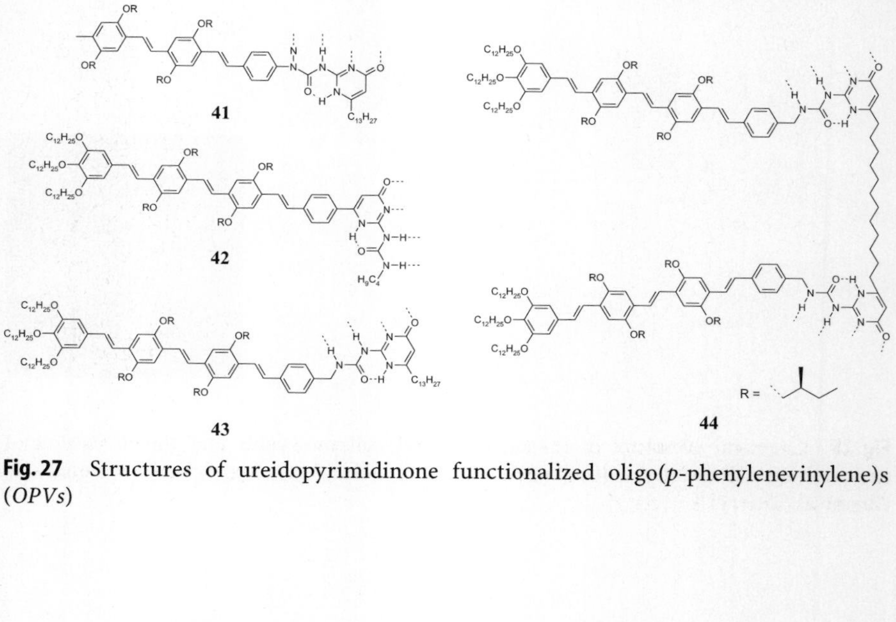

**Fig. 27** Structures of ureidopyrimidinone functionalized oligo(*p*-phenylenevinylene)s (*OPVs*)

(Fig. 26). For example, the OPV derivatives functionalized with ureidopyrimidinone at one end, **41**, **42** and **43** are shown to undergo dimerization in dilute chloroform solution (approximately $10^{-5}$ M), since the dimerization constant ($K_{\text{dim}}$) of the ureidopyrimidinone group is as high as $6 \times 10^7\ \text{M}^{-1}$ (Fig. 27). However, in dodecane solution the OPV dimers further aggregate into larger architectures, as evidenced from the optical and chiroptical properties [76]. Subsequently, El-ghayoury et al. [77] synthesized the bifunctional OPV **44** containing a ureidopyrimidinone group for the preparation of processable $\pi$-conjugated supramolecular polymers.

Mono- and bifunctional OPVs contaning ureido-*s*-triazine units have been shown to self-assemble in a distinctly different fashion [78]. The monofunctional OPVs **45–47** form dimers in chloroform solution ($K_{\text{dim}} = 2.1 \times 10^4\ \text{M}^{-1}$). In dodecane this compound is aggregated in helical columns, as could be concluded from the absorption, emission and CD spectral data. The absorption spectrum in dodecane showed an additional shoulder at longer wavelength, whereas the emission is quenched by approximately 1 order of magnitude with a shift to the red. The CD spectra showed a strong bisignated Cotton effect which is in agreement with the exciton model in which the OPV dimers aggregate in a chiral supramolecular stack as shown in Fig. 28 [79]. Interestingly, the absorption spectrum of the bifunctional OPV

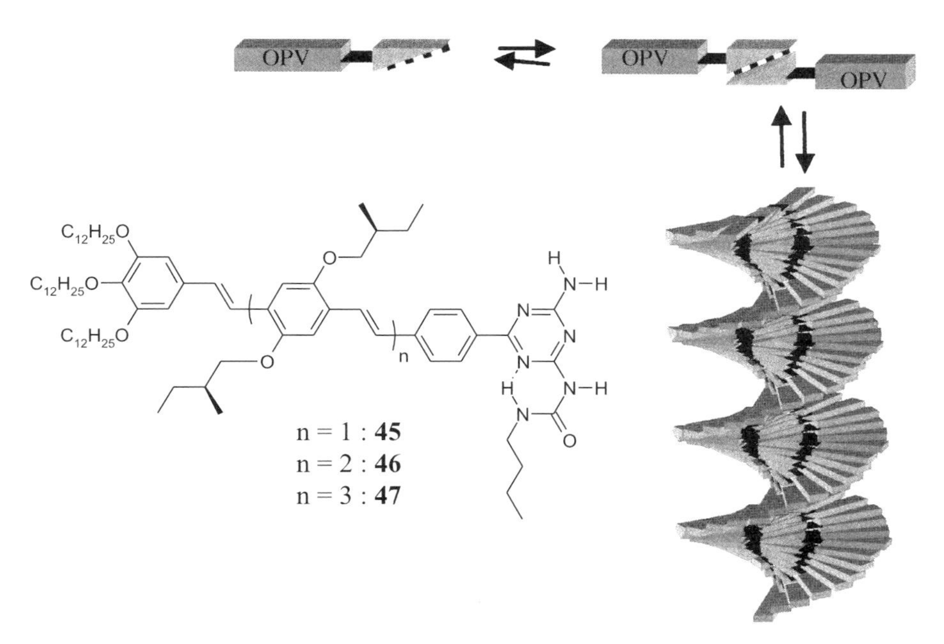


**Fig. 28** Schematic representation of the hierarchical organization of OPVs **45–47** into helical stacks in dodecane. (Reprinted with permission from Ref. [78]. Copyright 2001 American Chemical Society)

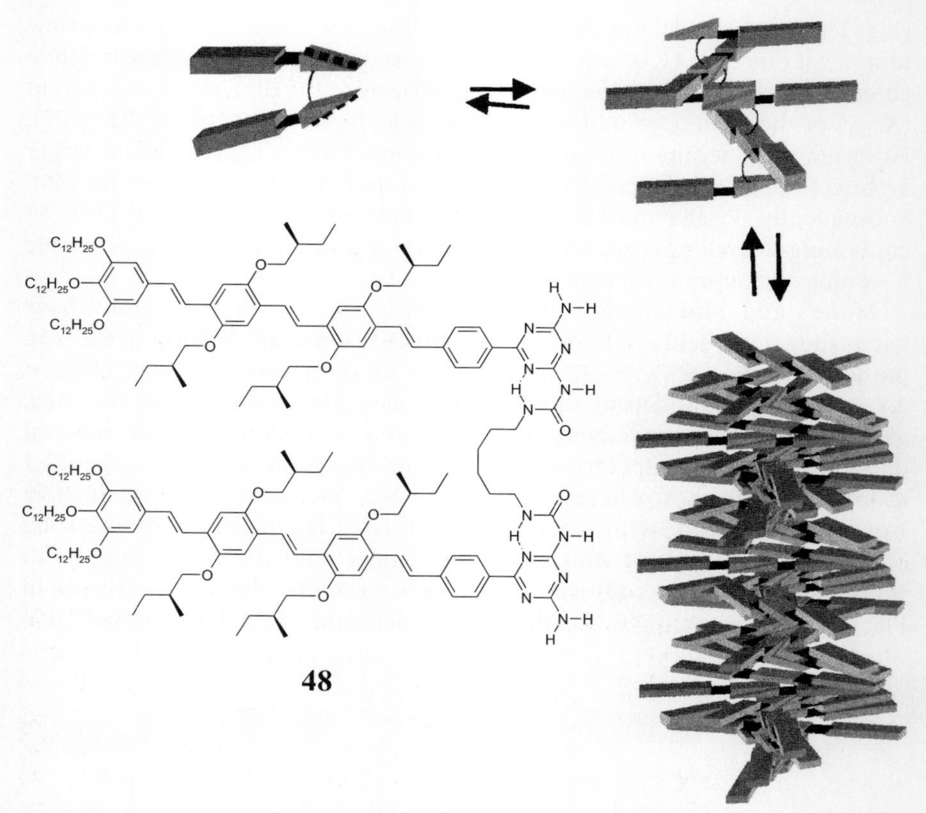


**Fig. 29** Structure of OPV **48** and the schematic picture of its organization into frustrated stacks in dodecane. (Reprinted with permission from Ref. [78]. Copyright 2001 American Chemical Society)

**48** is slightly redshifted in dodecane when compared with that in chloroform, which differs from the behavior of **47**. The emission spectrum of **48** in dodecane did not show much difference from that in chloroform. Surprisingly, in this case a weak Cotton effect was observed in the CD spectrum even at higher concentrations in dodecane. From these studies it was concluded that the aggregates of **48** are not as well organized as those of **47**; therefore, a supramolecular polymer in which the chromophores are present as frustrated stacks is formed (Fig. 29). The important point here is that it is possible to incorporate **47** as a chain stopper into the supramolecular polymers formed by **48**, thereby controlling the length and properties of these supramolecular polymers.

Beckers et al. [80] observed the preferential formation of a functional supramolecular heterodimer (**49**) between OPV **44** and a $C_{60}$ containing the two self-complementary ureidopyrimidinone moieties (Fig. 30). Formation of

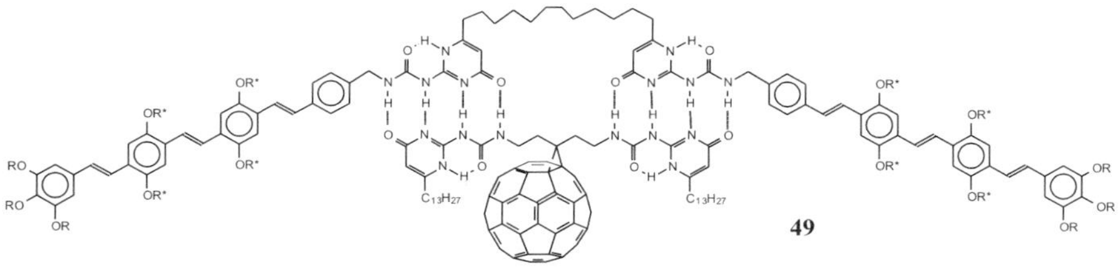

**Fig. 30** Quadruple hydrogen bonding in OPV–$C_{60}$ hetrodimer

the heterodimer was confirmed by $^{1}$H NMR studies and by changes in the emission spectra. During the mixing of the two components, the absorption bands remained unchanged, indicating a weak electronic interaction in the ground state. However, considerable quenching of the OPV emission could be observed, indicating singlet energy transfer from the excited OPV to the $C_{60}$. Electron transfer between OPV and $C_{60}$ within the heterodimer is ruled out owing to the unfavorable energy for the charge separation and the relatively long distance between the chromophores [81].

An increase in the conjugation length of the OPV backbone has a considerable influence on the self-assembling behavior of the ureido triazine functionalized OPVs **45–47** [82, 83]. Small-angle neutron scattering (SANS) studies show the hierarchical growth of rigid cylindrical objects, the length and diameter of which vary as a function of the conjugation length. For example, the stack of OPV **46** has a persistence length of 150 nm and a diameter of 6 nm, whereas OPV **45** forms cylindrical stacks of 60 nm in length and 5 nm in diameter. Studies of the temperature-dependence and the concentration-dependence revealed an increase in the stability of the self-assemblies with an increase in conjugation length. Atomic force microscopy (AFM) studies show that the transfer of the cylindrical objects to surfaces is specific to concentration and the nature of the solid surface. At higher concentrations, an intertwined network is formed, whereas at low concentrations, ill-defined globular objects are observed. Repulsive surfaces such as mica and glass induce clustering of stacks, while attractive surfaces like gold destroy the stacks. Transfer of a single fiber is successful only in the case of inert surfaces such as graphite and silicon oxide.

Recently, Hoeben et al. [84] demonstrated ultrafast energy transfer in mixed columnar aggregates of OPVs **45** and **46**. Fast and efficient energy transfer occurs from the short OPV **45** to the longer one **46** within the mixed assemblies when the latter is doped with the former. Incorporation of **46** within the stacks of **45** ensures an effective pathway for the transfer of the excitation energy, as illustrated by the loss of emission of the formers, above the melting temperature of the stacks. The energy transfer process is characterized by a very fast component in the rise of the acceptor luminescence, which is absent when the stacks are dissociated at high temperatures. This component is directly correlated with the high degree of order inside the OPV assemblies, which enables fast excitonic diffusion along the stacks before excitons are trapped inside the local potential minima. Coupling between chromophores is weak at longer time scales and therefore a direct Föster-type energy transfer to the guest molecules dominates. The authors of this study demonstrated that control of nanoscale order provides a strategy for tailoring macroscopic electronic properties of organic semiconductor systems. However, the main drawback of this system is that there is no possibility of controlling the length of the supramolecular stacks. Though the persistence length of the donor columns is approximately 60 nm, the system appears to be dynamic. In addition, the exact control of the energy trap position is not possible, which leads to the undesired clustering of the guest molecules owing to a phase separation between the donor and the acceptor molecules.

In a recent interesting report by Jonkheijm et al. [85] guidelines for programming $\pi$-conjugated molecules into self-assembled tubular aggregates in solution and in the solid surface are described. This is illustrated with the help of the STM analysis of OPVs **50** and **51** bearing a diamino triazine moiety. STM images showed chiral hexameric rosette structures lying flat on the surface with the diamino triazine moieties pointing to the center, forming a hydrogen-bonded cavity (Fig. 31). The rosettes are ordered in rows, which form a hexagonal two-dimensional crystal lattice with alkyl chains interdigitating with adjacent rosettes. The cavity of the rosette has an estimated diameter of approximately 0.7 nm. These cyclic hexameric rosettes are not fully planar, resulting in a propeller arrangement. SANS and AFM studies revealed the formation of tubules of 7 nm in diameter and 180 nm in length.

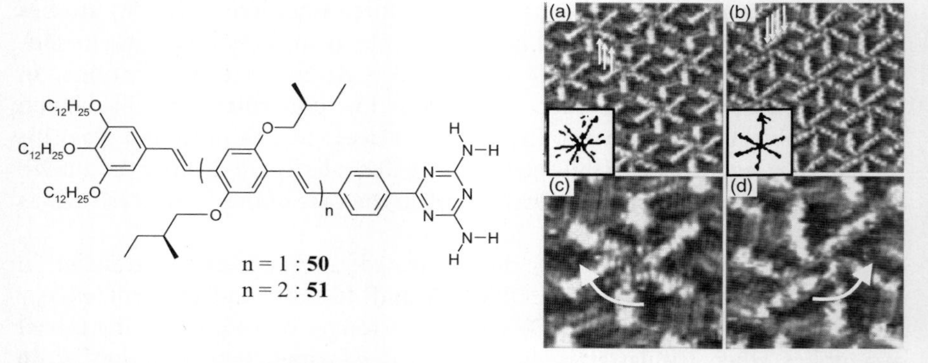


**Fig. 31** Structures of OPVs functionalized with triazine units that form hexameric rosettes. **a**, **b** Scanning tunneling microscope images of **50** and **51** monolayers at the solid–liquid interface using graphite as the substrate and 1-phenyloctane as the solvent. **c**, **d** Close-up images of the **50** and **51** rosettes. *Arrows* indicate the rotation direction of the rosettes. (Reprinted with permission from Ref. [85]. Copyright 2004 John Wiley & Sons)

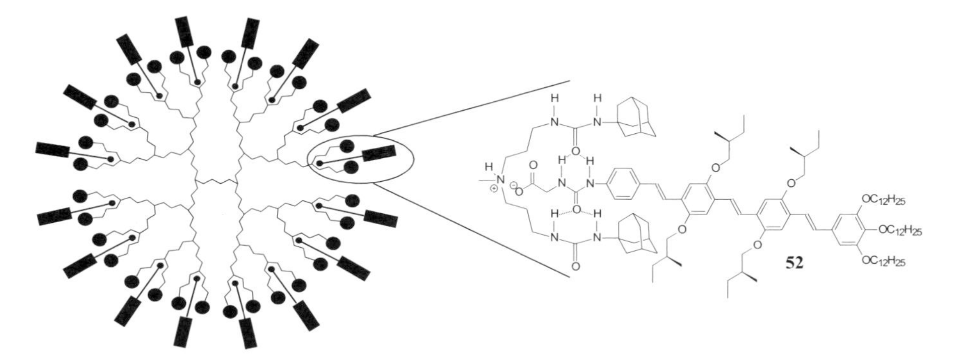


**Fig. 32** Schematic representation of the complexation of OPV chromophores to a dendritic scaffold via hydrogen-bond and ionic interactions

An interesting observation with respect to the supramolecular organization of **50** and **51** is the difference in the CD spectra in heptane solution. The bisignated CD spectrum of OPV **50** is opposite to that of **51** in sign and shape. However, in the case of **51** an inversion of the Cotton effect is observed with time (within 80 min), which is accompanied by a shift in the zero-crossing. The sign reversal of **51** shows first-order kinetics with a rate constant of $k = 5.6 \times 10^{-4}\ s^{-1}$. The initially formed helix may be kinetically controlled, whereas the finally formed helix may be the thermodynamically stable form. Thus a variety of nanoarchitectures of different shape, size and properties

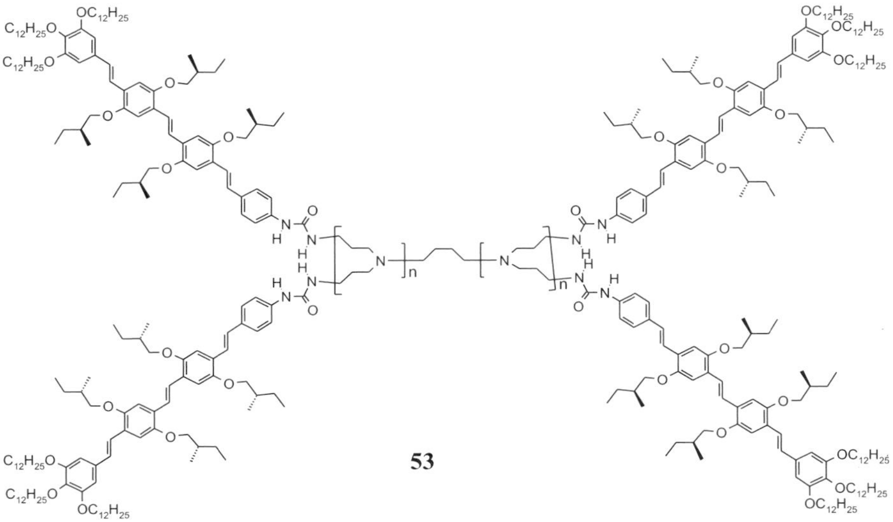

53

**Fig. 33** OPV-functionalized liquid-crystalline poly(propylene imine) dendrimers

could be achieved from functionalized OPVs by the rational supramolecular approach, taking advantage of the directional hydrogen-bonding interaction of ureidopyrimidinones, diaminotriazine and ureidotriazine moieties.

Dendrimers have been used as a spherical scaffold in which OPV chromophores (**52**) could be complexed via ionic and hydrogen-bond interactions, yeilding highly emissive amorphous films (Fig. 32) [86]. The $\pi$-oligomers are less aggregated in these assemblies because of the shielding effect of the bulky adamantyl units present in the dendritic host. Liquid-crystalline OPV carboxylic acid derivatives that dimerize via hydrogen bonds [87] and more exotic hydrogen-bonded liquid-crystalline OPV systems have been reported including different generations of poly(propylene imine) dendrimers modified peripherically with OPVs in which strong hydrogen bonds exist between the urea linkages (**53**, Fig. 33) [88].

## 6.1 Miscellaneous $\pi$-Conjugated Assemblies

Organized chromophore assemblies are crucial in biological electron and energy transfer processes. For example, hydrogen-bonded donor–acceptor dyads as well as complex architectures have been constructed in several cases to shed light on the biological light-harvesting phenomenon [89]. Such assemblies are not only important for addressing fundamental questions in chemical biology but are also of great interest in the design of photovoltaic devices. With this objective, Schenning, Wurthner et al. [90, 91] designed a hydrogen-bonded OPV–perylene bisimide chiral assembly **54** (Fig. 34). This

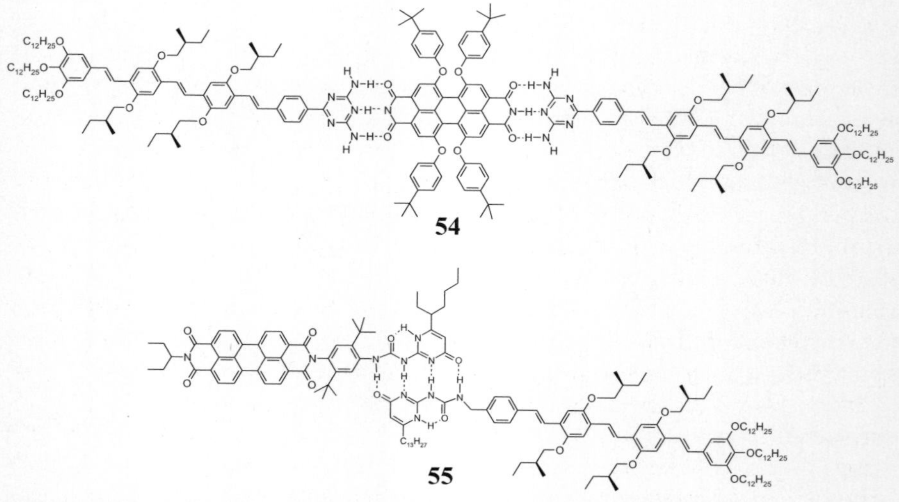

**Fig. 34** Perylene bisimide–OPV hydrogen-bonded triad (**54**) and dyad (**55**)

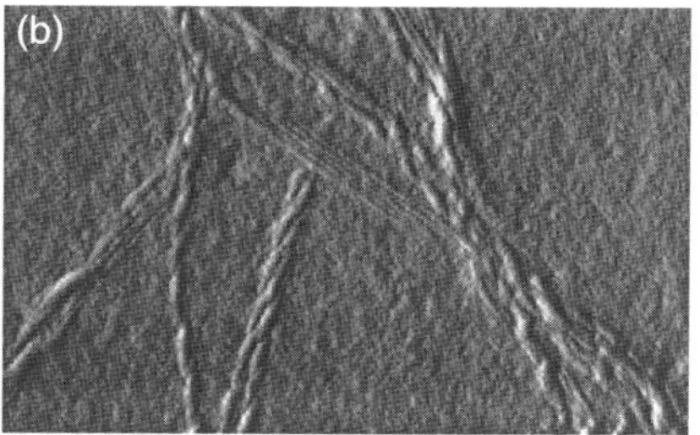


**Fig. 35** **a** Possible helical arrangement of **54** and **b** tapping mode atomic force microscope (*AFM*) image of the helical fibers of **54** upon spin-coating from methylcyclohexane solution. (Reprinted with permission from Ref. [90]. Copyright 2002 American Chemical Society)

is an example for a donor–acceptor–donor assembly where the acceptor perylene dye is placed between two OPV donors through the complementary triple hydrogen-bond interaction of diaminotriazine and imide moieties. UV–vis studies revealed the formation of J-type aggregates where the perylene bisimide units of the hydrogen-bonded triad pack tightly within the aggregates. Significant quenching of the perylene bisimide fluorescence is noticed, suggesting electron transfer from OPV to perylene bisimide. In addition, a strong negative Cotton effect was observed corresponding to the perylene absorption, indicating that the chirality of the OPV side chains is expressed in the perylene chromophore. Further insight into the self-assembly process was obtained from the temperature- and concentration-dependent optical and chiroptical properties. The large bathochromic shift of the perylene absorption and the strong negative CD of the dye assembly is rationalized on the basis of the winding of the complex in a chiral screw sense as shown in Fig. 35a. This is confirmed by the AFM images, which showed the formation of helical rodlike aggregates (Fig. 35b). Femtosecond pump–pulse spectroscopy measurements showed the presence of the OPV radical cation at 1450 nm, which is evidence for the electron transfer to the perylene moiety. In another example, singlet energy transfer has been demonstrated between the perylene–OPV hydrogen-bonded dyad **55** (Fig. 34) [92]. Owing to the high association constants of the quadruple hydrogen-bonded system used in this study, a mixture of heterodimers and homodimers is present even in dilute solutions. This feature allowed the authors to follow the temporal evolution of the singlet energy transfer process using subpicosecond transient absorption spectroscopy in the heterodimers on short time scales.

Interestingly perylene bisimides and a hydrogen-bonded OPV (p-type) derivative can also orthogonally self-assemble into separate nano-sized p-and n-type stacks in methylcyclohexane. In contrast, in toluene only molecularly dissolved species are present. In films deposited from methylcyclohexane as well as from toluene photoinduced electron transfer takes place from

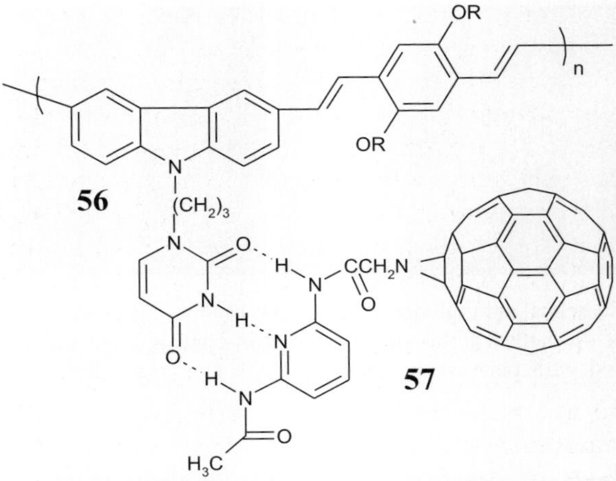

**Fig. 36** Hydrogen-bonding interaction between polyphenylenevinylene (**56**) and fullerene (**57**) derivatives functionalized with uracil and diaminopyridine units, respectively

the p-type material to the n-type material. As a result of the orthogonal self-assembly process, in films from methylcyclohexane an ordered network of fibers was formed, whereas in films from toluene no ordering was observed [93].

Emge and Bäuerle [94] also reported the functionalization of oligothiophenes and the corresponding polythiophenes with uracil units. Both thiophenes show a modification in the cyclic voltammogram in the presence of a complementary adenine hydrogen-bonding motive, indicating hydrogen-bond interactions. Polyphenylenvinylenes have also been functionalized with uracil units (**56**, Fig. 36), and are capable of forming hydrogen-bond interactions with a fullerene derivative equipped with a diaminopyridine unit (**57**) [95]. Fluorescence studies gave strong evidence of photoinduced electron transfer from the polymer backbone to the fullerene acceptor moiety.

## 6.2 Organogels of Oligo(*p*-phenylenevinylene)s and Related $\pi$-Conjugated Systems

Gelation is an interesting phenomenon shown by certain molecules in aqueous or organic solvents as a result of weak noncovalent interactions, leading to the formation of supramolecular structures of nanometer to micrometer dimensions [96–98]. The formation of H-bonded assemblies between molecules and the subsequent gelation is usually supported by other weak interactions such as $\pi$–$\pi$ stacking, van der Waals forces and electrostatic associations. Formation of one-dimensional stacks and their growth to three-dimensionally entangled architectures are responsible for the entrapment of large volumes of solvents within such self-assemblies. Though a variety of molecules are known to form gels, only a few examples are available for gels derived from dyes, and these are discussed in detail by Shinkai and coworkers in one of the chapters of this volume. Therefore, we will discuss briefly organogels based on extended $\pi$-conjugated systems, highlighting our contributions, for a continuity of the present review. Despite having intriguing optical and electronic properties, gels based on hydrogen-bonded conjugated systems are surprisingly few in number. Gelation has been observed in the case of certain $\pi$-conjugated polymers as a result of unwanted crosslinking and chain-branching reactions [99].

Ajayaghosh et al. [100] demonstrated the gelation of OPV units (**58**) when appropriately functionalized with hydrogen-bonding motifs and hydrocarbon side chains as a result of the cooperative hydrogen-bonding, $\pi$-stacking and van der Waals interactions. A number of OPVs with different functional groups have been prepared and have been shown to form gels of varying strength and morphology. Variable-temperature $^1$H NMR studies in benzene-$d_6$ revealed strong chromophore interaction owing to the self-assembly and gelation. X-ray diffraction patterns of the dried gel showed intense peaks in the small-angle region, which correspond to the lamellar packing distance of the side chains as well as the length of the molecules. In the wide-angle region, reflections corresponding to the $\pi$-stacking distance was predominant. The optical polarizing microscopy pictures of the gels obtained by cooling the isotropic solutions showed the presence of birefringent textures, the morphology of which was significantly influenced by the structure of the gelator as well as the rate of cooling. SEM analysis indicated the formation of entangled structures of 50–200 nm in width and several micrometers in length (Fig. 37a). TEM pictures of the film of the gelators from dilute solutions showed the presence of twisted tapes of 50 nm in width and several micrometers in length (Fig. 37b). The morphology of the gels obtained from AFM was in agreement with that obtained from the SEM analysis.

Optical properties of OPVs are useful tools for probing the self-assembly-induced gelation process. The absorption and emission spectra of the OPV

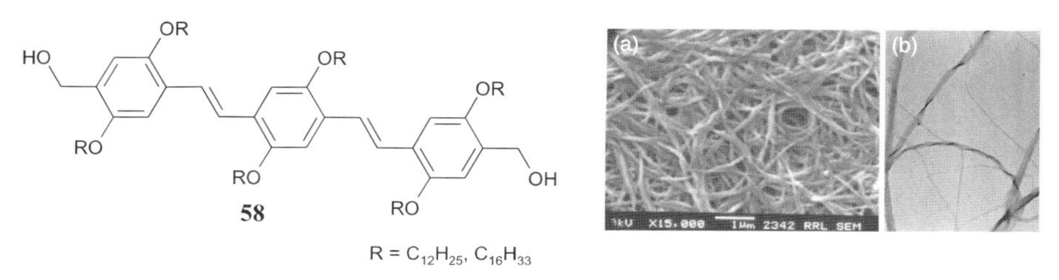


**Fig. 37** Structure of OPV organogelator **58** and the scanning electron microscope (*SEM*) (**a**) and transmission electron microscope (**b**) images of the resulting toluene organogels. (Reprinted with permission from Ref. [100]. Copyright 2001 American Chemical Society)

gelators in chloroform indicated that the molecules are in the unassembled state. In apolar solvents such as hexane, cyclohexane, decane and dodecane, an additional shoulder band is formed at the long-wavelength region. The emission spectra in these solvents were significantly quenched and new long-wavelength emission peaks were formed. These changes were reversible with respect to concentration and temperature, indicating the formation of non-covalent assemblies in apolar solvents. The stability of the self-assemblies could be determined from the plots of the fraction of the self-assembled species, which is obtained from the temperature-dependent changes in the absorption and emission spectra. The results obtained were in agreement with those obtained from the differential scanning colorimetric analysis of the gels. OPVs (**59**) bearing chiral side chains provided gels with helical morphology (Fig. 38). Detailed field-emission SEM and AFM analyses showed the hierarchical growth of helical nanotapes and coiled-coil ropes of approximately 90 nm in width and several micrometers in length [101] (Fig. 38). Temperature- and concentration-dependent CD spectra of **59** showed interesting changes during the gelation process. At low concentrations weak bisignated signals were observed which changed to intense bisignated signals at higher concentrations. Similar observations could be seen when the tem-

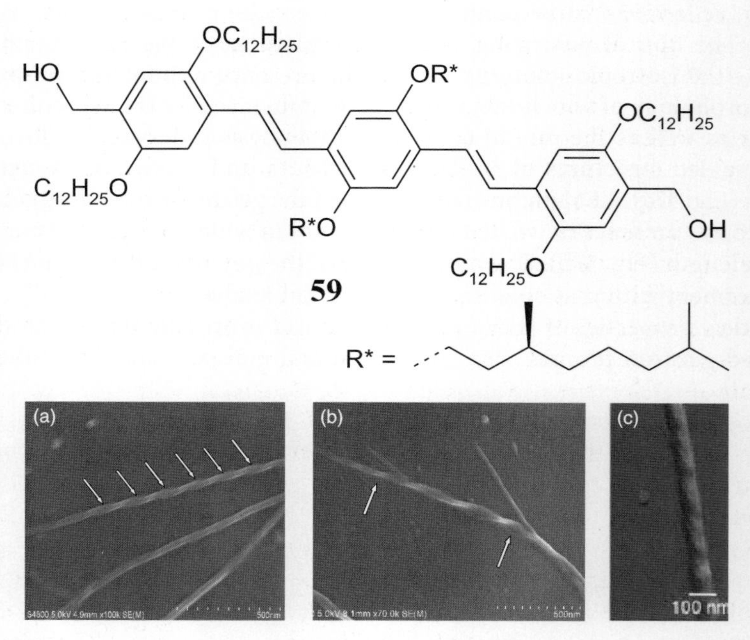


**Fig. 38** **a** SEM image of the left-handed helical fibers **b** filed-emission SEM image of the formation of coiled-coil ropes and **c** AFM image of the coiled-coil gel fibers of **59** in dodecane. (Reprinted with permission from Ref. [101]. Copyright 2004 John Wiley & Sons)

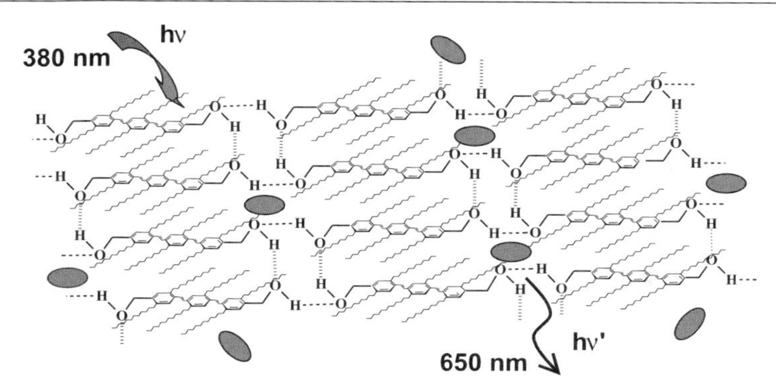


**Fig. 39** Schematic representation of the light-harvesting OPV gel network doped with rhodamine B (*ellipsoids*)

perature was increased from 10 to 35 °C. The complementary nature of the concentration- and temperature-dependent transitions of the CD signals indicates the involvement of similar chiral intermediates during the formation and breakdown of the supramolecular assembly.

Gelation-induced modulation of optical properties allows the use of OPV gels as a scaffold for fluorescence resonance energy transfer (FRET) selectively from the excited gel phase to a suitable acceptor as illustrated in Fig. 39 [102]. For this purpose, rhodamine B is a good acceptor since the emission of the self-assembled OPVs overlaps with the absorption of rhodamine B. When the OPV is excited at 380 nm in the presence of a slight excess of the dye, selective quenching of the emission of the self-assembled species and the concomitant emission of the dye could be observed. This experiment shows that in the present case FRET is not possible directly from the unassembled OPVs to the dye, which highlights the importance of chromophore orientation in excitation energy transfer processes. The weak emission of an optically matching dye solution when compared with that in the presence of OPV is a clear indication of the light harvesting. However, the main disadvantage of the system is the incompatibility between the hydrophobic gel and the cationic dye in apolar solvents. This could be solved to some extent in the case of rhodamine B doped xerogel film. In this case the FRET efficiency is significantly increased even in the presence of low dye concentrations.

Bisurea-appended oligo(thiophene)s (**60**) have also been prepared and in tetralin and 1,2-dichloroethane these compounds self-assemble into elongated fibers, resulting in the formation of organogels [103]. It was shown that the good electronic overlap between the thiophene rings due to the hydrogen-bond-assisted $\pi$-stacking of adjacent oligomers results in high charge carrier mobility. Lin and Tour [104] reported that the well-known $\pi$-stacking forces between the shape-persistent cyclophanes could be enhanced through

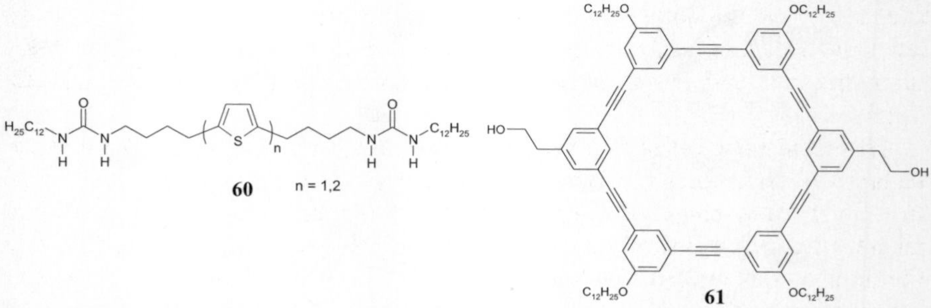

Fig. 40 Molecular structure of oligo(thiophene) (**60**) and cyclophane (**61**) based organogelators

hydrogen-bonding interactions by incorporating multiple hydrogen-bonding sites at the periphery of the cyclophane ring. The hydrogen-bond-assisted enhanced stacking interaction in **61** results in the formation of cyclophane dimers and gels at moderate concentrations in solvents such as $CH_2Cl_2$ and $CHCl_3$ (Fig. 40).

Survey of the current literature reveals that very little has been done with respect to hydrogen-bonded self-assembly induced gelation of $\pi$-conjugated molecules though significant changes can be expected to the optical and electronic properties of such systems. Hopefully in coming years we may witness more and more exciting results in this area.

# 7 Conclusions and Outlook

Hydrogen-bonded self-assembly offers an attractive tool to construct well-organized dye assemblies. The two most popular hydrogen-bonding motifs are the melamine–barbituric acid couple and the carboxylic dimer. The association constant for both couples is low and therefore additional secondary interactions are required to obtain stable assemblies. As can be deduced from this review, it is not only possible to study material properties at the supramolecular level, but also to tune the macroscopic properties to some extent. These tailor-made supramolecular assemblies will enormously influence the macroscopic properties. The combination of a supramolecular architecture with a dye molecule will not only give rise to emerging opportunities in materials science, but, in addition, will significantly contribute to bridging the gap between natural and artificial systems, in an effort to fully understand the guidelines used to assemble natural units in the different hierarchies of organization. However, there are still a number of very appealing targets that should be reached before these tailored supramolecular architectures are used. The lack of control to exactly position chromophores with nanometer precision and the difficulty in controlling the dimensions of supramolecular nanostructures will be the challenges. Another issue is the robustness of hydrogen-bond interactions, which are often temperature- and solvent-sensitive.

From the view point of advanced applications, it is a prerequisite to interface the disciplines of chemistry and materials where synthetic chemistry, functional dyes, chromophores, supramolecular chemistry and nanoscience all become important. Thus, traditional dye chemistry takes a new role in the design of advanced functional materials, when joined with supramolecular chemistry, where noncovalent interactions such as hydrogen-bonding are crucial. Taking into account the recent progress in obtaining full control over large three-dimensional dye architectures, we can safely conclude that in the near future a new generation of functional dye assemblies, with tailor-made properties, will be used as advanced materials in functional devices.

**Acknowledgements** The authors acknowledge the contributions of all coworkers whose names appeared along with them in the references cited. A.A is grateful to DST, New Delhi, for financial support. This is contribution no. RRLT-PPD-196 from the Regional Research Laboratory. A.S. is grateful to the Netherlands Organization for Scientific Research (NWO) and the Royal Netherlands Academy of Science (KNAW) for financial support.

## References

1. Hunger K (2003) Industrial dyes. Wiley-VCH, Weinheim
2. Steiner T (2002) Angew Chem Int Ed Engl 41:48
3. Prins L, Reinhoudt D, Timmerman P (2001) Angew Chem Int Ed Engl 40:2383
4. Sijbesma R, Meijer E (2003) Chem Commun 5
5. Hofmeier H, El-ghayoury A, Schenning A, Schubert U (2004) Chem Commun 318
6. Kadish K, Smith K, Guilard R (eds) (2000) The porphryin handbook. Academic, New York
7. Burell A, Officer D, Plieger P, Reid D (2001) Chem Rev 101:2751
8. van Rossum B, Steensgaard D, Mulder F, Boender G, Schaffner K, Holzwarth A, de Groot H (2001) Biochemistry 40:1587
9. Steensgaard D, Wackerbarth H, Hildebrandt P, Holzwarth A (2000) J Phys Chem 104:10379
10. Miyatake T, Tamiaki H, Holzwarth A, Schaffner K (1999) Helv Chim Acta 82:797
11. De Boer I, Matysik J, Amakava M, Yagai S, Tamiaki H, Holzwarth A, de Groot H (2003) J Am Chem Soc 125:13374
12. Prokhorenko V, Holzwarth A, Müller M, Schaffner K, Miyatake T, Tamiaki H (2002) J Phys Chem B 106:5671
13. Huang X, Nakanishi K, Berova N (2000) Chirality 12:237
14. Ogoshi H, Mizutani T, Hayashi T, Kuroda Y (2000) Porphyrins and metalloporphyrins as receptor models in molecular recognition. In: Kadish K, Smith K, Guilard R (eds) The porphryin handbook, chap 46. Academic, New York
15. Tecilla P, Dixon R, Slobodkin G, Alavi D, Waldeck D, Hamilton A (1990) J Am Chem Soc 112:9408

16. Myles A, Branda N (2001) J Am Chem Soc 123:177
17. Sessler J, Brown C, O'Connor D, Springs S, Wang R, Sathiosatham M, Hirose T (1998) J Org Chem 63:7370
18. Osuka A, Yoneshima R, Shiratori H, Okada T, Taniguchi S, Mataga N (1998) Chem Commun 1567
19. Ikeda C, Nagahara N, Motegi E, Yoshioka N, Inoue H (1999) Chem Commun 1759
20. Drain C, Shi X, Milic T, Nifiatis F (2001) Chem Comm 287; Ercolani G (2001) Chem Commun 1416
21. Shi X, Barkigia K, Fajer J, Drain C (2001) J Org Chem 66:6513
22. Elemans J, Nolte R, Rowan A (2003) J Porphyrins Phthalocyanines 7:249
23. Kuroda Y, Sugou K, Sasaki K (2000) J Am Chem Soc 122:7833
24. Sugou K, Sasaki K, Kitajima K, Iwaki T, Kuroda Y (2002) J Am Chem Soc 124:1182
25. Yamaguchi T, Ishii N, Tashiro K, Aida T (2003) J Am Chem Soc 125:13934
26. Kobuke Y, Ogawa K (2003) Bull Chem Soc Jpn 76:689
27. Nagata N, Kugimiya S-i, Kobuke Y (2000) Chem Commun 1389
28. Kobuke Y, Nagata N (2000) Mol Cryst Liq Cryst 342:51
29. Nagata N, Kugimiya S-i, Kobuke Y (2001) Chem Commun 689
30. Sagawa T, Fukugawa S, Yamada T, Ihara H (2002) Langmuir 18:7223
31. Král V, Schmidtchen F, Lang K, Berger M (2002) Org Lett 4:51
32. Drain C, Russell K, Lehn, J-M (1996) Chem Commun 337
33. Ni Y, Puthenkovilakom R, Huo Q (2004) Langmuir 20:2765
34. Würthner F (2004) Chem Commun 1564
35. Wang W, Wan W, Zhou H, Niu S, Li A (2003) J Am Chem Soc 125:5248
36. Bevers S, Schutte S, McLaughlin L (2000) J Am Chem Soc 122:5905
37. Abdalla M, Bayer J, Rädler J, Müllen K (2004) Angew Chem Int Ed Engl 43:3967
38. Würthner F, Thalacker C, Sautter A (1999) Adv Mater 11:754
39. Würthner F, Thalacker C, Sautter A, Schartl W, Ibach W, Hollricher O (2000) Chem Eur J 6:3871
40. Thalacker C, Würthner F (2002) Adv Funct Mater 12:209
41. Theobald J, Oxtoby N, Phillips M, Champness N, Beton P (2003) Nature 424:1029
42. Liu Y, Xiao S, Li H, Li Y, Liu H, Lu F, Zhuang J, Zhu D (2004) J Phys Chem B 108:6256
43. Liu Y, Zhuang J, Liu H, Li Y, Lu F, Gan H, Jiu T, Wang N, He X, Zhu D (2004) Eur J Chem Phys Chem 5:1210
44. Liu Y, Li Y, Jiang L, Gan H, Liu H, Li Y, Zhuang J, Lu F, Zhu D (2004) J Org Chem 69:9049
45. Wang X, Li X, Shao X, Zhao X, Deng P, Jiang X, Li Z, Chen Y (2003) Chem Eur J 9:2904
46. Katz H, Otsuki J, Yamzaki K, Suka A, Takido T, Lovinger A, Raghavachari K (2003) Chem Lett 32:508
47. Sautter A, Thalacker C, Heise B, Würthner F (2002) Proc Natl Acad Sci USA 99:4993
48. Würthner F, Sautter A, Thalacker C (2000) Angew Chem Int Ed Engl 39:1243
49. Sautter A, Thalacker C, Würthner F (2001) Angew Chem Int Ed Engl 40:4425
50. Gade L, Galka C, Williams R, De Cola L, McPartlin M, Dong B, Chi L (2003) Angew Chem Int Ed Engl 42:2677
51. Zhang Z, Achilefu S (2004) Org Lett 6:2067
52. Ye Y, Bloch S, Achilefu S (2004) J Am Chem Soc 126:7740
53. Würthner F, Yao S (2003) J Org Chem 68:8943
54. Würthner F, Yao S, Debaerdemaeker T, Wortmann R (2002) J Am Chem Soc 124:9431
55. Würthner F, Yao S, Heise B, Tschierske C (2001) Chem Commun 2260
56. Prins L, Thalacker C, Würthner F, Timmerman P, Reinhoudt D (2001) Proc Natl Acad

Sci USA 98:10042
57. Yang W, Chai X, Chi L, Liu X, Cao Y, Lu R, Jiang Y, Tang X, Fuchs H, Li T (1999) Chem Eur J 5:1144
58. Cao Y, Chai X, Smith J, Li D (1999) Chem Comm 1605
59. Bohanon T, Denzinger S, Fink R, Paulus W, Ringsdorf H, Weck M (1995) Angew Chem Int Ed Engl 34:58
60. Huang X, Li C, Jiang S, Wang X, Zhang B, Liu M (2003) J Am Chem Soc 126:1322
61. Bohanon T, Caruso P-L, Denzinger S, Fink R, Mobius D, Paulus W, Preece J, Ringsdorf H, Schollmeyer D (1999) Langmuir 15:174
62. Motesharei K, Myles D (1998) J Am Chem Soc 120:7328
63. Kimura M, Muto T, Takimoto H, Wada K, Ohta K, Hanabusa K, Shirai H, Kobayashi N (2000) Langmuir 16:2078
64. Kimura M, Kuroda T, Ohta K, Hanabusa K, Shirai H, Kobayashi N (2003) Langmuir 19:4825.
65. Kimizuka N, Kawasaki T, Hirata K, Kunitake T (1998) J Am Chem Soc 120:4094
66. Kawasaki T, Tokuhiro M, Kimizuka N, Kunitake T (2001) J Am Chem Soc 123:6792
67. Ariga K, Kunitake T (1998) Acc Chem Res 31:371
68. Yagai S, Karatsu T, Kitamura A (2003) Chem Commun 1844
69. Yagai S, Nakajima T, Karatsu T, Saitow K-I, Kitamura A (2004) J Am Chem Soc 126:11500
70. Rakotondradany F, Whitehead M, Lebuis A, Slieman H (2003) Chem Eur J 9:4771
71. Asanuma H, Shirasuka K, Takarada T, Kashida H, Komiyama M (2003) J Am Chem Soc 125:2217
72. Kashida H, Asanuma H, Komiyama M (2004) Angew Chem Int Ed Engl 43:6522
73. Tew G, Pralle M, Stupp S (1999) J Am Chem Soc 121:9852
74. Messmore B, Hulvat J, Sone E, Stupp S (2004) J Am Chem Soc 126:14452
75. Klok H-A, Rösler A, Götz G, Mena-Orsteritz E, Bäuerle P (2004) Org Biomol Chem 2:3541
76. El-ghayoury A, Peeters E, Schenning A, Meijer E (2000) Chem Commún 1969
77. El-ghayoury A, Schenning A, van Hal P, van Duren J, Janssen R, Meijer E (2001) Angew Chem Int Ed Engl 40:3660
78. Schenning A, Jonkheijm P, Peeters E, Meijer E (2001) J Am Chem Soc 123:409
79. Langeveld-Voss B, Beljonne D, Shuai Z, Janssen R, Meskers S, Meijer E, Brédas J (1998) Adv Mater 10:1343
80. Beckers E, Schenning A, van Hal P, El-ghayoury A, Sánchez L, Hummelen J, Meijer E, Janssen R (2002) Chem Commun 2888
81. Beckers E, van Hal P, Schenning A, El-ghayoury A, Peeters E, Rispens M, Hummelen J, Meijer E, Janssen R (2002) J Mater Chem 12:2054
82. Jonkheijm P, Hoeben F, Kleppinger R, van Herrikhuyzen J, Schenning A, Meijer E (2003) J Am Chem Soc 125:15941
83. Gesquière A, Jonkheijm P, Hoeben F, Schenning A, De Feyter S, De Schryver F, Meijer E (2004) Nano Lett 4:1175
84. Hoeben F, Herz L, Daniel C, Jonkheijm P, Schenning A, Silva C, Meskers S, Beljonne D, Phillips R, Friend R, Meijer E (2004) Angew Chem Int Ed Endl 43:1976
85. Jonkheijm P, Miura A, Zdanowska M, Hoeben F, De Feyter S, Schenning A, De Schryver F, Meijer E (2004) Angew Chem Int Ed Engl 43:74
86. Precup-Blaga F, Garcia-Martinez J, Schenning A, Meijer E (2003) J Am Chem Soc 125:12953
87. Eckers J-F, Nicoud J-F, Guillon D, Nierengarten J-F (2000) Tetrahedron Lett 41:6411
88. Precup-Blaga F, Schenning A, Meijer E (2003) Macromolecules 36:565

89. Ward M (1997) Chem Soc Rev 16:365
90. Schenning A, van Herrikhuyzen J, Jonkheijm P, Chen Z, Würthner F, Meijer E (2002) J Am Chem Soc 124:10252
91. Würthner F, Chen Z, Hoeben F, Osswald P, You C-C, Jonkheijm P, van Herrikhuyzen J, Schenning A, van der Schoot P, Meijer E, Beckers E, Meskers S, Janssen R (2004) J Am Chem Soc 126:10611
92. Neuteboom E, Beckers E, Meskers S, Meijer E, Janssen R (2003) Org Biomol Chem 1:198
93. van Herrikhuyzen J, Syamakumari A, Schenning A, Meijer E (2004) J Am Chem Soc 126:10021
94. Emge A, Bäuerle P (1997) Synth Met 84:213
95. Fang H, Wang S, Xiao S, Yang J, Li Y, Shi Z, Li H, Xiao S, Zhu D (2003) Chem Mater 15:1593
96. Terech P, Weiss R (1997) Chem Rev 97:3133
97. van Esch J, Feringa B (2000) Angew Chem Int Ed Engl 39:2263
98. Estorff L, Hamilton A (2004) Chem Rev 104:1201
99. Guenet J (1992) Thermoreversible gelation of polymers and biopolymers. Academic, London
100. Ajayaghosh A, George S (2001) J Am Chem Soc 123:5148
101. George S, Ajayaghosh A, Jonkheijm P, Schenning A, Meijer, E (2004) Angew Chem Int Ed Engl 43:3421
102. Ajayaghosh A, George S, Praveen V (2003) Angew Chem Int Ed Engl 42:332
103. Schoonbeek F, van Esch J, Wegewijs B, Rep D, de Hass M, Klapwijk T, Kellogg R, Feringa B (1999) Angew Chem Int Ed Engl 38:1393
104. Lin C, Tour J (2002) J Org Chem 67:7761

Top Curr Chem (2005) 258: 119–160
DOI 10.1007/b135554

Published online: 15 August 2005

# Dye-Based Organogels: Stimuli-Responsive Soft Materials Based on One-Dimensional Self-Assembling Aromatic Dyes

Tsutomu Ishi-i[1] · Seiji Shinkai[2] (✉)

[1]Institute for Materials Chemistry and Engineering (IMCE), Kyushu University, 6-1 Kasuga-kohen, 816-8580 Kasuga, Japan
*ishi-i@cm.kyushu-u.ac.jp*

[2]Department of Chemistry and Biochemistry, Graduate School of Engineering, Kyushu University, 6-10-1 Hakozaki, Higashi-ku, 812-8581 Fukuoka, Japan
*seijitcm@mbox.nc.kyushu-u.ac.jp*



**Abstract** In this review, low molecular-weight organogel based on aromatic dye molecules and its stimuli-responsiveness were summarized. Aromatic dyes from small and medium aromatics such as azobenzene, pyrene, bithiophene, tropone, and merocyanine to large aromatics such as porphyrin, phthalocyanine, fullerene, triphenylene, hexaazatriphenylene, and phenylenevinylene, are self-assembled and self-organized in organogel systems to afford a new class of dye-based supramolecular assemblies. In certain solvents, aromatic dye molecules are stacked one-dimensionally to form the elongated columnar-type fibrous aggregates, which are entangled into three-dimensional network structures to prevent the solvents from flowing, leading to viscoelastic fluid "organogel". $\pi$–$\pi$ stacking forces among core aromatic moieties and secondary noncovalent interactions such as hydrogen-bonding and van der Waals interactions cooperatively stabilize the supramolecular structures constructed in the organogel. The noncovalent assembling in the organogel organization is reflected in their reversible sol-gel phase transition controlled by physical or chemical stimuli such as thermo-, photo-, chemo-, metal-, proton-, and mechano-stimuli. By utilizing the columnar-type fibrous structure and the stimuli-responsiveness, the organogels can be engineered to show novel functions such as guest-recognizing, light-harvesting, carrier-transporting, and memory storage properties in soft materials.

**Keywords** Organogels · Supramolecular assemblies · Functional $\pi$-systems · Light-harvesting · Sol-gel transition

## 1 Introduction

In the 21st century, the self-organization of small molecules is one of the most important aims, because it is a fundamental concept of the bottom-up technique in nanotechnology to form well-defined supramolecular nanostructures. In supramolecular assembly systems, intermolecular noncovalent interactions such as hydrogen-bonding interactions, $\pi$–$\pi$ interactions, donor-acceptor interactions, van der Waals interactions, etc. are intensified so that they can control the supramolecular structures both in solid and liquid phases [1]. Recently, this concept has also been extended to low molecular-weight organogel systems, which is a useful strategy to create fibrous supramolecular structures in various organic solvents [2]. In the early stage, the discovery of low molecular-weight organogel depended largely on serendipity. In most cases, a viscoelastic fluid accidentally found during a recrystallization attempt did not attract the attention of many chemists, and such useful materials were probably discarded without any characterization. On the other hand, in the last decade, supramolecular design of the organogel has been proposed and demonstrated, although some aspects of the gelation mechanism are still unclear. The first step of the gelation is one-dimensional aggregation of the low molecular-weight organic molecules to form elongated strands. Subsequently, the strands assemble into fibers, which are entangled to form three-dimensional network structures. The solvent molecules are entrapped into the three-dimensional network to prevent their flowing, leading to the formation of viscoelastic fluid organogel. In the gelation process, a balance between gelator-gelator interaction and gelator-solvent interaction is important to control solubility and to prevent crystallization [3].

The low molecular-weight organogel has two fascinating properties. One is a physical gel property, which is attributed to the noncovalent interactions to maintain the organogel supramolecular structure. The physical organogel is responsive to external stimuli such as thermo-, photo-, chemo-, metal-, proton-, mechano-, by which the aggregate structure is stabilized or destabilized. In the stabilization process, the organogel structure is changed to a more ordered aggregation mode. On the other hand, in the destabilization process, the aggregate is dissociated to reversibly form a fluid liquid, indicating the gel-to-sol phase transition. The reversible system can be applied to a drug delivery system and a supramolecular switch system with memory function, etc. [4].

Another interesting property is their ambiguous supramolecular structure having contradictory flexibility and rigidity, because the organogel is regarded as a metastable state obtained during the crystallization process performed in solution. Morphologically, interconversion of organogel three-dimensional network structure is frozen to create rigid supramolecular structure even in solution state. On the other hand, the network structure maintained by noncovalent interactions is responsive to external stimuli such as additive, as described above, to permit the network interconversion. In one case of guest binding, for example, the binding functionalities in the organogel system can be efficiently frozen under the specific conditions of limited solvent diffusion in the three-dimensional network. The binding functionalities can interact flexibly with the guest molecule as assisted by an acceptable conformational change arising from soft noncovalent interactions. The facile interconversion between flexible and frozen supramolecular systems affords attractive specific media for molecular recognition, reaction, and energy- and electron-transfer fields, which cannot be created with the other supramolecular systems both in the solid and liquid phases. Combination of these flexibly-frozen organogel structures with the reversible organization phenomena will lead to the exploitation of very attractive nano-sized soft materials.

A number of organogels have been reported such as cholesterol [5–17], saccharide [18–20], and aliphatic amide and urea derivatives [21, 22]. The aromatic dye-based organogel is one of the new candidates for such soft materials because of its specific $\pi$-electron accumulated structure in the $\pi$-stacked fibrous aggregate and of the reversible change in the optical and electrochemical properties by external stimuli. In particular, large aromatic dye molecules are attractive self-assembling components due to their extended $\pi$-electron systems. The dye-based fibrous superstructures can be characterized by electron microscopy as well as UV/Vis absorption, fluorescence, and CD (circular dichroism) spectroscopy. The later spectroscopic methods afford useful information on how the dye molecules are assembled and how the organogel is formed. In this review, we first focus on the organogels based on self-assembling aromatic dye molecules, in particular large aromatics such as porphyrin, phthalocyanine, fullerene, triphenylene, phenylenevinylene, etc. Secondly, noteworthy topics related to the reversible functionality based on stimuli-responsiveness of the dye-based organogels are summarized.

# 2 Aromatic Dye-Based Gelators

## 2.1 Porphyrin-Based Gelators

### 2.1.1 Porphyrin-Appended Cholesterol Gelators

Cholesterol-based gelators have been extensively studied by Weiss's group [5–9], Terech's group [10, 11], Shinkai's group [12–15], and Whitten's group [16, 17]. A number of aromatic dyes such as anthracene [5–7, 10], anthraquinone [6–9, 11], azobenzene [12–14], nucleobase [15], and stilbene [16, 17], covalently-appended at 3-position of the cholesterol moiety are arranged in a highly-ordered manner in the organogel systems. When the cholesterol moieties constitute one-dimensional helical columnar stacking, the appended aromatic groups can enjoy the face-to-face type interaction and

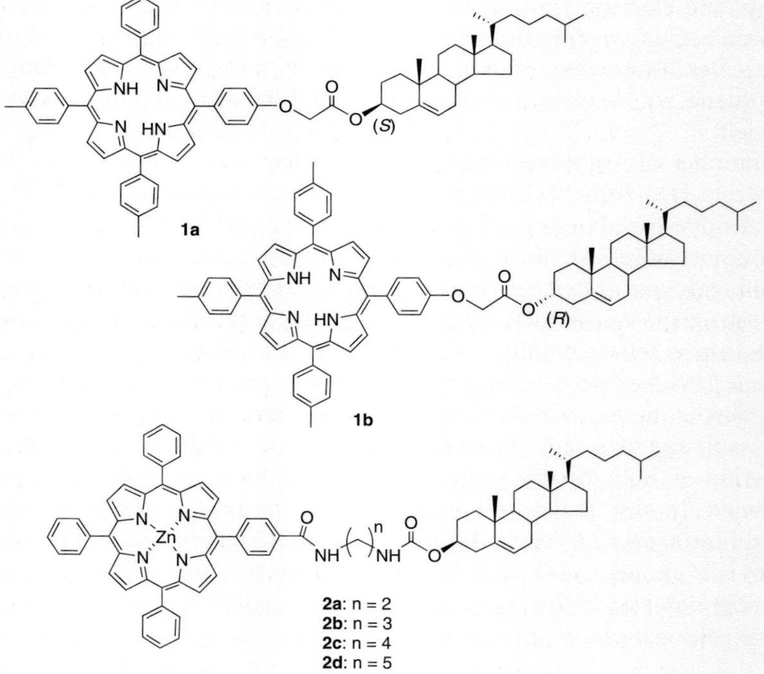

**Fig. 1** Porphyrin-appended cholesterol gelators

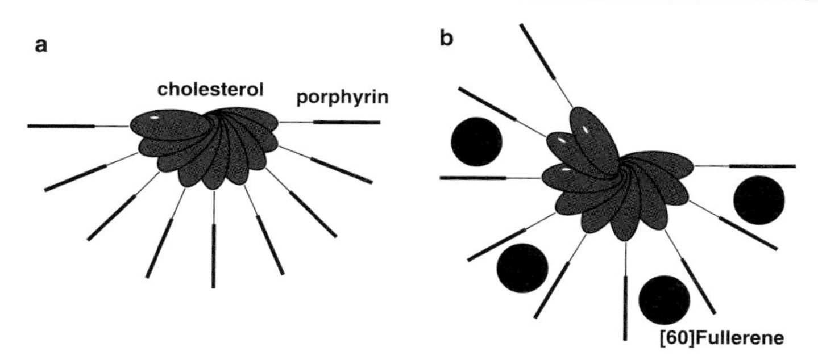


**Fig. 2** One-dimensional aggregate models of **2** (**a**) in the absence and (**b**) in the presence of [60]fullerene in the gel phase

stabilize the aggregate. The one-dimensional aggregate structure has been elucidated by UV/Vis and CD spectroscopy [13], electron-micrographic observation [13], XRD (X-ray diffraction) [8], AFM (atomic force microscopy) observation [17], and synchrotron small angle X-ray scattering [10, 11, 14].

Porphyrin-appended cholesterol gelators **1** and **2** were designed and prepared by Shinkai's group [23]. The first examples are **1a** with a natural (*S*)-configuration at 3-position and **1b** with an inverted (*R*)-configuration at 3-position (Fig. 1). Gelator **1a** can gelate cyclohexane and methylcyclohexane, whereas **1b** is soluble in these solvents. In the aggregate state, the porphyrin moieties in **1a** with (*S*)-configuration can interact with each other to stabilize the one-dimensional cholesterol columnar packing (Fig. 2a). On the other hand, such a packing mode is difficult in **1b** due to steric reasons arising from the inverted (*R*)-configuration [13]. The fact that the porphyrin rings in the gel phase are stacked according to the J-aggregation is confirmed by UV/Vis and CD spectroscopy [24].

According to a similar strategy as shown in **1a**, Zn(II) porphyrin-appended cholesterol gelators **2a–d** were designed and prepared (Fig. 1) [25, 26]. Interestingly, the gelation ability of **2a–d** is highly affected by odd or even carbon numbers in the spacer $(CH_2)_n$ group connecting the Zn(II) porphyrin moiety and the cholesterol moiety [26]. Gelators **2a** and **2c** with an even carbon number (2 and 4) in the spacer moiety gelated aromatic hydrocarbons such as benzene, toluene, and *p*-xylene. In contrast, **2b** and **2d** with an odd carbon number (3 and 5) could not gelate these solvents. As shown in Fig. 2a, when cholesterol-based gelators are packed into a one-dimensional helical column via the cholesterol-cholesterol interaction, the Zn(II) porphyrin moieties can interact with each other to stabilize the aggregate. The hydrogen-bonding interactions between amide and carbamate groups scarcely contribute to the construction of the helical column. The novel aggregate structure is very sensitive to the carbon number in the spacer $(CH_2)_n$ moiety in **2**. Only the even $(CH_2)_n$ spacer can arrange both the cholesterol and the Zn(II) porphyrin moieties cooperatively to generate the cholesterol-cholesterol interaction and the Zn(II) porphyrin-Zn(II) porphyrin interaction, leading to the construction of the one-dimensional helical column.

### 2.1.2
### Porphyrin Gelators with Peripheral Hydrogen-Bonding Functionalities

Hydrogen-bonding interaction has been used as a powerful driving force for the formation of organogel structures. Amide- and urea-based gelators

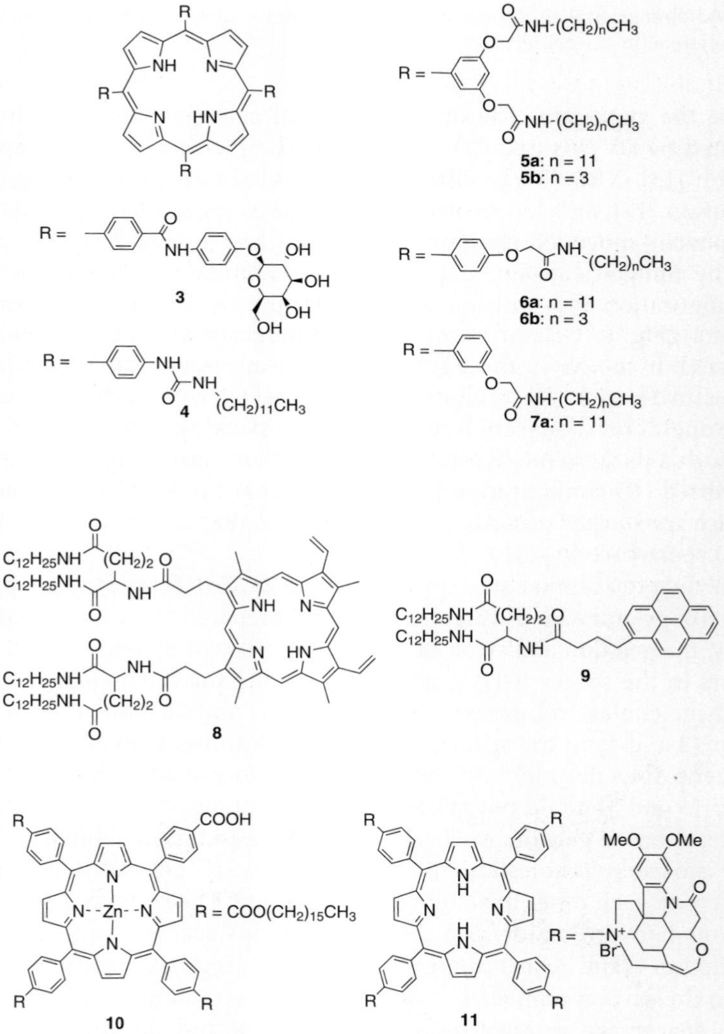

**Fig. 3** Porphyrin-based gelators having hydrogen-bonding functionalities

have been developed by Hanabusa's group [21], and Feringa and van Esch's group [22], independently. A natural library of carbohydrates has provided building blocks in hydrogen-bonded gelators [18–20]. These findings have offered a new idea that the porphyrin-based organogel would be further reinforced by the hydrogen-bonding interactions among urea and saccharide groups covalently-appended to the central porphyrin column. Thus, porphyrins with four $\beta$-D-galactopyranoside moieties **3** and with four urea groups **4** were designed and prepared as new porphyrin-based gelators (Fig. 3).

Gelator **3** resulted in gels in DMF mixed solvents with methanol, ethanol, *i*-propyl alcohol, and benzyl alcohol [27]. In this organogel structure, the $\pi$–$\pi$ stacking interaction among porphyrin moieties and the hydrogen-bonding interaction among the saccharide moieties cooperatively stabilized the one-dimensional aggregate structure, as indicated by UV/Vis and CD spectroscopic studies and electron-micrographic observations. In addition to the two different types of interactions, saccharide-solvent hydrogen-bonding interactions in the DMF/alcohol mixed solvents play an important role to control subtle balance in the gel formation.

Another porphyrin-based organogel was formed in anisole and diphenyl ether solutions of **4**, in which porphyrin-porphyrin $\pi$–$\pi$ stacking and urea-urea hydrogen-bonding interactions acted as driving forces for the formation of the one-dimensional aggregates [28]. Interestingly, the organogel structure was reinforced by addition of chiral urea "(*R*)- and (*S*)-enantiomers of *N*-(1-phenylethyl)-*N*'-dodecyl urea", which interacted with the urea groups in **4**. In the absence of the chiral additives, the porphyrin moieties in the aggregated **4** assemble into a one-dimensional aggregate in a face-to-face manner (Fig. 4). In contrast, the chiral additives twisted the one-dimensional porphyrin column in a chiral manner (Fig. 5). The formation of the chiral aggregate was reflected in an exciton-coupling Soret band found in the CD spectral observation.

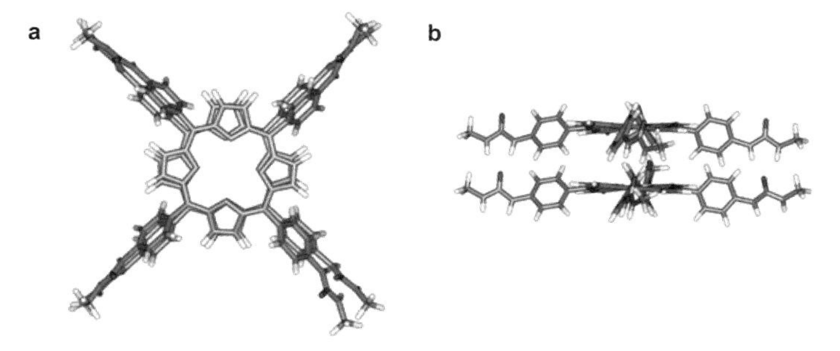


**Fig. 4** Energy-minimized structures of **4**: (**a**) top view, (**b**) side view

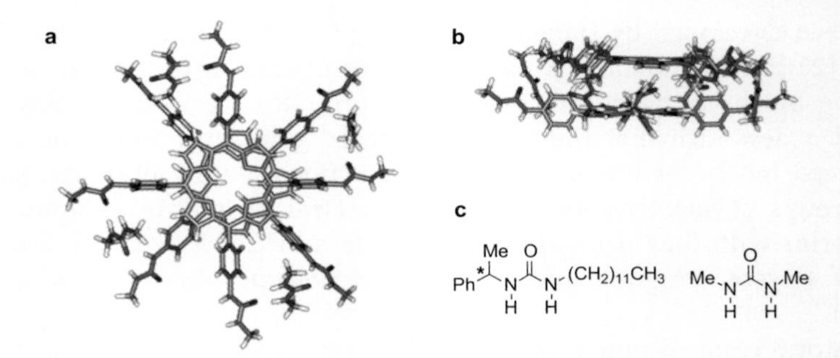


**Fig. 5** Energy-minimized structures of **4**: *N*,*N*′-dimethyl urea = 2 : 4 complex; (**a**) top view, (**b**) side view (to simplify the computational process, *N*,*N*′-dimethyl urea was used as additive instead of *N*-(1-phenylethyl)-*N*′-dodecyl urea). (**c**) The structures of *N*-(1-phenylethyl)-*N*′-dodecyl urea and *N*,*N*′-dimethyl urea used as additives

A systematic study of porphyrin-based gelators with peripheral hydrogen-bonding functionalities was performed in **5a**, **6a**, and **7a** with amide groups (Fig. 3) [29]. Gelators **6a** with the amide groups at the 4 position of the *meso*-phenyl groups acted as a versatile gelator. In contrast, gelators **5a** and **7a** with the amide groups at the 3,5-positions and 3-position, respectively, acted as poor gelators (Table 1). The difference in the gelation ability is explained by the aggregation mode. In the Soret band of UV/Vis spectral observations, the blue shift observed for **6a** indicates the H-aggregation mode of the porphyrin moieties, whereas the red shifts observed for **5a** and **7a** indicate the J-aggregation mode.

The J versus H-aggregation mode is supported by X-ray analysis of the single crystals in the model compounds **5b** and **6b** with terminal *n*-butyl groups. In **6b**, porphyrin rings form a one-dimensional stacking column with the H-aggregation mode, and the four columns are connected by amide-amide hydrogen-bonding interactions. The porphyrin-porphyrin stacking interaction and the amide–amide hydrogen-bonding interaction stabilize cooperatively the columnar superstructure, as found in **3** and **4**. On the other hand, in **5b** the porphyrin moieties are arranged two-dimensionally to form a layer structure coplanar to the porphyrin rings, and the amide groups are used to connect them within the layer. In each layer, the amide-amide hydrogen-bonding force plays a decisive role in determining the structure. SEM observations are also supportive of the H versus J aggregation mode, in which the one-dimensional fibrillar structure in **6a** is ascribed to the H-aggregate, whereas the two-dimensional sheet-like structure in **5a** to the J-aggregation (Fig. 6).

Microfibrous self-aggregation of porphyrins substituted by didodecyl L-glutamic acid **8** was investigated by Sagawa and Ihara's group (Fig. 3). The aggregation occurred even at a low concentration below the critical gel concentration. In the mixed system with the corresponding pyrene derivative

**Table 1** Organic solvents tested for gelation by **5a**, **6a**, **7a**, **5b**, and **6b**[a]

| Solvent | **5a** | **6a** | **7a** | **5b** | **6b** |
|---|---|---|---|---|---|
| benzene | P | P | S | G | G |
| toluene | P | P | S | G | G |
| *p*-xylene | G | P | S | P | G |
| pyridine | S | S | S | S | S |
| diphenyl ether | P | G | S | G | G |
| anisole | P | P | S | P | G |
| tetralin | G | S | S | P | G |
| cyclohexane | G | G | G | I | I |
| methylcyclohexane | G | G | G | I | P |
| hexane | P | P | P | I | I |
| octane | G | G | G | I | P |
| 1-octanol | S | P | S | G | G |
| 1-butanol | S | P | S | S | G |
| 2-propanol | S | G | S | S | G |
| 1-propanol | S | G | S | S | G |
| ethanol | S | G | S | S | G |
| methanol | P | G | P | S | P |
| acetonitrile | P | P | P | S | P |
| ethyl acetate | P | P | S | S | G |
| acetone | P | G | S | S | G |
| THF | S | S | S | S | S |
| dioxane | S | S | S | S | S |
| carbon tetrachloride | G | G | S | I | G |

[a] [gelator] = 10–50 g $dm^{-3}$; G = gel, S = solution, P = precipitation, I = insoluble

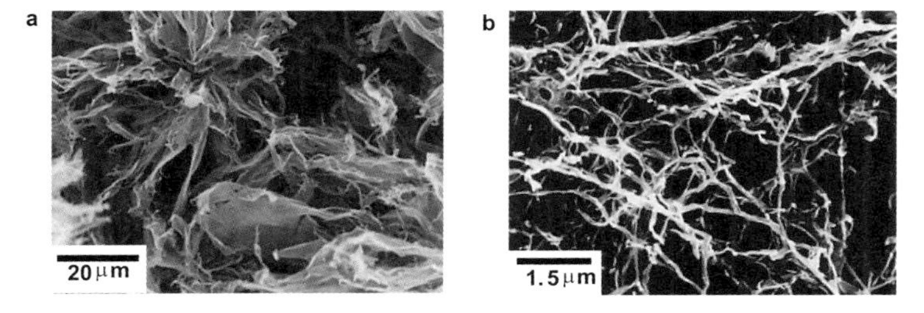


**Fig. 6** SEM images of (**a**) **5a** and (**b**) **6a** xerogels prepared from cyclohexane

**9**, a singlet-singlet energy-transfer from the pyrene to the porphyrin chromophore was detected, which suggests an advantage of the gel system as a medium for energy transfer [30]. Zn(II) porphyrin with a carboxyl group and three hexadecyloxycarbonyl groups **10** formed rod-like aggregates in the cyclohexane gel phase (Fig. 3). Small-angle neutron scattering and small-angle X-ray scattering experiments indicate that the aggregates are molecular threads with only one molecular per cross section [31].

Porphyrin-based gelator **11** with four chiral brucine pendants can aggregate in methanol and methanol/DMSO mixed solvents, leading to a robust dark red-brown transparent organogel with a highly ordered chiral supramolecular structure (Fig. 3) [32]. The $\pi$–$\pi$ stacking force among porphyrin moieties seems to be responsible for the chiral superstructure formation and indispensable for the organogel formation. The polar brucine units have crucial influence on the gelation abilities, although the details are not clear. The organogel of **11** was investigated by vibrational circular dichroism spectroscopy for the first time in the organogel system, which opens a new characterization method for the chiral organogels.

### 2.1.3 Porphyrin-Fullerene Interactions Working in Organogel Systems

Recently, porphyrin-fullerene conjugate systems have been of much concern as new, potential architectural photosynthetic models [33]. Because of the experimental convenience, the electronic interactions have so far been studied in covalently linked porphyrin-fullerene intramolecular systems. In contrast, the intermolecular system, in which fullerenes are used without chemical modification, has never been reported except only in special cases which found inclusion of [60]fullerene in cyclic porphyrin dimers [34] and acyclic porphyrin dimers having dendritic wedges [35]. The finding implies that, although the intermolecular porphyrin-fullerene interaction is considerably weaker, it becomes detectable by preorganization of porphyrin-based host molecules. As described in Sects. 2.1.1 and 2.1.2, even porphyrins can be assembled into fibrous aggregates in organogel systems. It thus occurred that added fullerenes would be intermolecularly bound to the aggregated porphyrin stacks due to the multipoint interaction and that such an interaction would reinforce the gel structure. The organogel supramolecular structures of porphyrin-based gelators can be reinforced by intermolecular porphyrin-fullerene interactions [25, 26, 36].

Firstly, the influence of added [60]fullerene on the gelation ability was studied in Zn(II) porphyrin-appended cholesterols **2a** and **2c**, which can gelate aromatic hydrocarbons [25, 26]. The gel-to-sol phase transition temperature of **2a** and **2c** significantly increased with increasing the equivalent of added [60]fullerene, indicating that the reinforcement of the organogel structure occurred by the intermolecular Zn(II) porphyrin-[60]fullerene interaction (Table 2). The interaction is detected with UV/Vis spectra, in which the porphyrin Soret absorption band shifts to a longer wavelength region. In the aggregate of **2a** and **2c**, the Zn(II) porphyrin moieties oriented outside the one-dimensional cholesterol stacking are preorganized to precisely interact with the [60]fullerene molecule (Fig. 2a). In the presence of [60]fullerene,

**Table 2** Influence of the added [60]fullerene on the sol-gel phase transition temperatures ($T_{gel}$) of **2a**/toluene and **2c**/toluene gels[a]

| [60]fullerene/**2a,c** (mol/mol) | $T_{gel}$(°C) **2a** | **2c** |
|---|---|---|
| 0.0 | 29 | < 5 |
| 0.1 | 36 | < 5 |
| 0.2 | 51 | < 5 |
| 0.3 | 64 | < 20 |
| 0.4 | 72 | 35 |
| 0.5 | 78 | 49 |
| 0.6 | 78 | 49 |

[a] [**2a,c**] = 0.20 mol $dm^{-3}$

the two Zn(II) porphyrin moieties interact with one [60]fullerene molecule to form the 2:1 Zn(II) porphyrin/fullerene sandwich complex (Fig. 2b). The helicity of the one-dimensional column is controlled to be an anti-clockwise mode as indicated by the Cotton effects found in the porphyrin Soret absorption band.

As described in section Sect. 2.1.2, **5a** takes a two-dimensional sheet structure in the organogel aggregate system but not a one-dimensional columnar one arising from the strong amide-amide hydrogen-bonding interactions (Fig. 7) [29]. Interestingly, the two-dimensional aggregate structure is transformed into a one-dimensional one by the intermolecular porphyrin-fullerene interaction (Fig. 7) [36]. The transformation can be visualized by SEM and TEM observation (Fig. 8). The striking morphological change is ascribed to the formation of 1 : 2 [60]fullerene/**5a** complex and a circular hydrogen-bonding array among two sets of four amide groups. In the newly generated structure, four amide groups in the [60]fullerene-including side form straight hydrogen-bonding bridges with those in another porphyrin, whereas four amide groups on the opposite side form bent hydrogen-bonding bridges with those in the next porphyrin (Fig. 7). The one-dimensional aggregate of **5a** thus provides cavities large enough to entrap [60]fullerene and cavities too small to accept it.

The intermolecular porphyrin-fullerene interaction found in the organogel system of **2a,c** and **5a** is very rare and valuable, because in the isotropic solution state the intermolecular interaction of porphyrin with fullerene has never been reported except only in special cases as described above [34, 35]. The supramolecular organogel system can provide a specific guest-recognition medium, which cannot be created in an isotropic solution system.

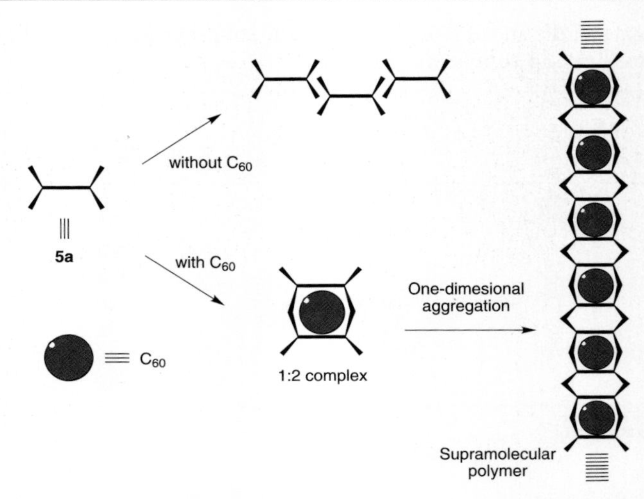


**Fig. 7** Aggregation mode of **5a** in the presence and the absence of [60]fullerene

**Fig. 8** SEM images of the xerogels prepared from (**a**) **5a**, (**b**) **5a** with 0.1 equiv. of [60]fullerene, and (**c**) **5a** with 0.5 equiv. of [60]fullerene

## 2.1.4
## Transcription of Organogel Structure into Silica Materials

Recently, Shinkai's group has found that organogel supramolecular structures can be elaborately transcribed into inorganic silica materials by utilizing the template-silanol interaction [37, 38]. Sol-gel polycondensation of tetraethoxysilane (TEOS) proceeds along the organogel aggregate used as a template fiber, leading to various types of hollow silica.

This concept called "sol-gel transcription" can be applied to a porphyrin-based organogel system of **3** [39]. The slow speed of the bundle formation of **3** in DMF/alcohol provided a greater chance to transcribe nano-sized incipient fibers into silica. Gelator **3** aggregates unimolecularly into a one-dimensional stack, along which sol-gel polycondensation of TEOS proceeds in the gel phase. The formed hollow fiber silica has a nearly monodispersed hollow size with 4–5 nm inner diameters, which is comparable to the mo-

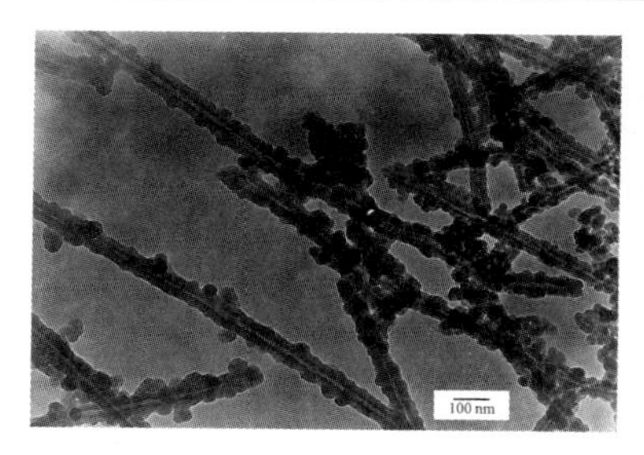

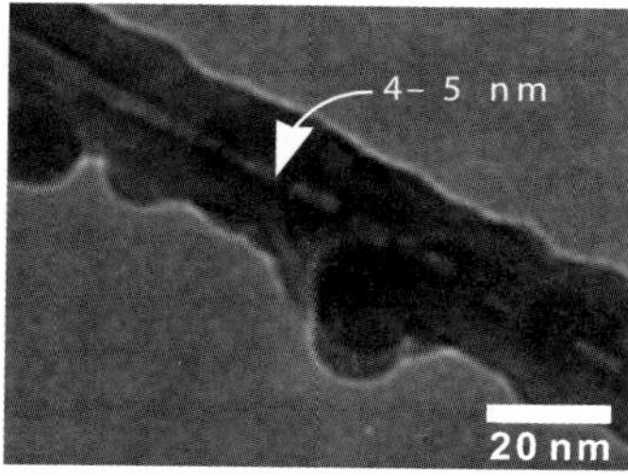


**Fig. 9** TEM images of the silica structure prepared by transcription of unimolecular aggregate of 3 as a template

lecular length of **3** (Fig. 9). This is a rare example of a unimolecular stack immobilized in an inorganic material.

## 2.2 Fullerene-Based Gelators

It is well known that [60]fullerene and its homologues tend to form three-dimensional aggregates without any specific orientation mode. In spite of this disadvantage, ordered fullerene assemblies have been successfully formed in some specific fields, e.g. (i) monolayers of fullerene and its derivatives with hydrophilic groups at the air-water interface [40–42]; (ii) rods and vesicles of fullerenes with hydrophilic groups [43]; (iii) encapsulated fullerenes in spherical aggregates of block copolymers [44]; and (iv) self-assembled monolayers of thiol-containing fullerenes on gold [40]. In general, however, the design of highly ordered aggregates from [60]fullerene has been considered to be difficult in solution.

An example of ordered [60]fullerene assembly in solution is found in an organogel system of [60]fullerene-containing amphiphile **12** [45]. A methanol solution of **12** was totally transformed into an organogel after prolonged standing for a few days (Fig. 10). SEM and XRD observations indicate that a molecular orientation change in **12** from the less ordered structure to the more ordered structure induces a morphological change from globular aggregates to fibrous aggregates.

[60]fullerene-based organogel was created on the basis of a similar strategy using a cholesterol unit [46]. Cholesterol-appended [60]fullerene gelator **13** can gelate dichloromethane leading to the formation of chirally ordered [60]fullerene aggregates (Fig. 10). In the gel state of **13**, the cholesterol-cholesterol interaction and the fullerene-fullerene cohesive force cooperatively act to construct the chiral assembly, as found in precedent cholesterol-based gelators [13]. The columnar one-dimensional packing of the cholesterol moieties constitutes the helical structure where the [60]fullerene moieties are chirally oriented outside the helical column (Fig. 2a). This

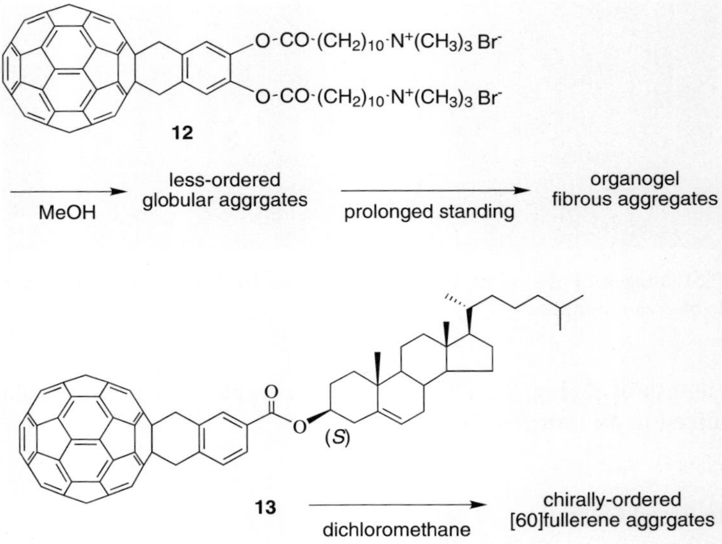

**Fig. 10** Fullerene-based gelators

is the first example for chirally ordered fullerene assemblies. Thus, the utilization of characteristic organogel supramolecular structures opens a new possibility to create ordered fullerene assemblies in solution.

## 2.3 Phenylenevinylene-Based Gelators

Phenylenevinylene derivatives have been of much concern because of their potential use in photovoltaic and light-emitting devices. Reports on phenylenevinylene-based aggregated structures were limited only to a few liquid crystal, solid state packing, and self-assembly systems. The first example on phenylenevinylene-based organogel supramolecular structure was reported by Ajayaghosh's group [47]. Phenylenevinylene **14a,b** with terminal two hydroxyl groups can gelate hexane, cyclohexane, benzene, and toluene (Fig. 11). Compared to **14a,b**, gelation by **15a,b** without the terminal hydroxyl groups occurred at a temperature below 0 °C as well as at much higher concentration. The results indicate that in addition to $\pi$-stacking forces of the phenylenevinylene backbone, hydrogen bonding interactions among the hydroxyl groups in **14a,b** play a crucial role in stabilizing the organogel structures. The XRD pattern and FT-IR spectrum of **14b** show a $\pi$-stacked lamellar packing structure assisted with hydrogen bonding interactions.

When chiral substituents were introduced in the phenylenevinylene-based gelator **14c**, a hierarchical self-organization into coiled-coil aggregates of nanometer dimensions was achieved [48]. AFM analysis of **14c** gel from do-

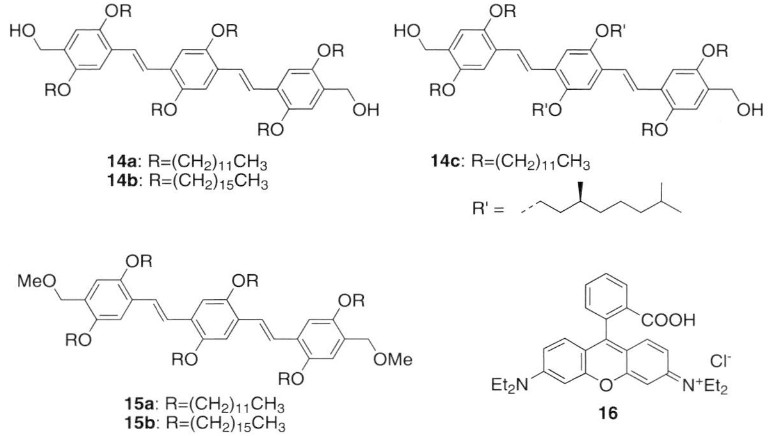

**14a**: $R=(CH_2)_{11}CH_3$
**14b**: $R=(CH_2)_{15}CH_3$

**14c**: $R=(CH_2)_{11}CH_3$

**15a**: $R=(CH_2)_{11}CH_3$
**15b**: $R=(CH_2)_{15}CH_3$

**16**

**Fig. 11** Phenylenevinylene-based gelators

decane showed the formation of entangled left-handed helical coiled-coil fibers of 50–100 nm in diameter. The morphology of a left-handed coiled-coil rope is approximately 90 nm in width and 1.1 nm in height, in which the pitch of each helix has an angle of about 40° with respect to the main fiber axis. Concentration- and temperature-dependent changes in UV/Vis and CD spectroscopic measurements of **14c** provided insight into the hierarchical helical self-assembly with two-stage transition. In the initial stage, the molecules of **14c** organize to form a left-handed chiral aggregate. Then, the chiral assemblies grow further into helical fibers and coiled-coil ropes.

An interesting light-harvesting phenomenon was found in **14b**-based organogel containing Rhodamine B dye **16** [49]. The dye molecules are trapped in the $\pi$-stacked lamellar packing structure. After irradiation of phenylenevinylene chromophore in **14b**, energy-transfer occurred to a significant level from **14b** to **16** to generate emission of **16** selectively. The emission from dye **16** was several times more intense than that resulting from the direct excitation of the dye, indicating the realization of an efficient light-harvesting system in this phenylenevinylene-dye organogel. The energy transfer preferably occurs to the gel-trapped dye molecules, which seem to be present within Förster radii.

## 2.4 Phthalocyanine-Based Gelators

Aggregation of phthalocyanine molecules in an organogel system was reported by Nolte's group [50, 51]. Phthalocyanine **17** with four crown ether rings and eight alkyl tails self-assembles in chloroform solution to form an organogel (Fig. 12) [50]. In the UV/Vis spectra, a blue shift of Q-bands

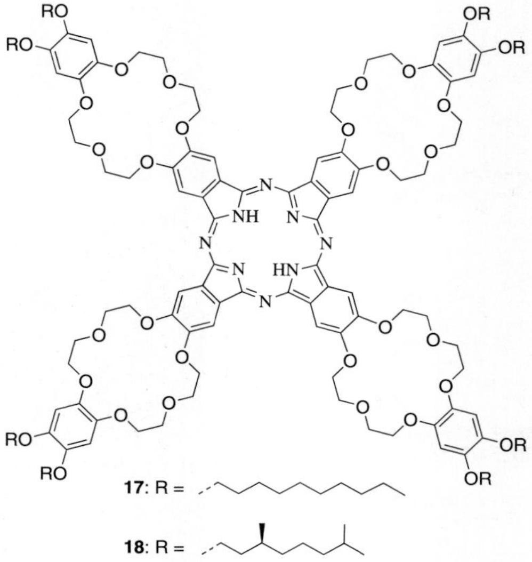

**Fig. 12** Phthalocyanine-based gelators

and a broadening of the signals were observed, which are characteristic of aggregated phthalocyanine species. TEM observation shows that in the three-dimensional network gel structure each fiber is built up of parallel aggregation of unimolecularly stacked strands. The aggregated fibers can be considered to be a multiwired molecular cable containing a central wire of stacked phthalocyanines as an electron channel, four molecular channels built up from stacked crown ether rings as ion channels, and a surrounding hydrocarbon mantle.

Phthalocyanine **18** with eight chiral tails also forms organogel having an aggregation mode similar to that found in **17** [51]. Interestingly, in the organogel system, the (*S*)-chiral centers in the tails induce a clockwise orientation of the molecular disks which leads to the formation of fibers with right-handed helicity (Fig. 12). The fibers, in turn, self-assemble to form coiled-coil aggregates with left-handed helicity.

## 2.5 Triphenylene- and Hexaazatriphenylene-Based Gelators

Triphenylene has attracted increasing attention because of its possible application to the materials science field arising from the high hole-transporting ability [52]. In this section, new organogels based on triphenylene and its hexaaza-analogue hexaazatriphenylene are introduced.

Triphenylene **19**, which has six long alkyl chains as tail groups and six amide groups as hydrogen-bonding sites, gelated some hydrocarbon solvents

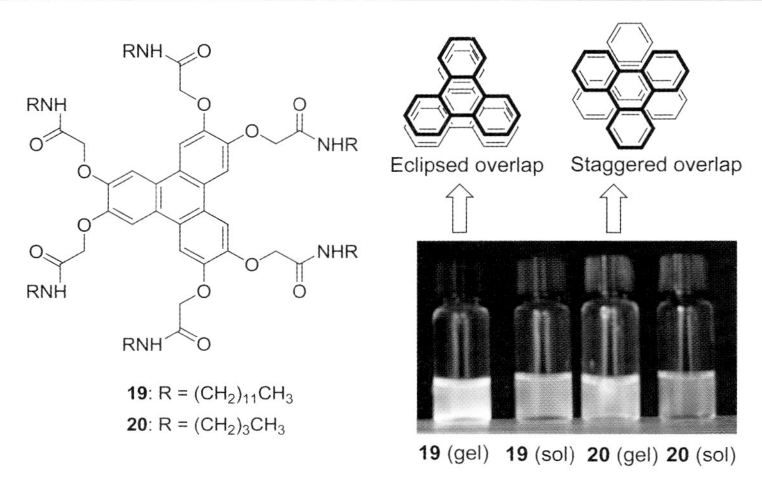


**Fig. 13** Triphenylene-based gelators and their fluorescent images in gel and sol state

such as *n*-hexane, *n*-octane, cyclohexane, and *p*-xylene [53]. In contrast, **20** with short alkyl chains was not able to gelate such hydrocarbon solvents except cyclohexane (Fig. 13). Interestingly, in the gel phase of **19**, triphenylene moieties adopt an unusual eclipsed overlap to yield excimer emission (Fig. 13). In the gel phase of **20** which did not show such excimer emission, the triphenylene moieties adopt a staggered or helical overlap as found in triphenylene-based liquid crystal phases reported so far (Fig. 13) [52]. The hydrogen-bonding interaction among the amide moieties as well as the van der Waals interaction among the long alkyl chains in **19** would play important roles in constructing the unique but unusual eclipsed overlap of triphenylenes moieties in the supramolecular assembly.

n-Type semiconducting hexaazatriphenylene, which acted as an electron-transporting material [54], can be developed as an organogel. Ishi-i's group reported that hexaazatriphenylene derivative **21a** can gelate 1,2-dichloroethane and aromatic hydrocarbons such as benzene, toluene, *p*-xylene, aniline, and nitrobenzene (Fig. 14) [55]. Interestingly, **21a** composed

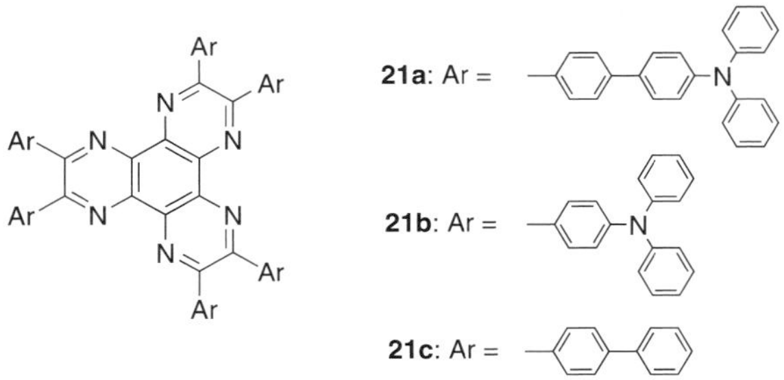

**Fig. 14** Hexaazatriphenylene-based gelators

**Table 3** Organic solvents tested for gelation by **21a-c**[a]

| Solvent | **21a** | **21b** | **21c** |
|---|---|---|---|
| hexane | I | I | I |
| cyclohexane | I | I | I |
| toluene | pG | S | R |
| *p*-xylene | pG | S | R |
| nitrobenzene | S (G)[b] | S | S |
| aniline | G | S | R |
| dichloromethane | S | S | R |
| chloroform | S | S | S |
| 1,2-dichloroethane | pG | S | R |
| ethanol | I | I | I |
| ether | I | I | I |
| THF | S | S | R |
| ethyl acetate | I | I | I |
| *N*-methylpyrrolidone | R | S | S |

[a] [**21a–c**] = $1.0 \times 10^{-2}$ M; at 20 °C, G = gel, *p*G = partial gel, R = recrystallization, S = solution, I = insoluble.
[b] The data in parenthesis is at 5 °C

of an aromatic hexaazatriphenylene core and flexible aromatic side chains acts as a gelator in spite of lacking alkyl side chains. In contrast to **21a**, no organogel phase could be formed from **21b** with a phenylene spacer, and **21c** without terminal diarylamino groups (Table 3). It can be concluded that in **21a** the flexibility of the terminal groups and the length of the biphenylene spacer are crucial for the formation of the organogel supramolecular structure. The aromatic side chains with terminal flexible groups make up a soft region, which cooperatively stabilize the supramolecular structure together with the hard region of the hexaazatriphenylene core. The created supramolecular aggregate would be considered as an attractive functional material because of its semiconducting property as well as high $\pi$-electron density and the presence of photochemically and electrochemically active aromatic side chains.

## 2.6 Azobenzene-Based Gelators

Azobenzene-based gelators have been studied extensively, because spectroscopic properties of the azobenzene moiety afford useful information on how the molecules are assembled and how the gel is formed. A typical sample was azobenzene-appended cholesterol gelators **22a–e**, in which azobenzene $\pi$–$\pi$ stacking and cholesterol-cholesterol interaction cooperatively stabilize the organogel structure as described in Sect. 2.1.1 (Fig. 15) [12, 13]. Hydrogen bond-assisted azobenzene gelators **23a,b** were created by introducing two urea functional groups to an azobenzene core (Fig. 15) [56]. Azobenzene gelator **24** containing a cyclic syn-carbonate moiety was self-assembled through the dipole-dipole interaction of the carbonate group to form a ribbon-like helical aggregate (Fig. 15) [57]. A hydrogel was obtained from azobenzene-based dye **25** containing two trifluoromethyl groups [58]. In the lower concentration region, the dye dimmer is formed, whereas the polyaggregates with several types of aggregation are formed in the higher concentration to afford a hydrogel (Fig. 15).

An interesting class of azobenzene-based gelators with two saccharide moieties was reported by Shinkai's group (Fig. 15) [59]. Bolaamphiphilic azobenzene derivative with two $\beta$-D-glucopyranoside moieties **26a** formed a gel in water even at concentrations as low as 0.05 wt% (0.65 mM). In a water-DMSO mixture the gelation ability of **26a** decreased with an increasing portion of DMSO (Table 4), showing that $\pi$–$\pi$ stacking of the azobenzene moieties is most probably the principal driving force for the gelation in water. In the hydrogel structure, the azobenzene moieties adopt an H-aggregation mode, as indicated by the blue shift of the UV/Vis absorption spectra. SEM

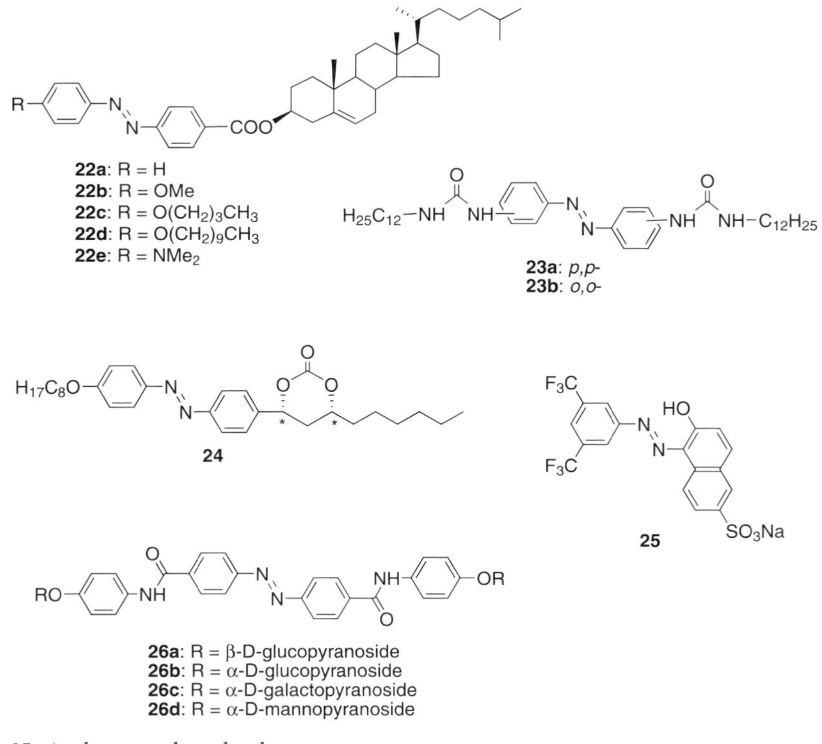


**Fig. 15** Azobenzene-based gelators

**Table 4** Gelation Test for **26a-d** in DMSO-water mixtures[a]

| DMSO:water (v/v) | **26a** | **26b** | **26c** | **26d** |
|---|---|---|---|---|
| 0:100 | G | P | I | I |
| 20:80 | G | P | | |
| 40:60 | G | G | | |
| 50:50 | G | G | P | P |
| 60:40 | G | G | | |
| 80:20 | S | G | | |
| 100:0 | S | S | P | P |

[a] [**26a–d**] = 3.0 (wt/v)%; G = gel, S = solution, P = precipitation, I = insoluble

and TEM observation and CD spectroscopic studies indicate that in the H-aggregation mode the azobenzene moieties are stacked in a clockwise direction with right-handed helicity (Fig. 16a). Thus, the chirality of the saccharide moieties at the molecular level is reflected in the hydrogel supramolecular level to form the unique helical aggregate. The aggregate structure of bolaamphiphilic azobenzene derivatives **26a–d** is affected by the structure of the terminal saccharide moieties (Figs. 15 and 16). $\alpha$-D-Glucopyranoside derivative **26b** formed a hydrogel, which is less stable than that of **26a** (Table 4). In contrast, $\alpha$-D-galactopyranoside derivative **26c** and $\alpha$-D-mannopyranoside derivative **26d** did not form the hydrogel (Table 4) and tended to aggregate into a two-dimensional lamellar structure (Fig. 16b). The difference is rationalized in terms of the difference in the absolute configuration in the saccharide moieties. The axial OH groups in **26c** and **26d** facilitate the two-dimensional aggregation (Fig. 16b), whereas the equatorial OH groups in **26a** and **26b** facilitate the one-dimensional fibrous aggregation (Fig. 16a). The foregoing studies indicate that in the bolaamphiphiles bearing two terminal

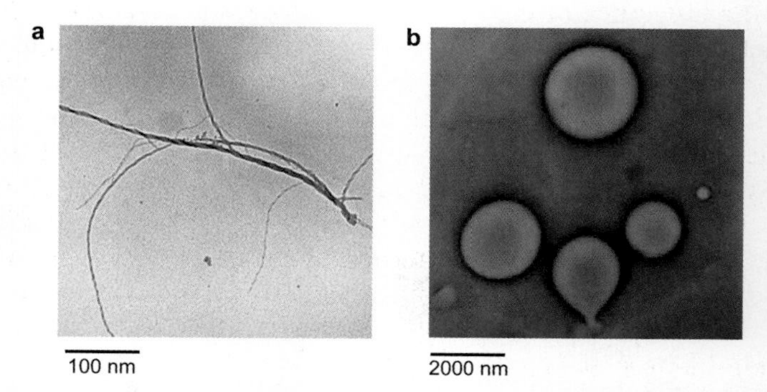


**Fig. 16** TEM images of (**a**) **26b** and (**b**) **26c** in 1 : 1 (v/v) DMSO-water mixtures. The TEM images of **26a** and **26d** (not shown here) were basically similar to those of **26b** and **26c**, respectively

saccharide moieties and one central azobenzene segment, excellent gelators can be found by combinatorial screening of the appropriate saccharide moieties from a carbohydrate family.

## 2.7
## Other Aromatic Dye-Based Gelators

Other aromatic dye-based gelators **27–31** were designed and prepared utilizing the concept of hydrogen-bonding functionality (Fig. 17). In the organogel phase thiophene-bisurea **27a** and bithiophene-bisurea **27b** form a lamellar structure, which is stabilized by $\pi$–$\pi$ interactions among the thiophene rings and hydrogen-bonding chains among the urea functional groups [60]. The thiophene moieties adopt a closely packed arrangement with a face-to-face stacking along the same direction of the infinite hydrogen-bonded chains (Fig. 17). Feringa and van Esch's group found that the extended aggregates of the $\pi$-stacked thiophene moieties provide an efficient path for charge-carrier transport within the organogel fibers. In a solid state, **27a** and **27b** indicated conductivity of $1 \times 10^{-3}$ cm $V^{-1}$ $s^{-1}$ and $5 \times 10^{-3}$ cm $V^{-1}$ $s^{-1}$, respectively, which are larger than that of their model compound [60].

A tropone-based gelator was reported by Mori's group. Gelator **28** takes a flat conformation arising from an intramolecular hydrogen-bonding interaction between the amide proton and the tropone carbonyl group (Fig. 17). The flat $\pi$-system in turn gives rise to strong $\pi$–$\pi$ interactions and provides the good gelation ability [61]. Meijer's group reported that $C_3$-symmetrical trisamides **29a,b** and trisureas **30a,b** act as gelators by contribution of the hydrogen-bonding and $\pi$–$\pi$ interactions (Fig. 17) [62]. In **29b** and **30b** with chiral substituents, elongated helical stacks of the $C_3$-symmetrical molecules are formed, as indicated by their strong Cotton effects. The urea stacks in **30a,b** are much more rigid than the corresponding amide ones in **29a,b.** Kinetic and thermodynamic stability of the organogel structures highly depend on the difference of the hydrogen-bonding functionalities. The amide disks **29a,b** immediately reach their thermodynamic equilibrium, whereas kinetic factors seem to govern the urea aggregation in **30a,b**. Similarly, $C_3$-symmetrical trisamide **31** containing three oxadiazole rings showed gelation capabilities for THF, 1,4-dioxane, and chlorinated solvents (Fig. 17) [63]. Interestingly, at the monomer state **31** was nonfluorescent, while the aggregate of **31** at the gel state emitted strong fluorescence. The nonradiative intersystem crossing process is inhibited by the aggregation of the **31** molecules through face-to-face intermolecular hydrogen-boding interactions among amide groups. This phenomenon is a novel supramolecular version of the aggregation-induced enhanced emission.

A similar aggregation-induced enhanced emission is reported by the same group of Park and co-workers (Fig. 17). Trifluoromethyl-based cyanostilbene **32** acts as a gelator for 1,2-dichlormethane. The formed gel is strongly fluor-

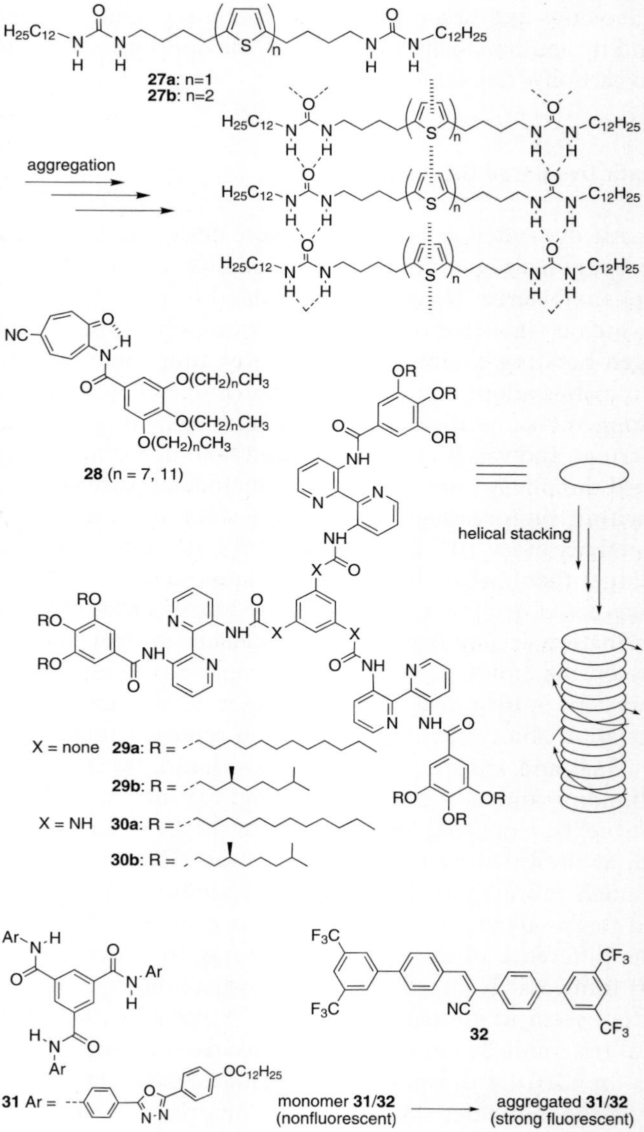

**Fig. 17** Oligothiophene-, tropone-, tris(bipyridylamide- and -urea)-, tris(oxadiazolylamide)-, and trifluoromethyl-containing cyanostilbene-based gelators

escent, while **32** itself is totally nonfluorescent in the 1,2-dichloromethan solution [64]. The gelation capability of **32** is attributed to the cooperative effect of the strong $\pi$–$\pi$ stacking interactions exerted by rigid rodlike aromatic segments in **32** and the supplementary intermolecular interactions induced by four trifluoromethyl groups. The gelator **32** is a unique and interesting example of a new class of gelators without long alkyl chains or steroidal substituents as found in hexaazatriphenylene-based gelator **21a** [55]. The trifluoromethyl group seems to acts as an important structural part for the effective gelation process [58].

Merocyanine dye has a highly dipolar structure with a large dipolar moment, by which the dye molecules can be self-assembled through electrostatic interactions to form tightly bound aggregates [65]. A merocyanine-based organogel was reported by Würthner's group [66]. Self-assembling of bis(merocyanine) dye **33** in nonpolar solvents such as cyclohexane results in a supramolecular polymer through the electrostatic interactions among the highly dipolar dye moieties (Fig. 18). The aggregation proceeds through hierarchical organization. The **33** molecules aggregated in a circular fashion to give rod-shaped fibers with ca. 5 nm and several micron length, in which six helically preorganized strands of the supramolecular polymer intertwine. At higher concentration, further association of the rod-shaped fibers takes place to form hexagonal columnar ordering by entanglement of the peripheral alkyl groups, leading to a significant increasing of viscosity and gelation.

Perylene diimide and its derivatives are well-known as photofunctional dyes with unique optical and electrochemical properties. They have excellent light fastness, high chemical stability, high photoluminescence yield, and

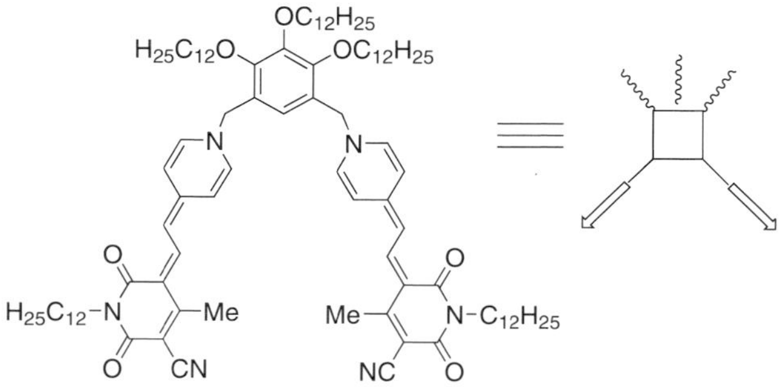

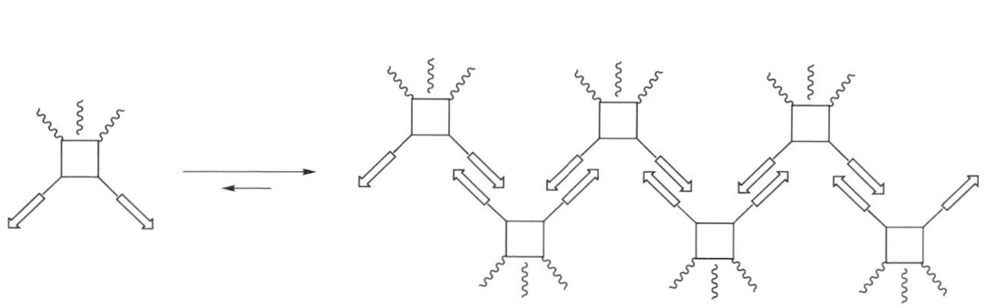

**Fig. 18** Bis(merocyanine)-based gelator and its supramolecular polymerization process

a wide range of colors that are tunable by various substituents at the bay position [67]. Perylene diimide-based gelators **34a–d**, in which a perylene diimide core and two cholesterol moieties are connected with urethane linkages, were designed and prepared by Shinkai's group for the first time [68]. In addition to $\pi$–$\pi$ stacking among perylene diimide moieties, the gel-forming ability of the cholesterol moieties is useful to tune the structure of the one-dimensional aggregate (Fig. 19). The aggregate network structure in **34a** can be visualized by SEM and AFM as well as confocal laser scanning microscopy (CLSM) (Fig. 19). In a mixture system where the perylene diimide moiety of the guest **34b–d** is intercalated into the gel fibrils of the host **34a**, an efficient energy-transfer can take place from the host **34a** to the guest **34b–d**, by which the fluorescence emitted from the host is quenched significantly whereas the strong fluorescence is emitted from the guest perylene diimide (Fig. 20). One can regard, therefore, that this is a new visible-light-harvesting system utilizing a supramolecular structure created in a fluorescent-dye-based organogel medium.

Anthracene- and anthraquinone-based gelators were extensively studied by Weiss's group as one of the pioneer works in the organogel field [5–9]. The dye moiety is appended to a steroid moiety such as cholesterol to show gelation ability. Other anthracene- [69] and anthraquinone-based gelators [70] were created by the Desvergne and Bouas-Laurent groups. Two long alkoxy chains were introduced to the 2 and 3 positions on the dye moiety to con-

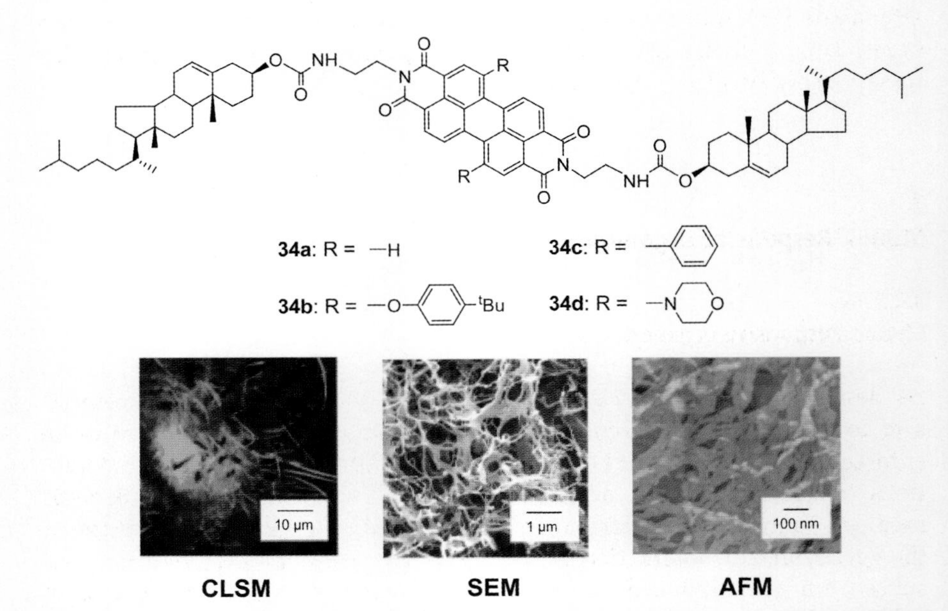


**Fig. 19** Perylene diimide-based gelators **34a-d**, and CLSM (ex 488 nm), SEM, and AFM images of a **34a** gel in *p*-xylene/1-propanol

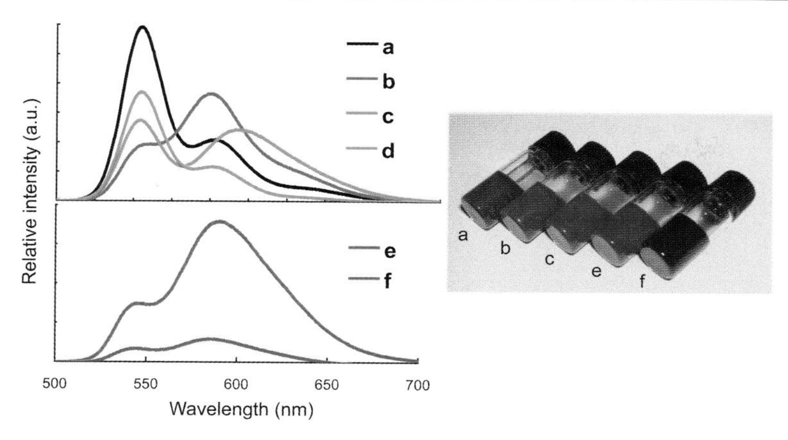


**Fig. 20** Fluorescence spectra and fluorescent images of the mixtures of perylene diimides **34a-d** ([**34a**] = 0.5 wt/vol % in *p*-xylene/1-propanol [3 : 1, (v/v)]: (**a**) **34a**, (**b**) **34a**/**34b** (10 : 3, mol/mol), (**c**) **34a**/**34c** (10:3, mol/mol), (**d**) **34a**/**34d** (10 : 3, mol/mol), (**e**) **34a**/**34b**/**34c** (10 : 3 : 3, mol/mol/mol), and (**f**) **34a**/**34b**/**34c**/**34d** (10 : 3 : 3 : 3, mol/mol/mol/mol)

trol solubility and to prevent crystallization, leading to the formation of organogels [69–71]. The alkoxy-chain-containing anthracene gelator can be applied to functional materials such as aerogel created from an organogel-supercritical fluid system [72], porous silica created by using the organogel as a template [73], and organogel electrolyte [74]. Terech's group analyzed the organogel structures of cholesterol-based and alkoxy-chain-containing gelators by using small angle X-ray scattering [10, 11, 75].

# 3 Stimuli-Responsive Organogels

## 3.1 Chemo-Responsive Organogels

As described in Sects. 2.1.2 and 2.1.3, dye-based organogels are responsive to added guest molecules to reinforce the organogel structure or to enforce the morphological change by intermolecular host-guest interactions [25, 26, 28, 36]. The organogel is formed only in a specific organic medium, so that the superstructure is maintained by a subtle balance between the gelator-gelator interaction and the gelator-solvent interaction. Thus, the subsequent guest-induced morphological change can be induced by a slight shift of this balance in the flexible organogel structure organized by soft noncovalent interactions. This section summarizes chemo-responsive dye-based

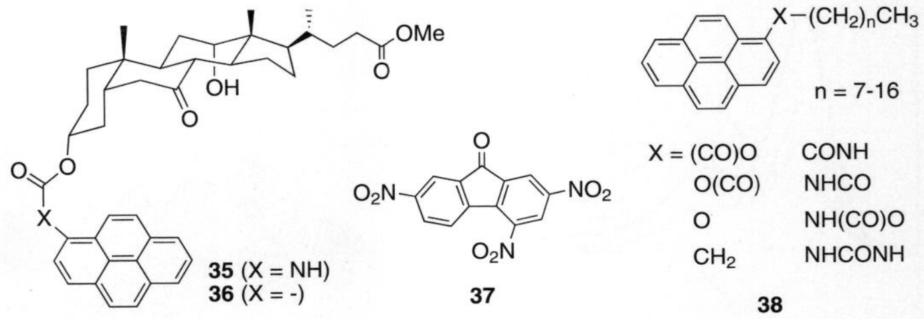

**Fig. 21** Pyrene-based organogel formation induced by trinitrofluorenone additive

organogels, which indicate either stabilization or destabilization of organogel structures.

The first example of organogel formation induced by host-guest interaction was reported by Maitra's group [76]. Bile acid derivatives with a pyrene unit **35** and **36**, which did not show the gelation ability, gelated certain organic solvents (alcohols and chloroform) in the presence of trinitrofluorenone **37** (Fig. 21). The main driving force is an intermolecular donor-acceptor interaction between the pyrene unit on **35**/**36** and the additive **37** with 1 : 1 stoichiometry, which was confirmed by the charge-transfer band in the UV/Vis spectra.

In the simple system of alkyl-chain-appended pyrene derivatives **38** (Fig. 21), the organogel formation induced by the donor-acceptor interaction was studied systematically [77]. Compounds with ester, ether, or alkyl linkages gelated a number of hydroxylic and hydrocarbon solvents only in the presence of **37**. This observation suggests that the formation of organogel aggregates is efficiently driven by a charge-transfer interaction. The long alkyl chain could play only a secondary role in modulating the solubility/crystallinity balance of the molecule in the given solvents. Since the stability of the **37**/**38** organogels was found to be at a maximum at a 1 : 1 molar ratio of the donor and the acceptor, the supramolecular aggregate is likely to be an alternative stack of donor and acceptor surface to form one-dimensional aggregates. On the other hand, pyrene derivatives **38** with hydrogen-bonding functional linkages (amide, urethane, and urea) formed organogels on their own in a variety of solvents by the cooperative action of $\pi$–$\pi$ stacking and hydrogen-bonding interactions. These compounds did not gelate the solvents in the presence of **37** because of incompatible hydrogen-bonding and donor-acceptor interaction geometries.

Kimizuka's group reported that aqueous dispersion of cationic L-glutamate derivatives **39** and **40** with two alkyl tails gives fibrous nanoassemblies with the bilayer structure (Fig. 22). When anionic naphthalene dye **41** was added to the aqueous bilayer of **40** with long undecyl tails, immediate precipitation occurred. Apparently, formation of the hydrophobic ion-pair destabilizes the long-chained ammonium bilayer of **40**. On the other hand, **39** with short octyl tails formed a transparent hydrogel in the presence of anionic dye **41**

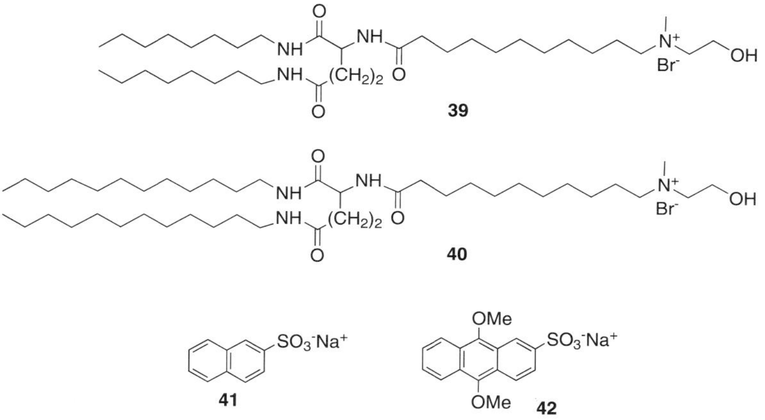

Fig. 22 Amino acid-based organogel formation induced by aromatic sulfonate additives

(Fig. 22) [78]. As **39** has less amphiphilicity arising from the short octyl tails, ion pairing with **41** must provide the moderate amphiphilicity to the complex that is required for ordered self-assembly. AFM observation indicates that the fibrous aggregate is comprised of bundles of bilayer membranes. Interestingly, when a small amount of anthracene chromophore **42** was added to the **39**/**41** hydrogel system, energy-transfer from bound **41** to **42** is facilitated significantly compared to that occurring in the aqueous dispersion system. It is considered, therefore, that in the hydrogel system, naphthalene chromophores **41** are accumulated in the fibrous nanoassemblies to make a light-harvesting supramolecular system.

It is known that 1,2-bis(ureido)cyclohexanes (*R*)-**43** and (*S*)-**43** act as highly effective gelators, and contain two chiral centers which are expressed in the formation of twisted fibers with a helicity that depends on the handedness of **43** [22b]. In contrast, 1,2-bis(ureido)cyclohexane (*R*)-**44** with azobenzene chromophores could not act as a gelator itself, however, it can be incorporated into the organogel aggregate structure composed of gelators (*R*)-**43** and (*S*)-**43**, as reported by Feringa and van Esch's group (Fig. 23). Interestingly, in the coaggregation systems, self-association among host **43** and hetero-association between host **43** and guest (*R*)-**44** are highly affected by the chirality of the host and guest [79]. The association constant between (*S*)-**43** and (*R*)-**44** is almost twice as large as that between (*R*)-**43** and (*R*)-**44**, which indicates that the incorporation of (*R*)-**44** in aggregates of (*S*)-**43** is preferred to the incorporation in aggregates of (*R*)-**43**. The difference in the coaggregation between (*R*)-**43**/(*R*)-**44** and (*S*)-**43**/(*R*)-**44** results in the difference in the hydrogen-bonded structure and strength. The difference is reflected by the photochromic property of the azobenzene moiety in (*R*)-**44**. After irradiation of (*R*)-**44** in the (*S*)-**43** gel, thermal *cis*-to-*trans* isomerization in the azobenzene moiety was found to be slower than that in the (*R*)-**43**

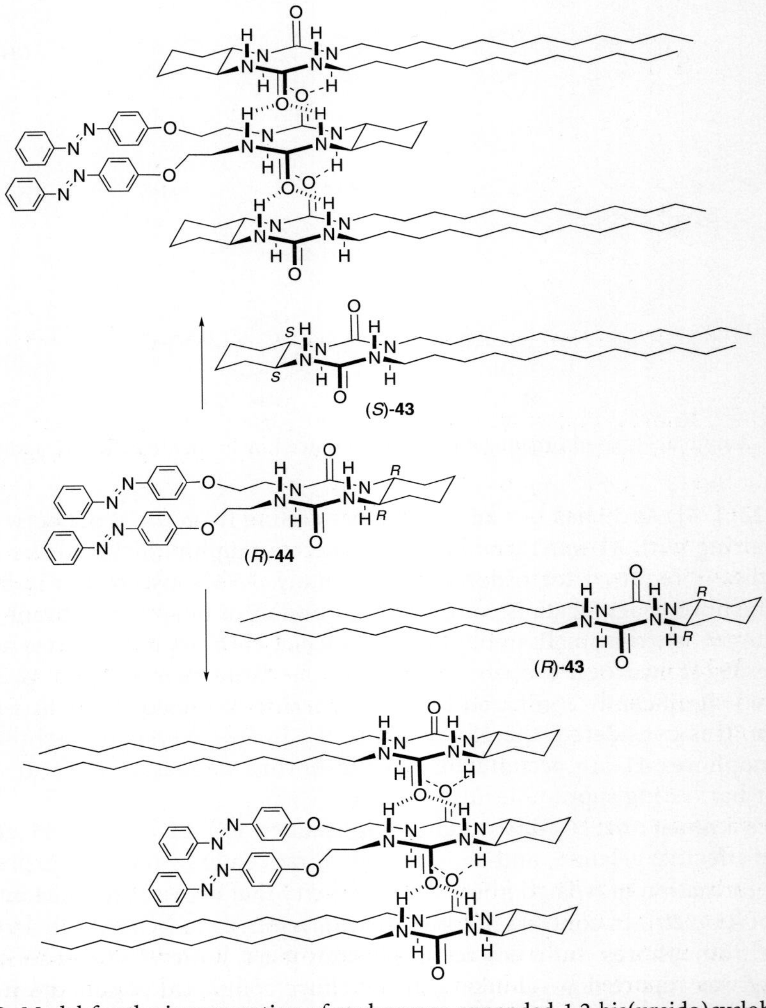

Fig. 23 Model for the incorporation of azobenzene-appended 1,2-bis(ureido)cyclohexane (*R*)-**44** in an aggregate of 1,2-bis(ureido)cyclohexane (*R*)-**43** with the same configuration and in an aggregate of 1,2-bis(ureido)cyclohexane (*S*)-**43** with the opposite configuration

gel. It is considered, therefore, that the azobenzene moieties in (*R*)-**44** are more exposed to solvent in the (*S*)-**43** gel than in the (*R*)-**43** gel.

Kimura and Shirai's group reported that a zinc porphyrin dimer (*R*)-**45**, in which two porphyrin moieties were appended to a 1,2-bis(amide)cyclohexane unit, is incorporated into fibers made of 1,2-bis(amide)cyclohexane (*R*)-**46** organogels (Fig. 24) [80]. The photochemical properties of (*R*)-**45** in a mixed organogel system were investigated by fluorescence quenching measurements using methyl viologen as an electron-transfer quenching agent. In the gel

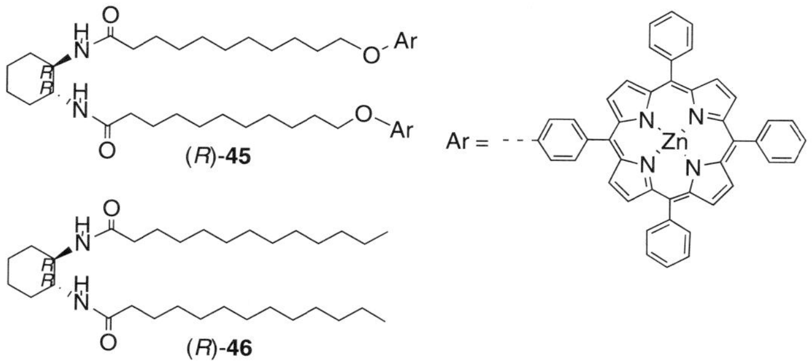

**Fig. 24** 1,2-Bis(amide)cyclohexane gelators with and without porphyrin moieties

phase the fluorescence intensity of zinc porphyrin moieties in (*R*)-**45** remained even in the presence of the quenching agent, although in the sol phase the fluorescent quenching by the quenching agent occurred easily. The collision between the zinc porphyrin moiety and methyl viologen molecule is depressed through incorporation into the organogel fibrous assemblies.

Similar incorporation was also examined in a cholesterol-based organogel. Porphyrin-cholesterol **2a** is incorporated into the gel fiber composed of nucleobase-appended cholesterol gelator (Fig. 1) [15]. In the mixed gel system, porphyrin moieties with large $\pi$-area are able to intercalate between the nucleobase moieties oriented outside the one-dimensional column of cholesterol stacking.

## 3.2 Photo-Responsive Organogels

As described in Sect. 2.6, azobenzene-appended cholesterol derivatives **22a–e** act as versatile gelators (Fig. 15). The sol-gel phase transition in 1-butanol gel of **22b** induced by *trans*-to-*cis* photoisomerization of the azobenzene moiety was reported by Shinkai's group (Fig. 25) [12, 13]. Upon UV irradiation ($330 < \lambda < 380$ nm) the gel turned into the sol. The formed *cis*-**22b** cannot aggregate efficiently because of the bent structure. Thermal *cis*-to-*trans* isomerization in the dark was very slow. On the other hand, *cis*-**22b** was rapidly isomerized to *trans*-**22b** upon Vis irradiation ($\lambda > 460$ nm), and the gel was reformed by cooling (Fig. 25). The sol-gel phase interconversion could be repeated reversibly by photoirradiation. Another interesting photo-induced isomerization of azobenzene chromophore was found in a mixed organogel system **43**/**44** as shown in Sect. 3.1 [79]. 3,3-Diphenyl-3H-naphtho[2,1-b]pyrans linked to sodium *N*-acyl-11-aminoundecanoate **47** showed a similar photo-responsive behavior [81]. Supramolecular gelating assemblies of **47** can be disrupted by the photo-induced ring opening of the chromene subunit (Fig. 26).

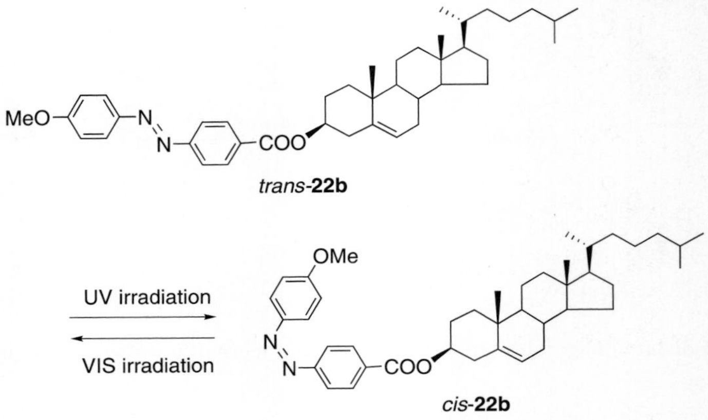

MeO N N COO trans-22b

UV irradiation VIS irradiation

OMe N N COO cis-22b

**Fig. 25** Photo-responsive azobenzene-cholesterol organogel

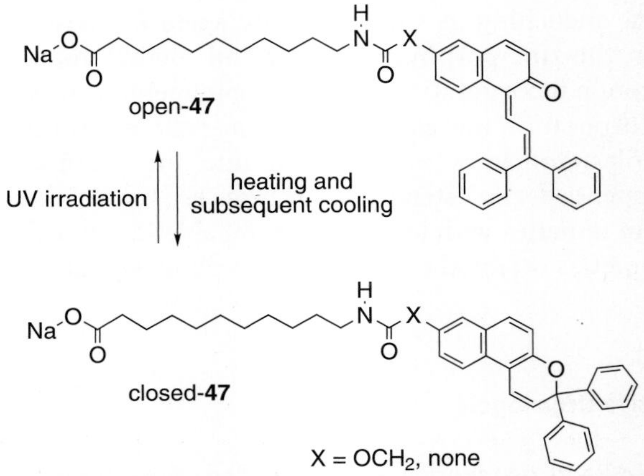

Na O O H N X O O open-47

UV irradiation heating and subsequent cooling

Na O O H N X O O closed-47

X = $OCH_2$, none

**Fig. 26** Photo-responsive 3,3-diphenyl-3*H*-naphtho[2,1-b]pyran organogel

Spiropyran-containing L-glutamic acid-based lipids SP-**48** form transparent organogels in benzene (Fig. 27). Upon UV irradiation, the photochromic spiropyran (SP) head group is converted to a zwitterionic merocyanine (MC) form. The thermal MC-to-SP isomerization process depends on the dispersion state. The isomerization rate constant for MC-**48** significantly decreased around the sol-to-gel transition point. The kinetic analysis can be applied to determine the critical gelation concentration [82]. Photochromic behavior of spiropyran dye SP-**49** with a succinyl ester functionality was studied in a organogel medium from 4-*tert*-butyl-1-phenylcyclohexanol gelator **50** (Fig. 28). In the thermal MC-to-SP isomerization process, a significant retardation of the decay rate was observed for MC-**49** in the organogel. In the gel phase, the succinyl functionality seems to interact with the gelator to retard the thermal isomerization [83].

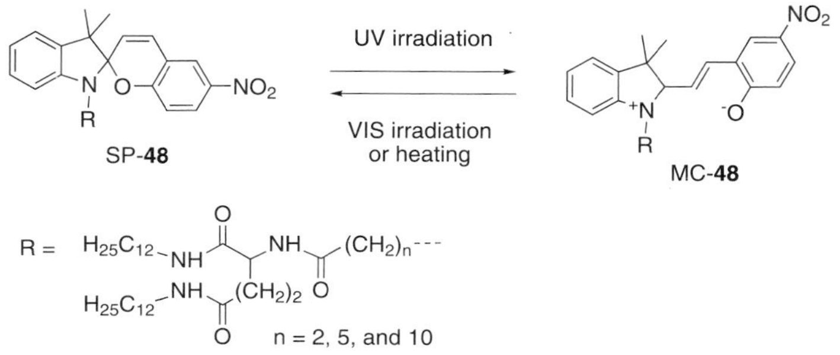

**Fig. 27** Photo-responsive L-glutamic acid-containing spiropyrane organogel

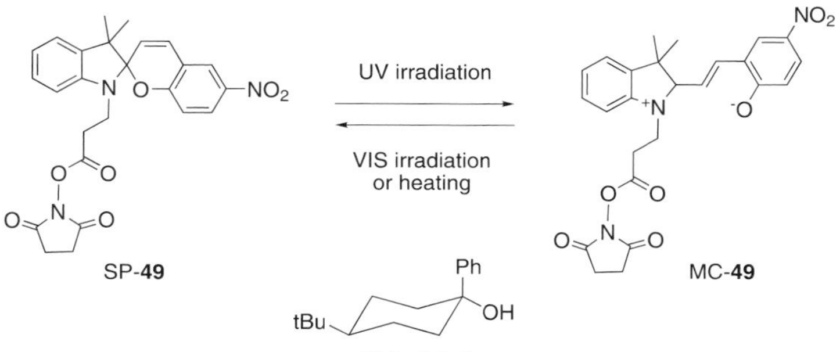

**Fig. 28** Photo-responsive spiropyrane in an organogel system of **50**

A novel photo-responsive organogel in a liquid crystalline state was reported by Kato's group [84]. An azobenzene dimer **51**, in which two azobenzene moieties were appended to a 1,2-bis(amide)cyclohexane unit, can gelate liquid crystal **52** to form a nematic liquid crystalline gel (Fig. 29). Upon UV irradiation of the nematic gel, *trans*-to-*cis* photoisomerization of the azobenzene moieties induces a transition to a cholesteric liquid crystalline phase. *cis*-**51** formed by UV irradiation is soluble in **52**, leading to the dissociation of its hydrogen bonding and the gel-to-sol phase transition is induced. When the cholesteric liquid crystalline state is kept at room temperature or irradiated with Vis light, *cis*-**51** gradually isomerizes to *trans*-**51**, which shows the gelation ability due to their hydrogen bond formation. The reaggregation of *trans*-**51** occurs along the fingerpoint structure of the cholesteric liquid crystal to form a cholesteric liquid crystal gel. Once the cholesteric gel is heated to an isotropic state and cooled to room temperature, the nematic liquid crystal gel is reproduced (Fig. 29). The photo-responsive liquid crystal gel will be applicable to new photon-mode re-writable information recording systems.

Another photo-responsive organogel in a liquid crystalline state was reported by Zhao's group [85]. A gelator **53** containing a photo-responsive azobenzene moiety and amide groups was dissolved in a cholesteric liquid

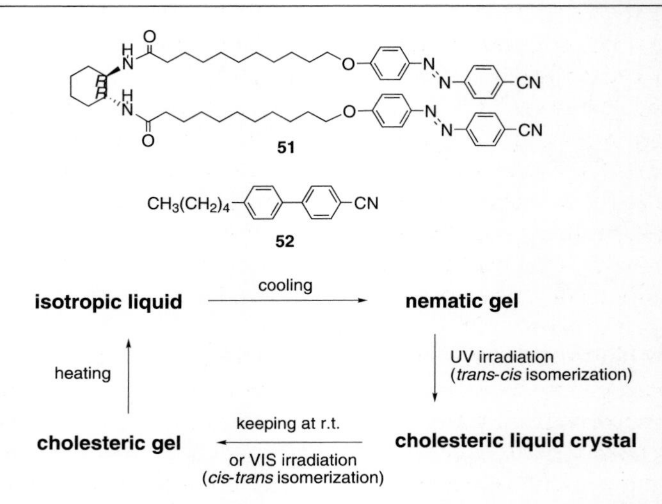


**Fig. 29** Photo-responsive liquid-crystalline physical gel based on azobenzene-containing bis(amide)cyclohexane gelator

crystal host. When the mixture was filled in a nonrubbed and ITO-coated electrooptic cell and rapidly cooled from 130 °C to room temperature, a liquid crystal gel was formed. Upon UV irradiation through photomask with parallel transparent strips, a light-induced reorganization in the gel takes place arising from the *trans*-to-*cis* photoisomerization of the azobenzene moiety in **53**, resulting in an apparent transport of the aggregates from the irradiated area to the nonirradiated area and, consequently, the formation of a grating (Fig. 30). The grating can be erased by heating the sample into the isotropic liquid phase to redissolve the gelator molecules in the liquid crystal, followed by cooling to ambient condition. The grating formed in the self-assembled cholesteric liquid crystal gel displays electrically switchable diffraction efficiency, which is determined by the response of the liquid crystal to the electric field.

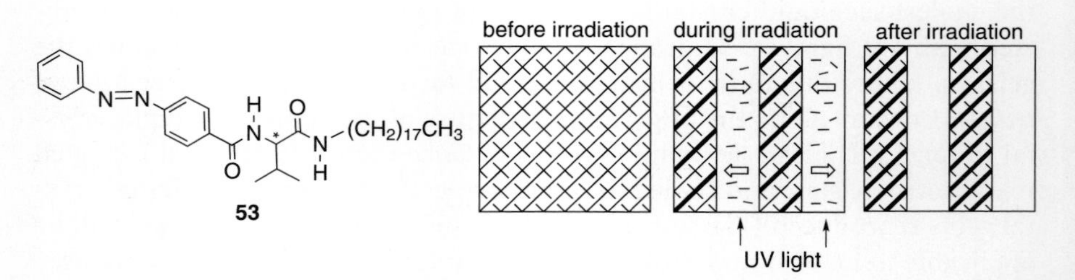


**Fig. 30** Photo-responsive liquid-crystalline physical gel based on azobenzene-containing bis(amide) gelator, and the schematic illustration of the mechanism for the light-induced reorganization in the liquid crystal gels and the grating formation

The examples of photo-induced *trans*-to-*cis* isomerization of azobenzene dyes described above are very rare and valuable in supramolecular assembled systems, because the difficulty of the photoisomerization of the azobenzene moiety has been recognized in micellar and monolayer supramolecular systems [86, 87]. In addition, other azobenzene-containing organogels, in which the *trans*-azobenzene moieties are tightly packed in the organogel supramolecular structure, did not show any photo-induced *trans*-to-*cis* isomerizations [59]. Presumably, in **22b** and **44** there is sufficient room for the isomerization around the azobenzene moiety. Thus, photo-induced isomerization is very sensitive to the organogel structure.

Another photo-responsive organogel was designed on the basis of a flexible binary system, as developed by Shinkai's group. The binary gelator **54** composed of alkylammonium and anthracene-9-carboxylate is not covalently-linked to the structure-forming component in order to acquire the more space necessary for isomerization [88]. In this binary organogel system, the aggregation mode seems to be mainly governed by the alkylammonium moiety and only partly by the anthracen-9-carboxylate moiety. Thus, the anthracen-9-carboxylates are not so well-ordered and not so tightly packed. They are just assembled as counteranions of the ordered alkylammonium assemblies. When the cyclohexane gel of **54** was photoirradiated ($\lambda > 300$ nm), absorption bands ascribed to monomeric anthracene-9-carboxylate disappeared and the physical state changed from the gel to the sol. The phase transition is ascribed to photo-induced dimerization of anthracene-9-carboxylate moieties (Fig. 31). At 30 °C in the dark, the gel phase could not be regenerated directly from the sol phase but the precipitation of **54** resulted, even though monomeric anthracene species were already regenerated. When the sol was once heated at boiling point and then cooled to 15 °C, the gel was in fact regenerated (Fig. 31). The findings suggest that the photo-induced monomer-to-dimer structural change eventually destroys the network gel superstructure. The alignment of the alkylammonium moieties is not stable as to suppress the structural change, because they are not linked by a covalent bond but just assembled by the weak van der Waals interaction.

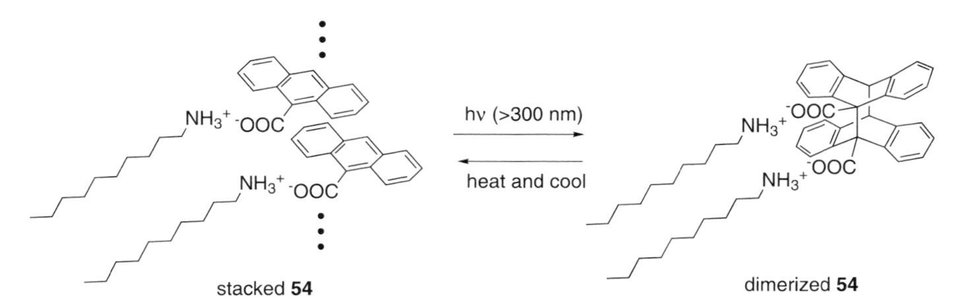


**Fig. 31** Photo and thermo-responsive phase changes in **54**

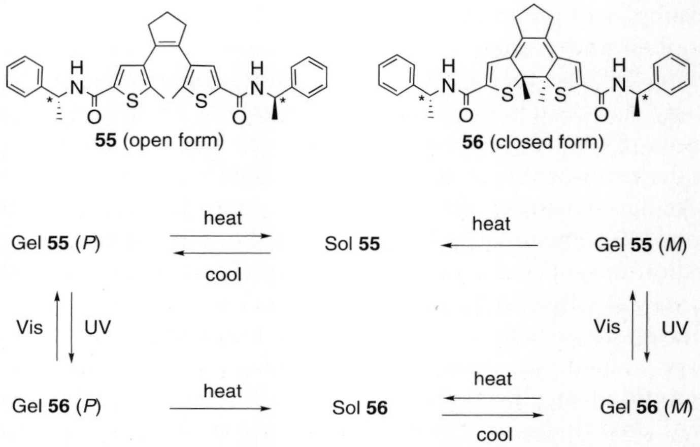

55 (open form)

56 (closed form)

Gel 55 (*P*) —heat→ Sol 55 ← heat — Gel 55 (*M*)

Gel 55 (*P*) ←cool— Sol 55

Gel 55 (*P*) ⇅ Vis / UV Gel 56 (*P*)

Gel 55 (*M*) ⇅ Vis / UV Gel 56 (*M*)

Gel 56 (*P*) —heat→ Sol 56 ←heat— Gel 56 (*M*)

Sol 56 —cool→ Gel 56 (*M*)

**Fig. 32** Photo-responsive diaryethene-bisamide gel having a chiral memory function

A novel photo-responsive organogel with a chiral memory function was created by Feringa and van Esch's group [89]. For design of gelator **55**, two urea groups are introduced into a diarylethene photochromic moiety to form an extended hydrogen-bonding network (Fig. 32) [90]. Upon irradiation of Vis light, the open form **55** photocyclizes to the closed-form **56**. Upon heating and subsequent cooling of an isotropic solution of the open form **55**, a stable gel **55** (*P*) is obtained, in which the aggregates exhibit supramolecular chirality owing to locking of the *P*-helical conformation of the open form **55** (Fig. 32). Photocyclization of the stable gel **55** (*P*) results in the metastable gel **56** (*P*) to preserve supramolecular chirality of *P*-helicity under a photostationary state. Heating of the metastable gel **56** (*P*) leads to an isotropic solution of **56**, which upon cooling results in the stable gel **56** (*M*) to invert supramolecular chirality to opposite *M*-helicity. Irradiation of the stable gel **56** (*M*) with Vis light results in the metastable gel **55** (*M*), which can be reconverted to the stable gel **56** (*M*) by UV irradiation. Finally, a heating-cooling cycle results in the transformation of the metastable gel **55** (*M*) to the original stable gel **55** (*P*) via an isotropic solution of **55**. In the photochemical steps, supramolecular chirality is locked to be preserved. However, the photochemical ring-opening or ring-closure changes the rigidity and chirality of the central unit and as a consequence the stability of the chiral aggregate. Since a change in the steric bulkiness in a diarylethene moiety is smaller than that in an azobenzene moiety, the photoinduced interconversion may be induced more easily.

## 3.3 Metal-Responsive Organogels

The first metal-responsive organogel through the metal-coordinating interaction was reported by Shinkai's group [12]. In the gel phase, $Li^+$, $Na^+$, $K^+$, and $Rb^+$ ($M^+(ClO_4)^-$ and $M^+(SCN)^-$) complexed with a 18-crown-6 ether moiety in azobenzene-cholesterol gelator **57**, even though the metals themselves are scarcely soluble in the gelation solvents of cyclohexane-benzene (Fig. 33). The sol-gel phase transition temperature in the presence of the metals increased with increasing metal concentration, indicating the gel stabilization stimulated by metal ions. Similarly, pyridine-containing cholesterol gelator **58** is responsive to metal ions to stabilize the gel structure through the pyridine-metal interaction (Fig. 33) [91].

Transition metal-responsive organogel was first reported by Sohna and Fages [92]. Trisbipyridine derivative **59** acts as a gelator using mainly the hydrogen-bonding interaction among amide groups (Fig. 33). When water-containing $Fe(SO_4){\cdot}7H_2O$ was deposited at the top of a **59** toluene gel, the water phase turned a pink color due to the extraction of some **59** molecules from the gel into the isotropic water phase derived by coordination of the Fe(II) ion with the trisbipyridyl moiety. By heating the system, the complexation was accelerated to extract **59** completely and collapse the gel structure eventually. On the other hand, in the blank test without $Fe(SO_4){\cdot}7H_2O$], **59** was not extracted into water at all.

In contrast to **59**, an organogel structure is reinforced by coordination with transition metals in a saccharide-based organogel system [93]. The organogel network structure of $\alpha$-D-glucopyranose-based **60** with an amino functionality is constructed by mainly hydrogen-bonding interactions among the saccharide moieties. In addition, the amino group is significantly involved in the gel network formation. When $CoCl_2$ or $CdCl_2$ was added to the ethanol

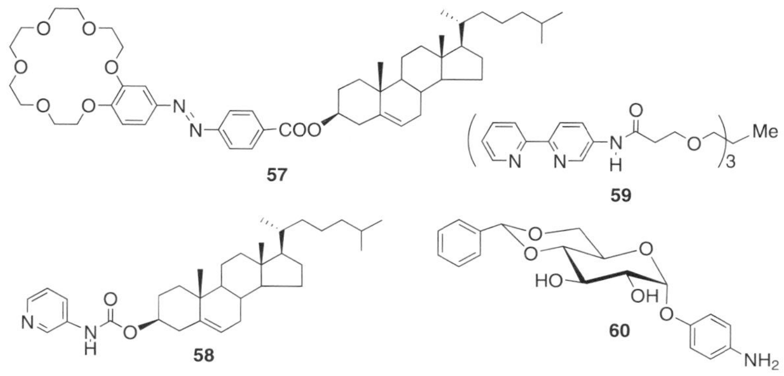

**Fig. 33** Metal-responsive organogels

gel of **60**, the system was markedly stabilized through amino-metal coordination (Fig. 33). SEM and TEM observation indicates that the added metal is homogeneously dispersed into the gel fibers and coordinated with the amino group without unfavorable destruction of the intermolecular aggregates. It is considered that the amino group, which is involved in the gel network in the absence of metal, interacts with the metal to form a better cross-linking network among **60** molecules.

## 3.4 Proton-Responsive Organogels

Proton-sensitive organogels were reported by Pozzo's group [94] and Shinkai's group [95]. A proton-sensitive organogel was formed from 2,3-bis(dodecyloxy) phenazine **61** in acetonitrile solution [94]. The gelation ability of **61** can be controlled by the reversible protonation of the phenazine moiety (Fig. 34). By addition of trifluoroacetic acid (TFA), a mono-protonated yellow species **61**·$H^+$ was formed and the organogel structure was reinforced through the hydrogen bonding interactions between nitrogen atoms and ammonium centers. Bubbling ammonia through the reinforced organogel provoked fading of the colored **61**·$H^+$ species and led to a transparent organogel.

A 1,10-phenanthroline-appended cholesterol **62** acts as a proton-sensitive fluorescent gelator (Fig. 35) [95]. Most gels of **62** exhibited violet emissions

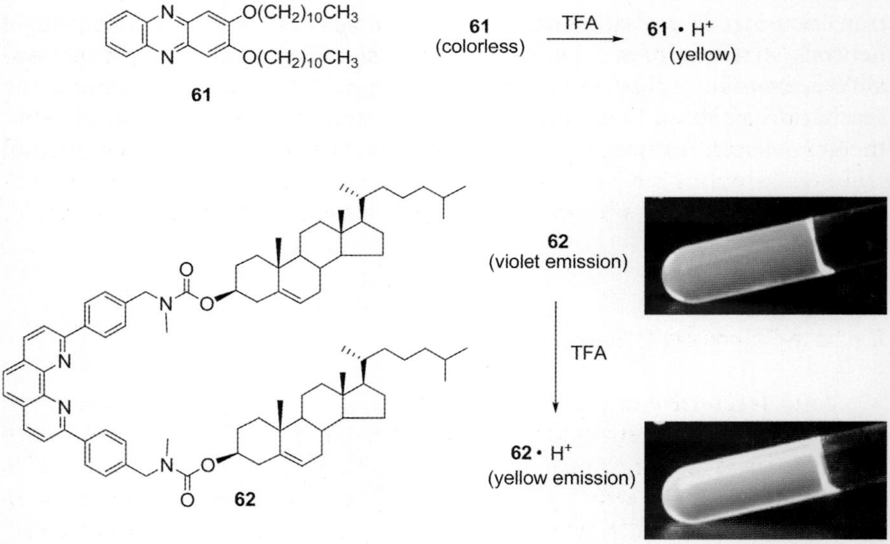

**Fig. 34** Proton-sensitive fluorescent gels based on 2,3-dialkoxyphenazine **61** and 1,9-phenanthroline-appended cholesterol **62**. The fluorescent images of **62** gels are shown in the right side

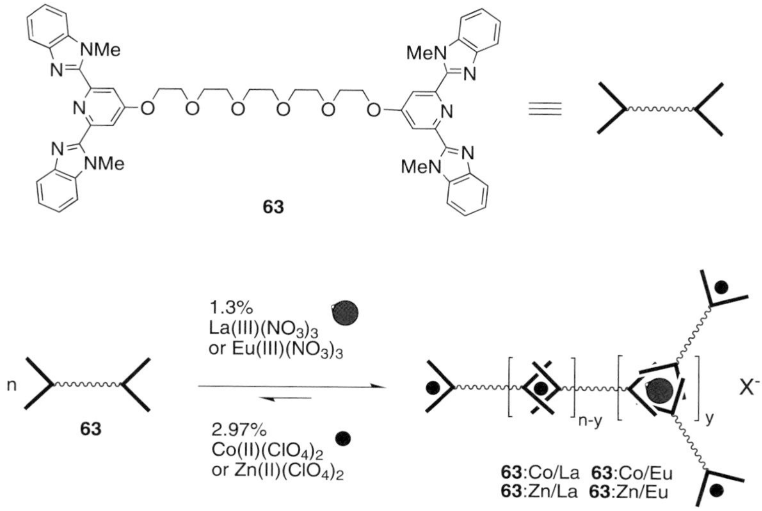

**Fig. 35** Metallo-supramolecular gel-like materials using a combination of lanthanoid and transition metal ions mixed with monomer **63**

at 396 nm arising from neutral **62**. By the addition of TFA the intensity of the violet emission decreased, whereas a yellow emission at 530 nm is newly generated, which is attributed to a $\pi-\pi^*$ transition of $\mathbf{62}\cdot H^+$ species (Fig. 34). Only in the presence of 0.2 equiv. of TFA does the violet emission disappear completely even though a significant amount of neutral **62** still remains. In contrast, the intensity of the newly generated yellow emission subsequently increases up to 2.0 equiv. with increasing TFA concentration. The findings suggest that energy-transfer from $\mathbf{62}^*$ to $\mathbf{62}\cdot H^+$ is taking place in the gel phase. The gel which is constructed from densely $\pi-\pi$ stacked 1,10-phenanthroline moieties thus becomes useful for designing a proton-sensitive gel system and an energy-transfer system.

## 3.5 Mechano-Responsive Organogels

Mechano-responsive organogels are reported by Beck and Rowan [96]. A solution of bis(methylbenzimidazolyl)pyridine-dimer **63** in chloroform/acetonitrile formed an organogel by coordination with either Co/La, Co/Eu, Zn/La, or Zn/Eu, in which either Co(II) or Zn(II) ions act as linear chain extension binding units, whereas either La(III) or Eu(III) ions act as cross-linking components (Fig. 35). These materials are mechano-responsive, exhibiting a thixotropic (shear-thinning) behavior. The reversible sol-gel formation by the mechanical shaking occurs for a short time of ca. 20 sec. Thus,

**63** can exhibit three different properties, i.e., thermo-, metal-, and mechano-responses.

## 4
## Conclusions

Low molecular-weight organogels have rapidly been developed from just construction of new-type fibrous supramolecular aggregates to the design of new-type self-organizing soft materials. Large aromatic dye molecules such as porphyrin, fullerene, phthalocyanine, triphenylene, phenylenevinylene, etc., possess, a priori, a natural tendency to assemble into one-dimensional aggregates due to the strong $\pi$–$\pi$ interactions, which is very advantageous to gel formation. In several special cases, the organogel supramolecular systems have opened a new possibility to program unique and unusually ordered aggregate structures. The typical example is a triphenylene-based organogel with eclipsed stacking [53], which cannot be created in other supramolecular systems in the solid and in the liquid phases.

The extended aggregates of the $\pi$-stacked aromatic moieties can provide an efficient path for charge-carrier transport within the organogel fibers [50, 51, 53, 55, 60]. The organogel fibrous structures can be elaborately transcribed into inorganic silica materials by utilizing template-silanol interactions [39]. Rare guest molecules can be recognized in the special binding sites preorganized in the organogel structure and induce morphological changes to thermodynamically more favored organogel structures [25, 26, 36]. The efficient guest recognition is ascribed to the frozen property characteristic of organogel under the limited solvent diffusion conditions, whereas the subsequent guest-induced morphological change is ascribed to the flexible structure organized by noncovalent interactions. When the guest molecule acts as an acceptor dopant, the doped organogel system shows unique and efficient energy-transfer from the $\pi$-stacked aggregate host to the guest, leading to new-type light-harvesting supramolecular systems [30, 49, 68, 78, 95]. Thus, by utilizing the unique and unusual dye-based self-assembling structures with both flexibility and rigidity, the organogels can be engineered to show novel functions such as guest-recognizing and organic-inorganic transcription properties in soft materials. Furthermore, in the $\pi$-electron accumulated structures of the $\pi$-stacked organogel aggregate, the optical and electrochemical properties of the aromatic dyes are changed and improved from the molecular level to the supramolecular level, leading to the generation of unique and efficient functions such as charge-carrier-transporting, light-harvesting, and energy-transfer properties.

The dye-based organogels also act as stimuli-responsive materials, because they are physically formed by noncovalent interactions, and are responsive to physical or chemical stimuli such as light, heat, proton, metal, shrinking, as well as guest molecules to induce reversible phase transition from a gel state to a solution state. According to the phase transition, the reversible interconversion in the optical and electrochemical properties of the aromatic dyes between the molecular level and the organogel supramolecular level can be realized by the external stimuli. It is expected that the reversible supramolecular systems can be applied to drug delivery, switching with memory functions, etc. [89]. The dye-based organogel has just started to be exploited in new applications for molecular recognition, catalysis, nanomaterials, etc. In this context, we believe that organogels will provide a broad foundation for new materials science.

## References

1. (a) Lehn JM, Atwood JL, Davies JED, MacNicol DD, Vögtle F (eds)(1996) Comprehensive Supramolecular Chemistry, Pergamon, New York (b) Conn MM, Rebek Jr J (1997) Chem Rev 97:1647 (c) Rebek Jr J (1999) Acc Chem Res 32:278 (d) Rebek Jr J (2000) Chem Commun 637 (e) Stang PJ, Olenyuk B (1997) Acc Chem Res 30:502 (f) Fujita M (1998) Chem Soc Rev 27:417 (g) Caulder DL, Raymond KN (1999) Acc Chem Res 32:975 (h) Leininger S, Olenyuk B, Stang PJ (2000) Chem Rev 100:853 (i) Fujita M, Umemoto K, Yoshizawa M, Fujita N, Kusukawa T, Biradha K (2001) Chem Commun 509 (j) Prins LJ, Reinhoudt DN, Timmerman P (2001) Angew Chem Int Ed 40:2382 (k) Reinhoudt DN, Crego-Calama M (2002) Science 295:2403
2. (a) Terech P, Weiss RG (1997) Chem Rev 97:3133 (b) Shinkai S, Murata K (1998) J Mater Chem 8:485 (c) van Esch JH, Schoonbeek F, de Loose M, Veen EM, Kellogg RM, Feringa BL (1999) Supramol Sci 6:233 (d) Abdallah DJ, Weiss RG (2000) Adv Mater 12:1237 (e) Melendez RE, Carr AJ, Linton BR, Hamilton AD (2000) Struct Bonding 96:31
3. van Esch JH, Feringa BL (2000) Angew Chem Int Ed 39:2263
4. (a) Oya T, Enoki T, Grosberg AY, Masamune S, Sakiyama T, Takeoka Y, Tanaka K, Wang G, Yilmaz Y, Feld MS, Dasari R, Tanaka T (1999) Science 286:1543 (b) Aggeli A, Bell M, Boden N, Keen JN, Knowles PF, McLeish TCB, Pitkeathly M, Radford SE (1997) Nature 386:259 (c) Inoue K, Ono Y, Kanekiyo Y, Ishi-i T, Yoshihara K, Shinkai S (1998) Tetrahedron Lett 39:2981
5. (a) Lin YC, Weiss RG (1987) Macromolecules 20:414 (b) Furman I, Weiss RG (1993) Langmuir 9:2084
6. Lin YC, Kachar B, Weiss RG (1989) J Am Chem Soc 111:5542
7. Lu L, Cocker TM, Bachman RE, Weiss RG (2000) Langmuir 16:20
8. Ostuni E, Kamaras P, Weiss RG (1996) Angew Chem Int Ed Engl 35:1324
9. (a) Mukkamale R, Weiss RG (1995) J Chem Soc Chem Commun 375 (b) Mukkamale R, Weiss RG (1996) Langmuir 12:1474
10. Terech T, Furman I, Weiss RD (1995) J Phys Chem 99:9558
11. Terech T, Otsuni E, Weiss RD (1996) J Phys Chem 100:3758
12. Murata K, Aoki K, Nishi T, Ikeda A, Shinkai S (1991) J Chem Soc Chem Commun 1715
13. Murata K, Aoki M, Suzuki T, Harada T, Kawabata H, Komori T, Ohseto F, Ueda K, Shinkai S (1994) J Am Chem Soc 116:6664
14. Sakurai K, Ono Y, Jung JH, Okamoto S, Sakurai S, Shinkai S (2001) Perkin Trans 2 108
15. Snip E, Shinkai S, Reinhoudt DN (2001) Tetrahedron Lett 42:2153

16. (a) Geiger C, Stanescu M, Chen L, Whitten DG (1999) Langmuir 15:2241 (b) Duncan DC, Whitten DG (2000) Langmuir 16:6445
17. Wang R, Geiger C, Chen L, Swanson B, Whitten DG (2000) J Am Chem Soc 122:2399
18. (a) Yoza K, Ono Y, Yoshihara K, Akao T, Shinmori H, Takeuchi M, Shinkai S, Reinhoudt DN (1998) Chem Commun 970 (b) Yoza K, Amanokura N, Ono Y, Akao T, Shinmori H, Takeuchi M, Shinkai S, Reinhoudt DN (1999) Chem Eur J 5:2722
19. Amanokura N, Yoza K, Shinmori H, Shinkai S, Reinhoudt DN (1998) Perkin Trans 2 2585
20. Gronwald O, Shinkai S (2001) Chem Eur J 7:4328
21. (a) Hanabusa K, Yamada M, Kimura M, Shirai H (1996) Angew Chem Int Ed Engl 35:1949 (b) Hanabusa K, Shimura K, Hirose K, Kimura M, Shirai H (1996) Chem Lett 885
22. (a) van Esch J, de Feyter S, Kellogg RM, de Schryver F, Feringa BL (1997) Chem Eur J 3:1238 (b) van Esch J, Schoonbeek F, de Loos M, Kooijman H, Spek AL, Kellogg RM, Feringa BL (1999) Chem Eur J 5:937
23. Tian HJ, Inoue K, Yoza K, Ishi-i T, Shinkai S (1998) Chem Lett 871
24. Kasha M, Rawls HR, El-Bayoumi MA (1965) Pure Appl Chem 11:371
25. Ishi-i T, Jung JH, Iguchi R, Shinkai S (2000) J Mater Chem 10:2238
26. Ishi-i T, Iguchi R, Snip E, Ikeda M, Shinkai S (2001) Langmuir 17:5825
27. Tamaru S, Nakamura M, Takeuchi M, Shinkai S (2001) Org Lett 3:3634
28. Tamaru S, Uchino S, Takeuchi M, Ikeda M, Hatano T, Shinkai S (2002) Tetrahedron Lett 43:3751
29. Shirakawa M, Kawano S, Fujita N, Sada K, Shinkai S (2003) J Org Chem 68:5037
30. Sagawa T, Fukugawa S, Yamada T, Ihara H (2002) Langmuir 18:7223
31. Terech P, Gebel G, Ramasseul R (1996) Langmuir 12:4312
32. Setnicka V, Urbanová M, Pataridis S, Král V, Volka K (2002) Tetrahedron Asymm 13:2661
33. (a) Imahori H, Sakata Y (1997) Adv Mater 9:537 (b) Martín N, Sánchez L, Illescas B, Pérez I (1989) Chem Rev 98:2527 (c) Guldi DM, Prato M (2000) Acc Chem Res 33:695
34. (a) Tashiro K, Aida T, Zheng JY, Kinbara K, Saigo K, Sakamoto S, Yamaguchi K (1999) J Am Chem Som 121:9477 (b) Zheng JY, Tashiro K, Hirabayashi Y, Kinbara K, Saigo K, Aida T, Sakamoto S, Yamaguchi K (2001) Angew Chem Int Ed 40:1858
35. Yamaguchi T, Ishii N, Tashiro K, Aida T (2003) J Am Chem Soc 125:13934
36. Shirakawa M, Fujita N, Shinkai S (2003) J Am Chem Soc 125:9902
37. Ono Y, Nakashima K, Sano M, Kanekiyo Y, Inoue K, Hojo J, Shinkai S (1998) Chem Commun 1477
38. van Bommel KJC, Friggeri A, Shinkai S (2003) Angew Chem Int Ed 42:980
39. (a) Tamaru S, Takeuchi M, Sano M, Shinkai S (2002) Angew Chem Int Ed 41:853 (b) Kawano S, Tamaru S, Fujita N, Shinkai S (2004) Chem Eur J 10:344
40. Mirkin CA, Caldwell WB (1996) Tetrahedron 52:5113
41. Jonas U, Cardullo F, Belik P, Diederich F, Gügel A, Harth E, Herrmann A, Isaacs L, Müllen K, Ringsdorf H, Thilgen H, Uhlmann P, Vasella A, Waldraff CAA, Walter M (1995) Chem Eur J 1:243
42. (a) Yanagida M, Kuri T, Kajiyama (1997) Chem Lett 911 (b) Oh-ishi K, Okamura J, Ishi-i T, Sano M, Shinkai S (1999) Langmuir 15:2224
43. Cassell AM, Asplund CL, Tour JM (1999) Angew Chem Int Ed 38:2403
44. Jenekhe SA, Chen XL (1998) Science 279:1903
45. Oishi K, Ishi-i T, Sano M, Shinkai S (1999) Chem Lett 1089
46. Ishi-i T, Ono Y, Shinkai S (2000) Chem Lett 808
47. Ajayaghosh A, George SJ (2001) J Am Chem Soc 123:5148

48. George SJ, Ajayaghosh A, Jonkheijm P, Schenning APHJ, Meijer EW (2004) Angew Chem Int Ed 43:3422
49. Ajayaghosh A, George SJ, Praveen VK (2003) Angew Chem Int Ed 42:332
50. (a) van Nostrun CF, Picken SJ, Nolte RJM (1994) Angew Chem Int Ed Eng 33:2173 (b) van Nostrun CF, Picken SJ, Schouten AJ, Nolte RJM (1995) J Am Chem Soc 117:9957
51. Engelkamp H, Middelbeek S, Nolte RJM (1999) Science 284:785
52. Adam D, Schuhmacher P, Simmerer J, Häussling L, Siemensmeyer K, Etzbach KH, Ringsdorf H, Haarer D (1994) Nature 371:141
53. Ikeda M, Takeuchi M, Shinkai S (2003) Chem Commun 1354
54. Gearba RI, Lehmann M, Levin J, Ivanov DA, Koch MHJ, Barberà J, Debije MG, Piris J, Geerts YH (2003) Adv Mater 15:1614
55. Ishi-i T, Hirayama T, Murakami K, Tashiro H, Thiemann T, Kubo K, Mori A, Yamasaki S, Akao T, Tsuboyama A, Mukaide T, Ueno K, Mataka S (2005) Langmuir 21:1261
56. van der Laan S, Feringa BL, Kellogg RM, van Esch J (2003) Langmuir 18:7136
57. Mamiya J, Kanie K, Hiyama T, Ikeda T, Kato T (2002) Chem Commun 1870
58. Hamada K, Yamada K, Mitsuishi M, Ohira M, Miyazaki K (1992) J Chem Soc Chem Commun 544
59. (a) Kobayashi H, Friggeri A, Koumoto K, Amaike M, Shinkai S, Reinhoudt DN (2002) Org Lett 4:1423 (b) Kobayashi H, Koumoto K, Jung JH, Shinkai S (2002) Perkin Trans 2 1930
60. Schoonbeek FS, van Esch JH, Wegewijs B, Rep DBA, de Haas MP, Klapwijk TM, Kellogg RM, Feringa BL (1999) Angew Chem Int Ed 38:1393
61. Hashimoto M, Ujiie S, Mori A (2003) Adv Mater 15:797
62. van Gorp JJ, Vekemans JAJM, Meijer EW (2002) J Am Chem Soc 124:14759
63. Ryu SY, Kim S, Seo J, Kim YW, Kwon OH, Jang DJ, Park SY (2004) Chem Commun 70
64. An BK, Lee DS, Lee JS, Park YS, Song HS, Park SY (2004) J Am Chem Soc 126:10232
65. (a) Würthner F, Yao S (2000) Angew Chem Int Ed 39:1978 (b) Würthner F, Yao S, Debaerdemaeker T, Wortmann R (2002) J Am Chem Soc 124:9431
66. (a) Würthner F, Yao S, Beginn U (2004) Angew Chem Int Ed 42:3247 (b)Yao S, Beginn U, Gress T, Lysetska M, Würthner F (2004) J Am Chem Soc 126:8336
67. Würthner F (2004) Chem Commun (Feature Article) 1564
68. Sugiyasu K, Fujita N, Shinkai S (2004) Angew Chem Int Ed 43:1229
69. (a) Brotin T, Utermöhlen R, Fages F, Bouas-Laurent H, Desvergne JP (1991) J Chem Soc Chem Commun 416 (b) Placin F, Colomès M, Desvergne JP (1997) Tetrahedron Lett 38:2665 (c) Pozzo JL, Desvergne JP, Clavin GM, Bouas-Laurent H, Jones PG, Perlstein J (2001) Perkin Trans 2 824 (d) Terech P, Meerschaut D, Desvergne JP, Colomes M, Bouas-Laurent H (2003) J Colloid Interface Sci 261:441
70. Clavier GM, Brugger JF, Bouas-Laurent H, Pzzo JL (1998) Perkin Trans 2 2527
71. Placin F, Desvergne JP, Belin C, Buffeteau T, Desbat B, Ducasse L, Lassegues JC (2003) Langmuir 19:4563
72. Placin F, Desvergne JP, Cansell F (2000) J Mater Chem 10:2147
73. Clavier GM, Pzzo JL, Bouas-Laurent H, Liere C, Roux C, Sanchez C (2000) J Mater Chem 10:1725
74. Placin F, Desvergne JP, Lassegues JC (2001) Chem Mater 13:117
75. Terech P, Bouas-Laurent H, Desvergne JP (1995) J Colloid Interface Sci 174:258
76. Maitra U, Kumar PV, Chandra N, D'Souza LJ, Prasanna MD, Raju AR (1999) Chem Commum 595
77. Babu P, Sangeetha NM, Vijaykumar P, Maitra U, Rissanen K, Raju AR (2003) Chem Eur J 9:1922

78. Nakashima T, Kimizuka N (2002) Adv Mater 14:1113
79. de Loos M, van Esch J, Kellogg RM, Feringa BJ (2001) Angew Chem Int Ed Eng 40:613
80. Kimura M, Kitamura T, Muto T, Hanabusa K, Shirai H, Kobayashi N (2000) Chem Lett 1088
81. Ahmed SA, Sallenave X, Fages F, Mieden-Gundert G, Müller WM, Müller U, Vögtle F, Pozzo JL (2002) Langmuir 18:7096
82. Hachisako H, Ihara H, Kamiya T, Hirayama C, Yamada K (1997) Chem Commun 19
83. Shumburo A, Biewer MC (2002) Chem Mater 14:3745
84. Moriyama M, Mizoshita N, Yokota T, Kishimoto K, Kato T (2003) Adv Mater 15:1335
85. Zhao Y, Tong X (2003) Adv Mater 15:1431
86. (a) Shinkai S, Matsuo K, Sano M, Sone T, Manabe O (1981) Tetrahedron Lett 22:1409 (b) Shinkai S, Matsuo K, Harada A, Manabe O (1982) J Chem Soc Perkin Trans 2 1261
87. Perlstein J, Whitten DG (1997) J Am Chem Soc 119:9144
88. Ayabe M, Kishida T, Fujita N, Sada K, Shinkai S (2003) Org Biomol Chem 2744
89. de Jong JJD, Lucas JN, Kellogg RM, van Esch JH, Feringa BL (2004) Science 304:278
90. Lucas LN, van Esch J, Kellogg RM, Feringa BJ (2001) Chem Commun 759
91. Kawano S, Fujita N, van Bommel KJC, Shinkai S (2003) Chem Lett 12
92. Sohna JE, Fages F (1997) Chem Commun 327
93. Amanokura N, Kanekiyo Y, Shinkai S, Reinhoudt DN (1999) Perkin Trans 2 1995
94. Pozzo JL, Clavier GM, Desvergne JP (1998) J Chem Mater 8:2575
95. Sugiyasu K, Fujita N, Takeuchi M, Yamada S, Shinkai S (2003) Org Biomol Chem 895
96. Beck JB, Rowan SJ (2003) J Am Chem Soc 125:13922

Top Curr Chem (2005) 258: 161–204
DOI 10.1007/b135804

Published online: 11 August 2005

# Intercalation of Organic Dye Molecules into Double-Stranded DNA – General Principles and Recent Developments

Heiko Ihmels (✉) · Daniela Otto

Organic Chemistry II, University of Siegen, Adolf-Reichwein-Str. 2, 57068 Siegen, Germany
*ihmels@chemie.uni-siegen.de*



**Abstract** Several aspects of the intercalation of organic dyes into DNA are presented. After a description of general features of intercalation and the analytical evaluation of this binding mode, recent applications of the intercalation process in chemistry, biology, pharmacy, and medicine are described, with examples from representative dye classes, namely acridines, anthraquinones, cyanines, and phenanthridinium dyes. Along with references to classical work in this research field, examples are presented which demonstrate the influence of the substitution pattern of the intercalator on the selectivity and efficiency of intercalation. In addition, a survey of the recent literature is given that covers significant developments in the application of intercalating dyes as chemotherapeutic drugs, as probe or sensor molecules in biophysical chemistry and molecular biology, as photosensitizers in DNA-damage reactions, and as fluorescence stains.

**Keywords** Drug design · Fluorescence probes · Intercalation · Nucleic acids · Organic dyes

### Abbreviations

| | |
|---|---|
| A | Adenine |
| Ala | Alanine |
| BER | Base Excision Repair |
| C | Cytosine |

| | |
|---|---|
| CD | Circular dichroism |
| ct DNA | Calf thymus DNA |
| dA | Deoxyadenosinephosphate |
| dC | Deoxycytidinephosphate |
| dG | Deoxyguanosinephosphate |
| dsDNA | Double-stranded DNA |
| dT | Deoxythymidinephosphate |
| ET | Electron transfer |
| Gln | Glutamine |
| Gly | Glycine |
| G | Guanine |
| LD | Linear dichroism |
| Lys | Lysine |
| SASA | Solvent-accessible surface area |
| Ser | Serine |
| ssDNA | Single-stranded DNA |
| T | Thymine |
| Trp | Tryptophane |
| Tyr | Tyrosine |
| Val | Valine |

## 1 Introduction

DNA is an important biomacromolecule [1, 2] that offers several binding sites for a variety of guest molecules [3]. The binding interaction between external molecules and nucleic acids often leads to a significant change in their structures and may have an important influence on their physiological functions. Thus, DNA-binding reagents exhibit a high potential as chemotherapeutic drugs which may suppress the gene replication or transcription in tumor cells [4–9]. In addition, such a host–guest interaction may be used for the detection of nucleic acids when the physical properties of the guest molecule change upon binding and may be easily monitored. Especially useful along these lines is DNA staining, which is based on the color change of a dye upon binding to the macromolecule [9, 10]. Because of these important applications of DNA binders, one of the most challenging goals in this area is the design of molecules which bind to DNA with high selectivity and large association constants. Thus, several classes of DNA-binding molecules have been established and investigated in detail in recent years [3, 11, 12]. For example, oligo- and polyamide [13] derivatives and oligosaccharides [14–16] are important biomolecules whose interaction with DNA is a physiologically relevant process; synthetic derivatives thereof also represent a promising lead structure for functional DNA-binding drugs. Furthermore, organometallic complexes have been shown to associate with DNA [17, 18] and been used to investigate the charge transfer in DNA [19]. Moreover, numerous organic dyes have been found to bind to DNA; this complex formation and its application in DNA visualization or intercalator-based drug design has been reviewed several times [20–25].

In this review we will focus on representative classes of dyes which intercalate into double-stranded DNA (dsDNA), namely acridines, anthraquinones, cyanine dyes, and phenanthridinium ions. Although other important polynucleotide structures, such as guanine quartets [26], RNA [27], and triplex DNA [28] are known as host structures for dye molecules, these host–guest interactions will not be covered in this review. We will try to cover the general principles and novel aspects of the intercalation process on the basis of the chosen well-studied representatives. In particular, we wish to focus on new approaches and developments in biochemistry, biology, and medicine, in which the particular dye properties have been used for analytical and therapeutic purposes. Please note that there also exist other important dye classes, such as naphthylimides and naphthyldiimide [29–35], azine dyes such as methylene blue [24, 36–42], perylenes [43–47], annelated quinolizinium ions such as coralyne [48–50], and porphyrins [51–55], whose members also exhibit pronounced intercalation properties and which are also important in analytical biochemistry or medicine. We may have neglected to cover a particularly important dye, but the overwhelming number of DNA-binding dye molecules explored and established during the past years in this field, along with an intention to limit this review to a reasonable length, required a subjective selection of representative examples.

## 2 Binding Modes

Prior to the presentation of DNA-binding dyes, a brief description of the possible binding modes will be given. In general, guest molecules may associate to DNA by (a) minor or major groove-binding or (b) intercalation. In Fig. 1, these binding modes are shown in a simplified illustration with representative examples, i.e., ethidium bromide (intercalation) and Hoechst 33258 (groove-binding). Additionally, a third binding mode, namely external binding by attractive electrostatic interactions, should be considered. Although single dye molecules usually bind by intercalation or, in fewer cases, by groove-binding, most dye aggregates are too large to fit into the corresponding binding pockets and bind to the phosphate backbone in a so-called outside-stacking [11]. Recently, it has been observed that in a cooperative aggregation reaction, selected cyanine dyes form helical aggregates in the minor groove by using the DNA as a template [56]. In contrast to the outside-stacking, which is mainly governed by electrostatic interactions, groove-binding and intercalation result from a supramolecular assembly based on $\pi$-stacking, hydrogen bonding, van der Waals, or hydrophobic interactions. Each bind-

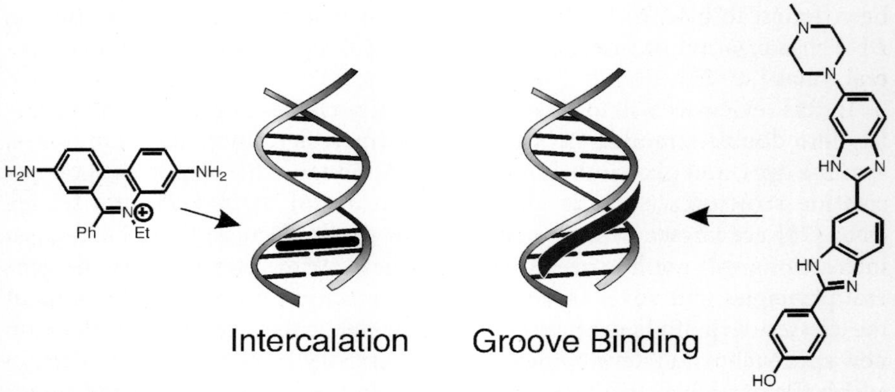

Fig. 1 Intercalation and groove binding of organic dyes

ing interaction is usually initiated by a hydrophobic transfer of the lipophilic DNA binder from the polar aqueous solution into the less polar environment of the DNA. Recently, the change of DNA hydration upon intercalation or groove-binding and its influence on the overall energy of the binding process has been discussed [57]. The dynamics of the binding processes have been monitored by time-resolved methods, and these experiments have revealed a crucial role of the rearrangement of the hydration layer on the DNA surface upon association of a DNA binder with respect to the enthalpic and entropic interactions.

It should be noted that in many cases all three binding modes may take place at the same time. Thus, it has been established that, for example, cationic metalloporphyrins, especially derivatives of *meso*-tetra(*N*-methylpyridinium-4-yl)porphyrin, bind to DNA by intercalation, groove-binding or by association of $\pi$-stacked aggregates with the DNA backbone [58]. In general, a combination of coulombic, hydrophobic, and steric forces influence the mode of binding which depends on the structure of the porphyrin as well as on the DNA sequence. Thus, the peripheral pyridinium groups of the porphyrin provide an electrostatic attraction to DNA, but increase the steric repulsion as well. Thus, a delicate balance between these forces is required to prevent mixed binding modes. Interestingly, the first porphyrin binding to DNA exclusively by intercalation was reported only recently [59].

## 2.1 Groove-Binding

The DNA helix exhibits two grooves of different size, namely the minor and the major grooves, which may serve as binding sites for guest molecules [60]. Whereas relatively large molecules such as proteins bind preferentially to the major groove [12], the minor groove is the preferred binding site for smaller ligands [61]. Typically, groove binders show a binding selectivity towards AT-rich areas [61], but some groove binders are known that bind preferentially in GC-rich grooves [62, 63]. The binding pocket of a DNA groove may be defined by two different regions, namely the "bottom", formed by the edges of the nucleic bases that face into the groove, and the "walls", which are formed from the deoxyribose–phosphate backbone of the DNA. Groove binders usually consist of at least two aromatic or heteroaromatic rings whose connection allows conformational flexibility such that a crescent-shaped conformation may be achieved and the molecule fits perfectly into the groove. In addition, functional groups are required to form hydrogen bonds with the nucleic bases at the bottom of the groove. Typical minor-groove binders are polyamides [13], Hoechst 33258 (**1**) [64], netropsin (**2**) [61], and furamidines such as **3** [65]. Many minor-groove binders exhibit a high potential as therapeutic reagents [66].

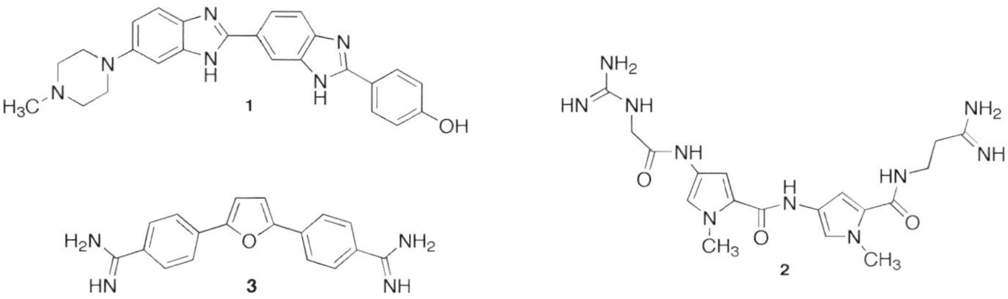

**Scheme 1**

## 2.2 Intercalation

In a DNA helix, the nucleic bases are located in an almost coplanar arrangement, which allows planar polycylic aromatic molecules to intercalate between two base pairs [67]. Important driving forces for this binding mode are electrostatic factors [68], dipole–dipole interactions, dispersive interactions and $\pi$-stacking [69] of the guest molecule with the aromatic nucleic bases. The analysis of calorimetric data led to the conclusion that hydrophobic effects and van der Waals contacts within the intercalation site are most important for efficient intercalation [70]. Extensive theoretical studies on the complexes of DNA with several intercalators reveal that the dipole–dipole interactions are not as important as the dispersion energy, which contributes most to the overall energy of the intercalation complex [71]. Nevertheless, an external positive charge appears to be essential for intercalation of small aromatic compounds such as benzene, naphthalene, quinoline, or even *N*-alkylquinolinium, since these compounds only intercalate into dsDNA if there are positive charges in a substituent side chain [72]. The delicate balance between forces that govern the propensity of a compound to intercalate in dsDNA is obvious, if one compares the latter results with the observation that additional benzo-annelation of naphthalene or quinolinium derivatives results in strong intercalators that no longer require positively charged substituents.

Other than groove-binding, intercalation has a significant influence on the DNA structure, because the DNA needs to unwind so that the intercalator fits between the two base pairs. This unwinding leads to a lengthening of the helix by approximately 3.4 Å, which causes a significant conformational change of some involved deoxyribose moieties from C2′-endo to C3′-endo [73, 74]. In addition, the unwinding and lengthening of DNA increases the phosphate spacing and decreases the charge density along the helix axis. This change in the electrostatic potential of the DNA surface leads to a release of counterions from the grooves and provides an energetically favorable contribution to the free energy of binding [75]. If the DNA binder carries a positive charge, the association with DNA leads to a release of additional counter cations from the DNA grooves, so that the association of cationic molecules is usually more favorable than the one of uncharged compounds. On the other hand, it has been observed that the binding of one intercalator between two base pairs hinders the access of another intercalator to the binding site next to the neighboring intercalation pocket ("neighbor exclusion principle") [76]. Presumably, this principle is a result of the structural changes of the DNA upon intercalation, which lead to limited access to the neighboring binding pocket for steric reasons. Furthermore, the intercalation process may reduce the negative electrostatic potential at the intercalation site, so that attractive electrostatic interactions no longer take place close to this site.

Dye molecules may be intercalated with their long molecular axis parallel or perpendicular to the binding pocket (Fig. 2), i.e., relative to the direction of hydrogen bonds in the base pair. The latter binding mode has been observed for anthracycline derivatives **4a** [77], whereas derivatives of acridine (**5a**) [78], methylene blue (**6**) [79], and cryptolepine (**7**) bind parallel to the binding pocket, with cryptolepine being the first molecule to intercalate in non-alternating binding sites [d(CpC)-d(GpG)] [80].

Typical binding constants for intercalation complexes between organic dyes and DNA range from $10^4$ to $10^6$ $M^{-1}$ and are usually significantly smaller than the binding constants of groove binders ($10^5$ to $10^9$ $M^{-1}$). Moreoever, the rate of the association/dissociation of intercalators is high: for example, se-

**Fig. 2** Schematic representation of intercalators with long molecular axis parallel (**a**) and perpendicular (**b**) to the binding pocket. *Grey* Nucleic bases, *black* intercalator

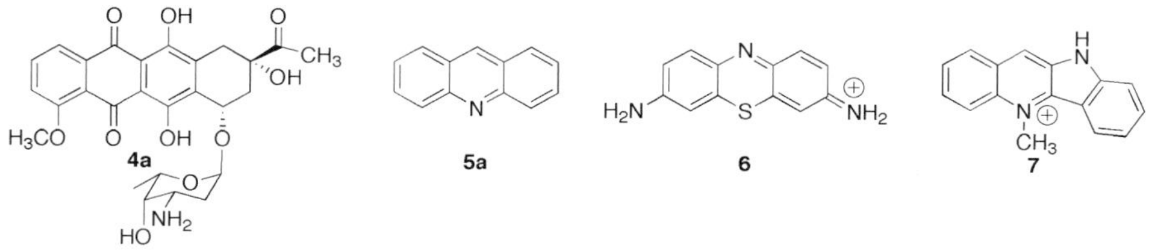

**Scheme 2**

lected anthraquinone derivatives have been shown to have residence times within the intercalation site of less than 1 ms [81]. This condition constitutes a major drawback for the application of intercalators as drugs that are supposed to suppress the proliferation of cells by the occupation of the binding sites of enzymes. To overcome this problem, derivatives were designed that carry two or more intercalation functionalities in a single molecule. It was early shown that the association constants of organic molecules to DNA may be enhanced significantly when one molecule carries two or more intercalating functionalities, as realized in bisintercalators, trisintercalators, and tetrakisintercalators etc. The design of DNA bisintercalators on the basis of alkyl-linked acridine functionalities has been introduced by Le Pecq et al. [82] and Cain et al. [83]. This strategy was subsequently established with numerous bisacridine derivatives such as **8**, in which the intercalating 9-aminoacridine moieties are connected by alkyl linkers [84–86] and extended to bisintercalators with other intercalating functionalities [87–97]. Notably, only recently a sequence-selective bisintercalator, i.e., **9**, was reported [98].

The structure of a complex between DNA and bisintercalators depends on the linker length [86]. With a short connection between the intercalators (< 9 Å), solely monointercalation takes place (Fig. 3A). With a longer alkyl chain between the intercalating parts, i.e., with a chain length of > 10.2 Å, bis-

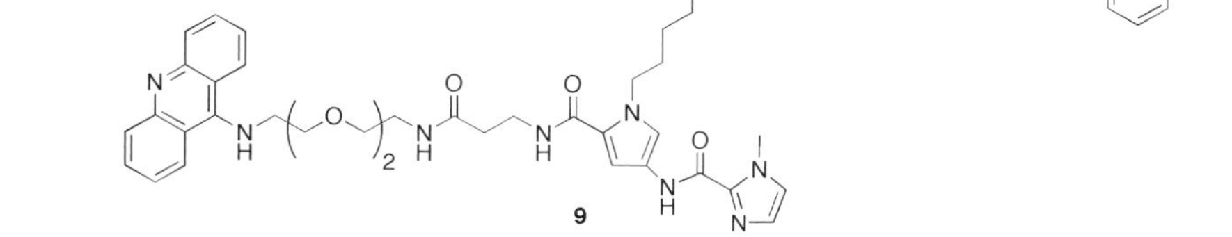

**Scheme 3**

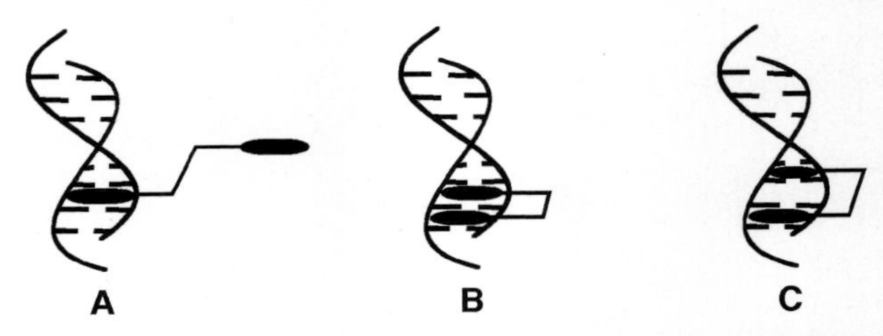


**Fig. 3** Binding mode of bisintercalators depending on the length of the alkyl chain between the intercalating moieties (**A**: Monointercalation. **B**: Bisintercalation with violation of neighbor-exclusion principle. **C**: Bisintercalation)

intercalation may be achieved without a violation of the neighbor-exclusion principle (Fig. 3C). It has been proposed that the neighbor-exclusion principle may be violated by bisintercalators, whose intercalating parts are separated by 9–10 Å (Fig. 3B) [99]. In contrast, $^{1}$H-NMR studies showed that such compounds only monointercalate [100].

Theoretical considerations predict that in the absence of significant steric and entropic factors, the binding constants of bisintercalators with DNA should be the square of that of the corresponding monomer [101]. Nevertheless, the experimental data are usually much smaller than expected, presumably due to an unfavorable geometric effect. Recently, the complex formation between DNA and a naturally occuring bisintercalator, namely echinomycin **10**, was evaluated on the basis of thermodynamic data from calorimetric data (differential scanning calorimetry and UV thermal denaturation) [102]. These results revealed that at 20 °C the entropy contributes mostly to the overall binding ($T\Delta S = -11.4\,\mathrm{kcal\,mol^{-1}}$), whereas the enthalpic term is slightly positive ($\Delta H = +3.8\,\mathrm{kcal\,mol^{-1}}$). Moreover, the overall free energy ($\Delta G = -7.6\,\mathrm{kcal\,mol^{-1}}$) was divided into different contributions, which reflect the different forces which may contribute to the overall binding and whose energetic contribution was estimated and calculated, respectively. Thus, $\Delta G_{\mathrm{conf}}$ (8 kcal mol$^{-1}$) considers the conformational changes of the DNA

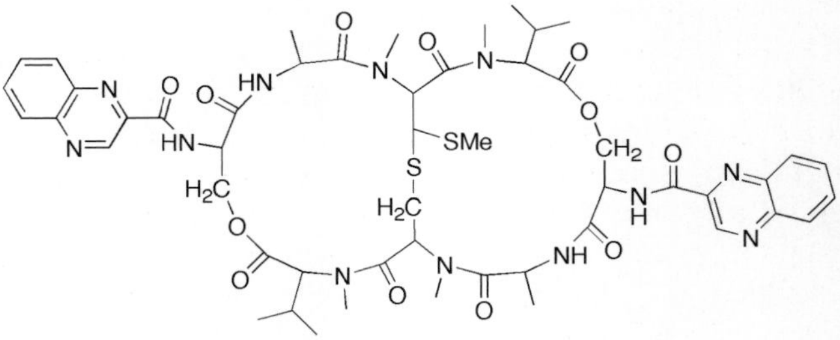

**10**

**Scheme 4**

and the drug upon complexation; $\Delta G_{t+r}$ (14.9 kcal mol$^{-1}$) is the contribution resulting from the suppressed translational and rotational freedom of both components within the DNA–drug complex, $\Delta G_{hyd}$ ( – 17.4 kcal mol$^{-1}$) is the free energy for the hydrophobic transfer of the echinomycin molecule from solution to the DNA binding sites, $\Delta G_{pe}$ (– 1 kcal mol$^{-1}$) is the electrostatic free energy contribution resulting from counterion release from the DNA backbone, and $\Delta G_{mol}$ (– 12 kcal mol$^{-1}$) is the term which considers weak non-covalent bonds including hydrogen bonds, van der Waals interactions and other weak forces upon complex formation. These results suggest that one major driving force for the association of echinomycin and DNA is the hydrophobic effect, i.e., the transfer of the hydrophobic echinomycin molecule from the hydrophilic solution into the hydrophobic intercalation pocket. Moreover, the direct contacts (hydrogen bonding, van der Waals interactions and base-stacking) between the echinomycin molecule and DNA are another key feature of the complex structure and contribute significantly to the overall free energy of binding. Most likely these contributions are the result of the binding of the peptide chain within the DNA groove. This interpretation is in agreement with results obtained for the minor-groove binder, Hoechst 33258. The binding of this compound is also assisted by a favorable entropic effect [103]. In contrast, the entropic contribution to the binding of classical intercalators such as ethidium bromide is unfavorable, i.e., $\Delta S < 0$ kcal mol$^{-1}$ [104]. Unfortunately, calorimetric data for bisintercalators with alkyl chains rather than peptide linkers are not available, so far.

## 2.3 Determination of the Binding Mode

Since the physical properties of DNA and the dye molecules change significantly upon complex formation, the determination of these properties is a useful tool for the qualitative and quantitative evaluation of the association process and to confirm the binding mode of a given DNA binder. Among the properties that may be used for this purpose, are the viscosity or the sedimentation coefficient, the melting temperature of the DNA, mass-spectrometric data [105], the NMR shifts, and the absorption and emission properties [106–108]. In the last case, variations are possible. For example, the absorption of circularly or linearly polarized light may be used in circular dichroism (CD) and linear dichroism (LD) spectroscopy to deduce the orientation of the dye molecule relative to the DNA [109, 110]. Also, steady-state fluorescence polarization measurements [111] as well as fluorescence energy transfer from the DNA bases to the bound dye have been used as criteria to elucidate the binding mode; however, it has also been suggested that this method should be used with caution [112]. Binding constants and binding selectivities may also be obtained by the DNA footprinting experiments [113, 114], by the equilibrium dialysis method [115], or by a well-designed fluorescent-intercalator displacement assay [116]. X-ray diffraction analyses of crystalline complexes between short double-stranded oligonucleotides and dyes provide the most detailed data on the binding, but it is often difficult to grow the required single crystals from the complexes.

In critical articles, it has been pointed out that the choice of the methods to determine the binding mode must be made with caution and that only a combination of selected methods provides sufficient information to determine the binding mode [117–119]. Thus, for a detailed analysis of criteria for intercalation, Long and Barton divided the aforementioned experimental techniques into three groups (Table 1) [118]:

(a) The evaluation of changes in the DNA helix by the determination of the sedimentation coefficient, the solution viscosity, the electrophoretic mobility of circular DNA, or the $^{31}$P-NMR shift of phosphate phosphorous atoms.
(b) Experiments that reflect the electronic interactions between intercalator and the nucleic acid. Usually these changes are monitored by absorption spectroscopy, as the absorption maxima of intercalated molecules are significantly red-shifted compared to those of the unbound molecule. In addition, a hypsochromic effect is observed upon DNA addition. If the intercalator is a fluorescent compound, fluorescence quenching or fluorescence enhancement may occur upon intercalation. Electronic interactions may be observed by high-resolution $^{1}$H-NMR spectroscopy, especially with the signals of aromatic protons in the binding pocket, which are usually high-field-shifted due to anisotropic cones and the aromatic ring current from the nucleic bases.
(c) The evaluation of molecular orientation or rigidity by CD spectroscopy and linear-flow or electric LD spectroscopy. If the direction of the transition dipole moment of the DNA binder is known, these experiments provide information on the relative orientation of the guest molecule. Fluorescence polarization techniques allow the determination of rotational diffusion of fluorophores and may thus be used to monitor the immobilization of molecules when bound to DNA.

In addition to these three groups, the particular structural features of each molecule must be considered, especially when complex structures may give rise to more than one binding geometry, or when the DNA binder may approach more than one binding site.

With these criteria in mind, the authors used the data of molecules whose DNA-binding mode is known and checked which of these methods is sufficient to deduce the corresponding binding mode [118]. Surprisingly, the authors found that "the proposal or dismissal of an intercalative interaction cannot be concluded on the basis of any of the above criteria". For example, it was observed that groove binders exhibit absorption spectra upon DNA addition which are rather characteristic for intercalators. Based upon these results it was strongly suggested that experiments from all three above mentioned groups, (a)–(c), should be applied along with a detailed evaluation of the molecular shape and structure of the DNA binder in order to deduce the binding mode.

Suh and Chaires also conducted a study to evaluate the methods for the determination of the binding mode with a known intercalator (ethidium bromide) and a groove binder (Hoechst 33258) [119]. These studies revealed that spectrophotometric and spectrofluorimetric titrations, or fluorescence polarization measurements may be used to deduce a general binding interaction between guest molecules and DNA, but they cannot be used for the unambiguous determination of the binding mode. Nevertheless, it was demonstrated that a combination of viscosimetric titrations and the determination of a fluorescence resonance energy transfer may serve as a reliable tool to determine the binding mode [119]. In the latter experiments the energy transfer from the excited DNA bases to a fluorescent ligand is monitored. Excitation energy transfer exhibits a distance dependence of $R^{-6}$ where $R$ is the distance between the species involved in the transfer of energy. This distance dependence, along with the relatively low fluorescence quantum yield of the DNA bases results in a short Förster critical distance $R_0$ (4–7 Å) at which half of the energy of the excited DNA bases is transferred to the bound molecule [120]. Intercalation of the molecule between stacked bases within the DNA host duplex is associated with small host–guest distances (about 2–4 Å). Thus, the observation of efficient energy transfer from a host DNA to a bound molecule is consistent with an intercalative mode of binding. This energy transfer may be determined by excitation spectra, i.e., when the emission wavelength is fixed on the DNA-bound molecule, the observed excitation spectrum should resemble the absorption bands of the nucleic bases (note that the excitation spectrum needs to be normalized with respect to the fluorescence quantum yield resulting from direct excitation of the intercalator). By contrast, binding of a molecule to the minor groove and/or drug-stacking along the surface of the helix are associated with larger base-to-molecule distances and minimal base–molecule stacking interactions. Consequently, such binding modes usually suppress the energy transfer from the host DNA to a bound molecule. Thus, it was shown that a fluorescence energy transfer between DNA-bound ethidium bromide and the nucleic bases takes places, whereas with Hoechst 33258 this effect was not observed. However, it should be noted that in a related study fluorescence energy transfer has been observed between nucleic bases and groove-bound molecules [112].

Although the sole detection of fluorescence energy transfer is not a sufficient criterium for intercalation, it was claimed that additional viscosimetric studies provide complementary information which allows unambiguous determination of the binding mode [119]. As already demonstrated by Lerman [67] intercalation leads to a significant lengthening of the DNA, which in turn has an influence on the colligative properties of solutions of the macromolecule. The determination of the viscosity changes upon DNA association of an intercalator turned out to be an especially useful tool to follow the lengthening and unwinding of the nucleic acid. Thus, the relative specific viscosity of ct DNA increased upon addition of ethidium bromide, whereas the titration with Hoechst 33258 did not change the viscosity. Based upon these results, it was proposed that the determination of the fluorescence energy transfer in combination with the determination of viscosity changes may be a useful combination to determine whether an external molecule intercalates into DNA.

## 2.4 Energetics of the Intercalation Process

Recent articles on thermodynamic [115, 121] and volumetric [122] analyis of drug–DNA interactions are also available. Although numerous binding constants are available, from which $\Delta G$ can be calculated, very few attempts have been made to determine $\Delta H$, $\Delta S$, or the change of heat capacity $C_p$ of the intercalation reaction. Isothermal titration calorimetry was used to obtain reliable data for $\Delta C_p$ [70]. Thus, the change of heat capacity upon association of ethidium bromide with DNA was estimated to be $-139 \pm 30$ cal mol$^{-1}$ K$^{-1}$ and claimed to be more accurate than the data reported earlier [123, 124]. Moreover, it was demonstrated that the experimental data for $\Delta C_p$ obtained for ethidium bromide, propidium bromide, daunorubicin, adriamycin and actinomycin D correlate very well with theoretically derived changes of solvent-accessible surface areas (SASA) upon intercalation. Nevertheless, a recent theoretical study on small-cluster hydrophobic interactions showed that the evaluation of hydrophobic interactions with the help of SASA may not be applied in general [125].

The free energy $\Delta G$ of the intercalation reaction may be obtained from the binding constants (for a brief overview over relationships between thermodynamic parameters see [126]) and attempts were made to parse the overall free energy into contributions of different intermolecular interactions. As already mentioned in Sect. 2.2, a simplified additive contribution of at least five different terms may be considered (Eq. 1). In the following the origins and simplified approaches for the estimation of the energetic contributions of these terms will be briefly discussed.

$$\Delta G_{\mathrm{obs}} = \Delta G_{\mathrm{conf}} + \Delta G_{\mathrm{t+r}} + \Delta G_{\mathrm{hyd}} + \Delta G_{\mathrm{pe}} + \Delta G_{\mathrm{mol}} \quad (1)$$

The DNA helix needs to undergo significant conformational changes in order to accommodate an intercalator. Since in B-form DNA the nucleic bases are in close contact due to $\pi$-stacking, they need to be separated from each other before an aromatic compound may intercalate between these base pairs. In general, the required space for a planar aromatic compound is approximately 3.4 Å, which corresponds to the average thickness of aromatic molecules.

Thus, the DNA needs to unwind for 3.4 Å prior to complex formation; however, the net lengthening is often somewhat smaller due to additional bending of the helix at the intercalation site. The lengthening of the nucleic acid is accompanied by an unwinding of approximately 10–30° depending on the nature of the intercalator and the DNA sequence. Several calculations revealed that these structural changes are energetically unfavorable and represent an endergonic contribution to $\Delta G_{obs}$ [104]. Moreover, detailed kinetic studies on the dependence of the intercalation rate allowed an estimation of $\Delta G_{conf} \approx +4\,\text{kcal mol}^{-1}$ [127]. It should be noted that this energy contribution depends significantly on the bases which constitute the intercalation pocket, and it may therefore be assumed that the sequence selectivities of particular intercalators may be governed by this energetic factor.

When two molecules associate, three rotational and three translational degrees of freedom are lost, which usually leads to a loss of entropy; however, some of this energetic loss is compensated for by six internal vibrational modes [128]. Because of the impossibility of obtaining experimental data along these lines, the magnitude of the loss of translational and rotational entropy is still under debate [129–131]. Thus, only empirical and theoretical data are available, all of which have particular advantages and disadvantages. At present, the best estimate for the energetic contribution of $\Delta G_{r+t}$ to the overall free energy, $\Delta G$, of bimolecular complex formation is presumably $\Delta G_{r+t} \approx +15\,\text{kcal mol}^{-1}$ [129].

The hydrophobic transfer, i.e., the energetically favorable change of the microenvironment of a hydrophobic guest molecule from hydrophilic hydrous solution to the hydrophobic binding sites of the nucleic acids, is considered to be one of the important driving forces for intercalation. It was reported that the change of heat capacity upon intercalation correlates with the contribution of the hydrophobic effect to the overall free energy of binding ($\Delta G_{hyd} \approx 80 \times \Delta C_p$) [132]. With this correlation, a hydrophobic contribution to the overall free binding energy of ethidium was estimated to be $\Delta G_{hyd} = -11.2\,\text{kcal mol}^{-1}$ [70]. Unfortunately, rather few calorimetric data are available for other intercalators, so that in these cases theoretical estimates are necessary to obtain $\Delta G_{hyd}$. Thus, the change of the SASA, which may be calculated by standard protocols, may be correlated with the free energy of the hydrophobic transfer ($\Delta G_{hyd} \approx -22(\pm 5) \times \Delta\,\text{SASA}$) [132, 133].

The contribution of $\Delta G_{pe}$ to the overall free energy may be experimentally determined as the binding constant changes as a function of metal-ion concentration $[M^+]$, most often sodium cations, according to Eq. 2 [75].

$$\Delta G_{pe} = -RT(m\psi)\ln[M^+] \qquad (2)$$

Thus, the number of counter ions released from the DNA backbone upon binding of another charged molecule may be determined from the plot of $\log K$ versus $\log[M^+]$, where the slope equals $m\psi$. The parameter $m$ represents the charge of the binding molecule and $\psi$ is the fraction of counter ions associated with each DNA phosphate moiety ($\psi = 0.88$ for the double-stranded B-DNA) [134]. From the dependence of the binding constant on the salt concentration, the polyelectrolyte contribution, $\Delta G_{pe}$, arising from coupled polyelectrolyte effects may be calculated according to Eq. 2. For example, $\Delta G_{pe}$ of ethidium bromide is approximately $-1.2$ kcal/mol. This energy is mostly determined by the release of condensed counter ions from the DNA helix upon binding of the charged ligand. It should be noted that $\Delta G_{pe}$ may be negative even with uncharged intercalators, since the lengthening of the DNA upon complex formation leads to a decrease of charge density along the backbone and subsequently to a release of counter ions.

Noncovalent bonding interactions such as hydrogen bonding, $\pi$-stacking, and Lewis acid–base interactions are general features of most supramolecular complexes, and in most cases the sum of the attractive interactions contributes significantly to the overall binding. The energy of noncovalent molecular interactions, $\Delta G_{mol}$, in a complex is difficult to quantify either from experiments or from calculations [135]. The most efficient approach applies the variation of the substitution pattern of a known intercalator and the performance of detailed binding studies. The results will give a structure–property correlation that allows the determination of $\Delta G_{mol}$. Since such studies are tedious and time-consuming, few data for $\Delta G_{mol}$ of intercalation complexes are known at present. On the other hand, $\Delta G_{mol}$ may be estimated if reasonable data exist for the other contributions to the overall energy by simple subtraction of the latter from $\Delta G_{obs}$. With this method, $\Delta G_{mol}$ of ethidium binding to DNA was estimated to be approximately $-13$ kcal mol$^{-1}$.

# 3 Representative Dye Classes

## 3.1 Acridines

Acridine (**5a**) [136] and derivatives thereof [137] are a dye class whose interactions with DNA have been extensively studied [138–142]. Acridines are protonated at the nitrogen atom even under neutral conditions; and the

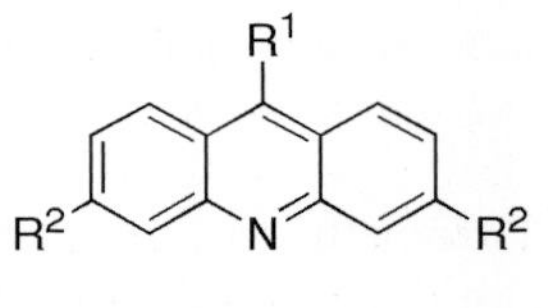

R$^1$ R$^2$ N R$^2$

**5a–e**

**5a**: $R^1 = R^2 = H$
**5b**: $R^1 = 4'\text{-}(NH_2)C_6H_4$; $R^2 = NH_2$
**5c**: $R^1 = H$; $R^2 = NH_2$
**5d**: $R^1 = H$; $R^2 = N(CH_3)_2$
**5e**: $R^1 = NH_2$; $R^2 = H$

**Scheme 5**

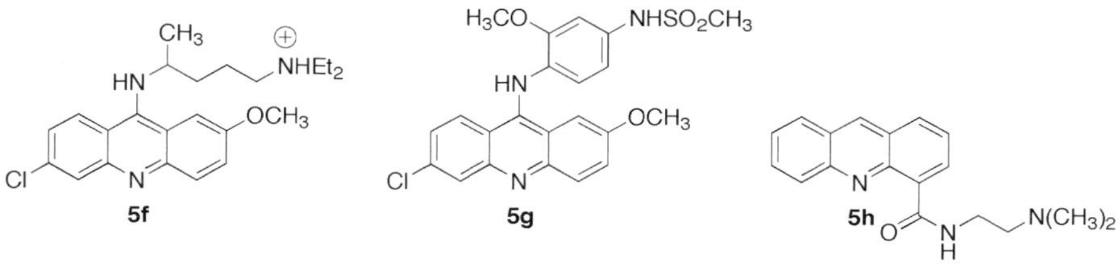

**Scheme 6**

resulting cationic aromatic compounds represent a promising class of intercalators. Most notably, acridine derivatives exhibit a high potential for their application in chemotherapy [143]. The first report on the use of acridines in medicine goes back to the late nineteenth century, when it was observed that anilinoacridine (**5b**) is a weakly active anti-malaria drug [144]. Later it was shown that, for instance, proflavine (**5c**) and derivatives thereof exhibited antibacterial activity [145, 146]. Moreover, 9-aminoacridine (**5e**) has been used clinically as an antiseptic drug, since its intercalation into DNA suppresses DNA replication at the intercalation site [147, 148]. Up to now, numerous derivatives have been synthesized and evaluated with respect to their biological and potential clinical activities [149–152]. Among these derivatives, the quinacrines (**5f**) [153–156], used clinically as antimalarial drugs, the antileukemic amsacrine derivatives (**5g**) [157–159], and derivatives of acridine carboxamides (**5h**) are especially promising lead structures. The cytotoxicity of acridines has been evaluated by an assay that monitored the influence of acridines on the luminol-dependent chemiluminescence of polymorphonuclear leucocytes during phagocytosis [160]. It has also been reported that several acridine derivatives induce tacrine-like subcellular changes in hepatocytes which subsequently cause mitochondrial dysfunction [161]. Recently, it has been shown that 9-aminoacridine (**5e**) and its derivatives promote the binding of the RecA protein to DNA [162]. The effect of the binding affinity of the intercalator on cytotoxic potential has been discussed for acridine-based drugs [138]: although one should expect that intercalators with high binding constants have the highest cytotoxic potential, it has been pointed out that this strong binding is counterproductive with respect to the diffusion of the drug within the extravascular medium [163]. According to these studies, the extravascular diffusion decreases with an increasing binding constant, which significantly limits the distribution of the drug in the medium and, thus, its cytotoxic potential. As a consequence, it has been proposed that an efficient drug needs to exhibit binding affinities sufficiently large as to provide for intercalation, but small enough to allow diffusion of the drug within the whole volume of the target tissue (concept of the "minimal intercalator") [138].

The cytotoxicity of most acridine-based drugs is founded on their ability to suppress topoisomerase activity. Topoisomerases are essential enzymes which bind to DNA and, within this complex, initially induce a DNA-strand cleavage. Subsequently, both single strands are reorganized within the enzyme–DNA complex and finally reconnected. The final result of this process is the relaxation of the DNA structure [164, 165]. Topoisomerases induce a single-strand break (topoisomerase I) or a double-strand break (topoisomerase II) of the double helix. The resulting change in DNA structure is an important requirement for replication, transcription or recombination of the genetic material. There exist two possibilities for an intercalator to influence the topoisomerase activity and therewith suppress the proliferation of the cell [166]: (a) by the intercalation of a drug, the binding site of the topoisomerase is occupied, and the complex formation between the enzyme and the DNA is hindered; and on the other hand, (b) a ternary complex between DNA, intercalator and topoisomerase may be formed which is significantly more stable than the DNA–topoisomerase complex. The stability of the ternary complex may lead to an enhanced lifetime of the cleaved DNA, i.e., the re-ligation of the strands cannot take place and the strand breaks remain permanent. Thus, the topoisomerase acts as an endogeneous poison and may induce apoptosis. Such a simultaneous formation of ternary complexes between DNA, intercalator and topoisomerase has been studied systematically with benzimidazole derivatives [167].

Acridine rings are often used as the intercalating moiety in more complex molecules. In most cases, the intercalation properties of the acridine are used to bring particular functionalities, which do not bind to DNA themselves, into the close vicinity of the nucleic acid. Also, the intercalation of an acridine moiety may be used to enhance the binding constant and/or binding selectivity when the other part of the molecule also binds to nucleic acids. Thus, acridines have been connected to, for instance, polyamides [98, 168], peptide nucleic acids (PNA) [169], purine derivatives [170–172], oligonucleotides [173, 174], *cis*-platin derivatives [175–177], bis-phenanthroline derivatives [178], polyamines [179], viologens [180], curcurbituril [181], and porphyrins [97]. Acridines may also be connected to one or more other acridine rings [182, 183] to give so-called bis-, tris-, or tetrakisintercalators such as **5i** and **5j** [86, 139, 184].

These compounds exhibit significantly enhanced binding affinities towards dsDNA [144]. Two acridine rings may also be connected in cylophane-type macrocylces, e.g. **5k**, which were found to exhibit large binding affinities towards hairpins in ssDNA, G-quartets, and abasic sites [185–189].

**5i,j**

**5i**: R = H

**5j**: R =

**Scheme 7**

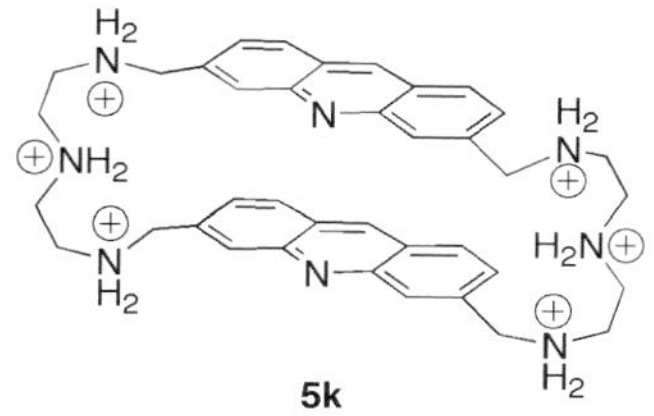

**Scheme 8**

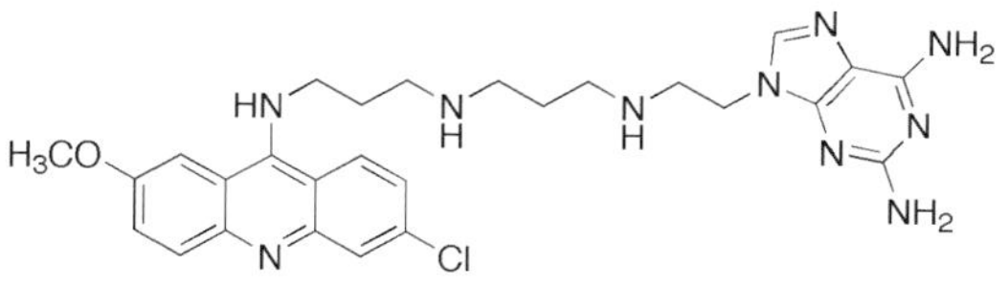

Recently, particular interest has been focused on the interaction of nucleic base–intercalator conjugates such as **5l** with abasic sites, because the latter play a crucial role in DNA damage and DNA repair mechanisms in the cell. Abasic positions, i.e., sites in which the deoxyribose does not carry a nucleic base, usually occur during enzymatic repair of alkylated or dimerized nucleic bases. During this base-excission repair (BER) [190], the damaged base and the corresponding deoxyribosephosphate are removed by an AP endonuclease, and with the help of DNA polymerase and DNA ligase the correct nucleotide is replaced. Unfortunately, this repair mechanism also takes place in tumor cells whose DNA bases are damaged by anti-tumor drugs, so that the efficiency of the drug is limited. In one novel approach in chemotherapy, the anti-tumor reagent is assisted by an intercalator which associates at abasic sites with high selectivity [191]. Binding in abasic sites is usually established in nucleic base–intercalator conjugates, mostly purine–intercalator combinations, in which the purine binds in the abasic site and the intercalator in a binding pocket nearby (Fig. 4). The linker between the intercalator and the purine carries a secondary amine, which induces an elimination in the deoxyribose moiety, which finally leads to a DNA strand cleavage. Usually, the latter cannot be repaired by BER.

The diamino-substituted acridine derivatives proflavine (**5c**) and acridine orange (**5d**) are classical representatives of this dye class and their DNA intercalating properties have been studied in detail. Both dyes were shown early to intercalate into DNA [192–194], and in both cases the dye properties, i.e., absorption and emission in the UV/vis area, could be used to

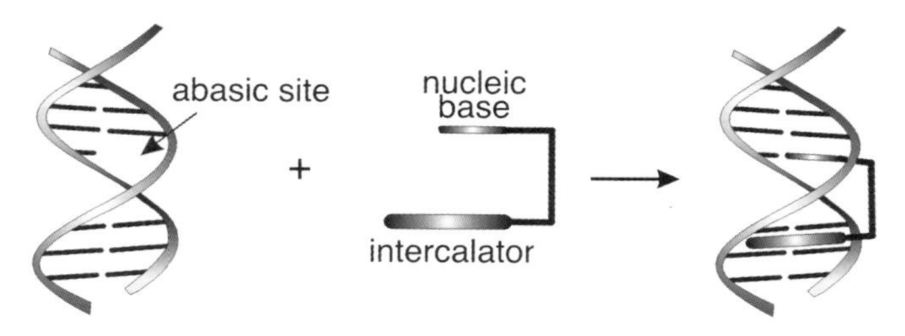


**Fig. 4** Association of a nucleic base–intercalator conjugate with an abasic site

evaluate the binding by spectrophotometric and spectrofluorimetric titrations [195]. In general, upon DNA binding the dye molecules are positioned in an environment which is different from that of the uncomplexed molecule in solution, because the interior of DNA is proposed to exhibit a low dielectric constant [196], and the pH in the grooves is significantly lower than that of the surrounding solution [197]. Also, the electron distribution of the intercalator is distorted upon $\pi$-stacking with the bases. All these factors contribute to significantly different dye absorption properties in the complexed and uncomplexed forms. Thus, the addition of DNA to a solution of an intercalator results in a characteristic shift of the absorption maximum to longer wavelengths (bathochromic shift or red shift) and a decrease of the absorbance (hypochromicity). Consequently, spectrophotometric titrations of DNA to acridine orange and proflavine have been used to prove the binding of the acridine dyes to the nucleic acid. In combination with equilibrium dialysis and low-shear viscosimetry it has been demonstrated that proflavine (**5c**) and acridine orange (**5d**) are intercalating dyes; however, at ratios between 0.2 and 0.3 mol of dye per 1 mol of DNA phosphate, a binding mode other than intercalation takes place [193]. Acridine orange intercalates with its long molecular axis parallel to the binding pocket [198]. In the case of proflavine, these observations have been verified by kinetic measurements which revealed that, along with the intercalative binding mode, a so-called "pre-intercalative binding" with an even larger binding constant takes place [199, 200]. Detailed data for a complex of the intercalated acridine orange [198, 201] and proflavine [202–204] in double-stranded oligonucleotides are also available from X-ray diffraction analyses. Notably, the intercalation process of acridine orange may be reversed or inhibited in the presence of xanthine derivatives, which exhibit a high propensity for complex formation with acridine orange [205].

The intercalation of proflavine (**5c**) and acridine (**5d**) into DNA also has a significant influence on their emission properties. Thus, the emission of **5c** is quenched upon DNA addition [206], which is mainly the result of a photoinduced electron-transfer (ET) reaction between excited proflavine and the nucleic bases [207]. This ET reaction is energetically more favorable with guanine and adenine, which exhibit lower oxidation potentials than the pyrimidine bases [208, 209]. In contrast, the fluorescence of acridine orange (**5d**) is enhanced upon DNA addition [210]. Most notably, the emission enhancement takes place only at high DNA-to-dye ratios, whereas at low DNA-to-dye ratios the emission is also quenched [211]. Because of this fluorescence enhancement upon DNA intercalation of acridine orange it is used as a DNA stain, e.g., in gel electrophoresis or for the detection of cellular DNA [212]. Moreover, acridine orange exhibits metachromism [213], i.e., it is yellowish-green fluorescent when intercalated in dsDNA ($\lambda_{ex}$ = 488 nm), but when it is associated with single-stranded DNA or RNA, it exhibits red fluorescence at the same excitation wavelength [214]. This behavior is used in dual flow cytometry, for example, where the relative content of red and green fluorescence is used for the simultaneous measurement of cellular nucleic acids, i.e., RNA and DNA [215].

Acridine orange exhibits a pronounced Stokes shift which is dependent on the solvent. Thus, the determination of the Stokes shift in several solvents at different temperatures, subsequent calibration of the data towards the solvent polarity and comparison with the Stokes shift of DNA-intercalated acridine orange has been used to characterize the interior of the DNA helix [216]. These studies have revealed that the inside of DNA is highly polarizable, like solvent cages in aromatic solvents such as aniline or acetophenone. Also, temperature-dependent determination of the Stokes shift showed that the diffusive and viscous dynamics inside the DNA helix are comparable to those of fluid solutions rather than to solid-state dynamics.

## 3.2 Anthraquinones

The DNA-binding properties of anthraquinones have been studied intensively over the past 25 years because of their clinical potential as anticancer drugs. The anthraquinone system is often found in anti-tumor drugs such as anthracyclines, mitoxantrone (**11a**), ametantrone (**11b**) and derivatives thereof [217–220]. Mitoxantrone (**11a**) and ametantrone (**11b**) have attracted much interest because of their lower risks of cardiotoxic effects compared with the naturally occurring anthracyclines doxorubicine (adriamycin) (**4b**) and daunorubicin [221].

The cytotoxic properties of anthraquinones result from their binding to cellular DNA [222, 223]. The planarity of the dye allows an intercalation between the DNA base pairs while its redox properties may be used for the formation of reactive radical or radical-ion intermediates in biological systems. The chemical and biological activity exhibited by anthraquinone derivatives is significantly affected by the different substituents at the planar ring system [224–227]. Thus, side chains usually carry one or two positive charges, in order to establish an electrostatic interaction with the phosphate backbone

**11a**: R = OH

**11b**: R = H

**4b**

**Scheme 9**

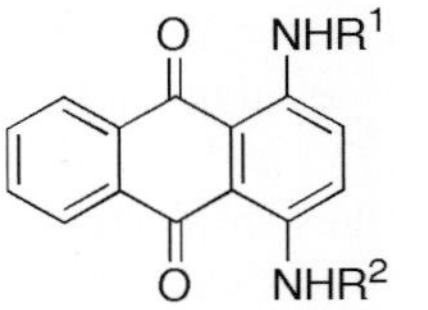

O NHR$^1$

O NHR$^2$

**12a–f**

**12a**: $R^1 = R^2$ = Gly
**12b**: $R^1 = R^2$ = Gly-Gly
**12c**: $R^1 = R^2$ = Gly-Gly-Gly
**12d**: $R^1 = R^2$ = Gly-Lys
**12e**: $R^1 = R^2$ = Gly-Trp
**12f**: $R^1$ = Gly-Gly; $R^2$ = H

**Scheme 10**

of the polynucleotide. It appears that the substitution pattern plays a major role for the binding selectivity and also for the chemical interactions with the physiological system. The substituent effects on the physical and biological properties in anthraquinone diamides have been explored by several groups. For example, the anthraquinone derivatives **12a–f**, in which one or two peptide chains are connected to amino functionalities at positions 1 and/or 4 by amide bond formation, have been synthesized and systematically investigated [228]. The introduction of peptidyl side chains into the anthraquinone structure has important effects on the physicochemical, DNA-binding and biological properties when compared to the anthraquinones **11a** and **11b**. The binding constants of mono-, di-, and triglycyl derivatives are comparable to those found for **11b**, but 5–10 times lower than those reported for **11a** [228]. On the other hand, compounds with glycyl–lysyl side chains, e.g. **12d**, bind DNA to the same extent as **11a** (L-isomer) or even better (D-isomer).

The anthraquinones **11a** and **11b** as well as the derivatives with peptidyl substituents **12** exhibit binding selectivities for alternating CG binding sites, although to different extents. The bis-Gly-Lys-substituted derivatives, e.g. **12d**, are the least sensitive towards DNA base composition, which may be due to the large unselective electrostatic interactions with the polynucleotide backbone. As far as redox properties are concerned, all peptidyl-substituted anthraquinones show a reduction potential which is very close to that of ametantrone (**11b**) and 60–80 mV less negative than that of mitoxantrone (**11a**); hence, they may lead to free-radical-damaging species with similar yields as the parent drugs. Other peptidyl-substituted anthraquinones **13a–f** have been synthesized as potential transcription-factor inhibitors [229]. These 1-[*N*-{2-succinamidylethyl}amino]-substituted pep-

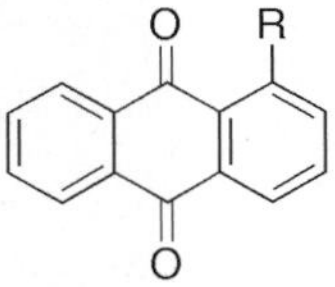

O R

O

**13a–d**

**13a**: R = $NH(CH_2)_2NHC(O)(CH_2)_2C(O)$-Ala-Arg-Cys-Lys-Ala
**13b**: R = $NH(CH_2)_2NHC(O)(CH_2)_2C(O)$-Ala-Lys-Cys-Arg-Ala
**13c**: R = $NH(CH_2)_2NHC(O)(CH_2)_2C(O)$-Ala-Lys-Ser-Arg-Ala
**13d**: R = $NH(CH_2)_2NHC(O)(CH_2)_2C(O)$-Ala-Lys-Cys-Arg-Asp-Ala

**Scheme 11**

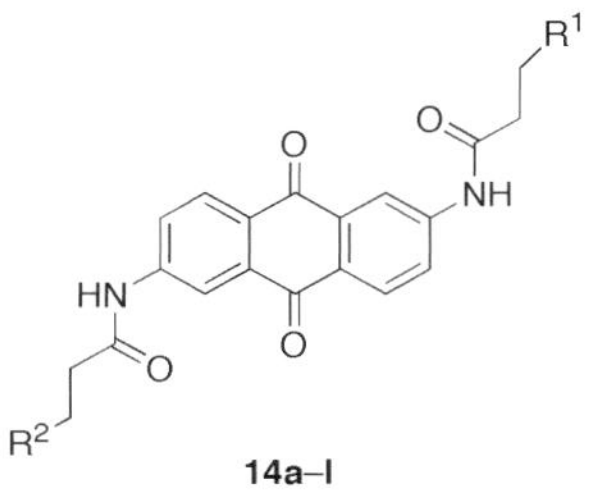

14a–l

**14a**: $R^1 = NH(CH_2)_2O(CH_2)_2OH$; $R^2 = NH(CH_2)_2O(CH_2)_2OH$
**14b**: $R^1 = NH(CH_2)_3NCH_3(CH_2)_3NH_2$; $R^2 = NH(CH_2)_3NCH_3(CH_2)_3NH_2$
**14c**: $R^1 = NH(CH_2)_2NH_2$; $R^2 = NH(CH_2)_2NH_2$
**14d**: $R^1 = NH(CH_2)_4NH_2$; $R^2 = NH(CH_2)_4NH_2$
**14e**: $R^1 = NH(CH_2)_3N(CH_3)_2$; $R^2 = NH(CH_2)_3N(CH_3)_2$
**14f** : $R^1 = NH(CH_2)_4N(CH_3)_2$; $R^2 = NH(CH_2)_4N(CH_3)_2$
**14g**: $R^1 = NH(CH_2)_2O(CH_2)_2OH$; $R^2 = H$
**14h**: $R^1 = NH(CH_2)_3NCH_3(CH_2)_3NH_2$; $R^2 = H$
**14i** : $R^1 = NH(CH_2)_2NH_2$; $R^2 = H$
**14j** : $R^1 = NH(CH_2)_4NH_2$; $R^2 = H$
**14k**: $R^1 = NH(CH_2)_3N(CH_3)_2$; $R^2 = H$
**14l** : $R^1 = NH(CH_2)_4N(CH_3)_2$; $R^2 = H$

**Scheme 12**

tidyl anthraquinones **13a–d** contain five to seven amino acid residues including the lysine–cysteine–arginine motif (also named KCR motif), which is important in AP-1 protein binding to DNA [230]. The anthraquinone–peptide conjugates **13a–d** intercalate into DNA and inhibit the association of AP-1 protein to its DNA consensus sequence [230].

Anthraquinone intercalators **14a–f**, which are substituted with amine-containing side chains at positions 2 and 6, are able to cleave plasmid DNA at abasic sites (apurinic or apyrimidinic sites) [231]. It has been proposed that the cleavage is induced by deprotonation in the $\alpha$ position to the aldehyde functionality of the open-ring glycoside of the abasic site followed by $\beta$ elimination of the 3′-phosphate. A following $\beta$-elimination leads to the loss of the sugar from the 5′-phosphate as well [232–234]. Since the DNA cleavage is the result of an initial base-induced $\beta$-elimination on the deoxyribose moiety, the position of the amino functionality with respect to the ribose hydrogen atoms is crucial for the cleavage yield. Thus, the structure of the amino side chain, especially the length of the alkyl chain between the amino functionalities, determines the cleavage efficiency of the molecule. The yield of strand breaks at abasic sites in plasmid DNA is higher with intercalators containing two amino functionalities (**14c–f**) than for anthraquinone derivatives with only one amino group at each substituent (**14a**). Side chains with terminal tertiary amines (**14e–f**) are slightly more effective than those with terminal primary amines (**14b–d**). Examples with a propylene bridge between neighboring nitrogen atoms cleave more efficiently than those with more or fewer methylene groups, presumably due to the most favorable conformation for proton abstraction in this configuration. Comparison of the disubstituted anthraquinones (**14a–f**) with their monosubstituted analogues (**14g–l**) have indicated that only one side chain is required for single-strand cleavage. Studies on DNA damage with the polyamines, which lack the substitution with an intercalator, have shown that these compounds are less potent than amino-substituted intercalators. Nevertheless, DNA cleavage with the simple polyamines has also been observed at low concentrations [231].

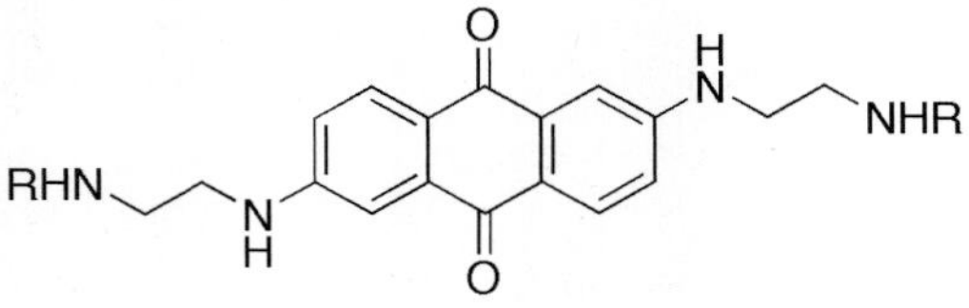

**15a–d**

**15a**: R = $(CH_2)_3OCH_3$
**15b**: R = $(CH_2)_3O(CH_2)_2OCH_3$
**15c**: R = $(CH_2)_3O[(CH_2)_2O]_2CH_3$
**15d**: R = $(CH_2)_3O[(CH_2)_2O]_3CH_3$

**Scheme 13**

The DNA binding characteristics of a series of homologous 2,6-disubstituted anthraquinone intercalators **15a–d** with varying ethylene glycol units in their amino-linked side chains have also been studied [235]. The side chains vary with respect to the polyethylene–glycol unit which allows a significant increase of the side chain length with a relatively small change in the hydrophobicity of the molecule. The latter has been designed such that the cationic charge (protonated secondary amine) is in a constant position in the series. These 2,6-disubstituted anthraquinone intercalators **15** bind with higher affinity to AT-rich hairpins than to GC-rich hairpins. The binding constants decrease with the elongation of the side chain and are not influenced by the decreasing hydrophobicity with increasing side chain length.

DNA intercalating agents with DNA cross-linking potential are expected to be ideal anticancer drugs and should exhibit improved selectivity in the therapeutic process [236–238]. For example, mitomycin C (**16**) [239] and its analogue, indoloquinone (**17**) [240] are the prototypes of compounds with improved selectivity toward major solid tumors, with identified biochemical and functional differences. Mitomycin C (**16**) may serve as a bifunctional alkylating agent that forms covalent bonds with two DNA strands with a resulting cross-link of the two independent strands [241], but **16** is not able to intercalate into dsDNA. Thus, a series of cyclopenta-annelated anthraquinone derivatives, e.g. **18**, with an additional aziridine moiety and a carbamate substituent has been synthesized. The anthraquinone moiety intercalates into dsDNA and covalently crosslinks to the macromolecule after bioreductive activation of the aziridine function. In fact, one of these derivatives, **18b**, ex-

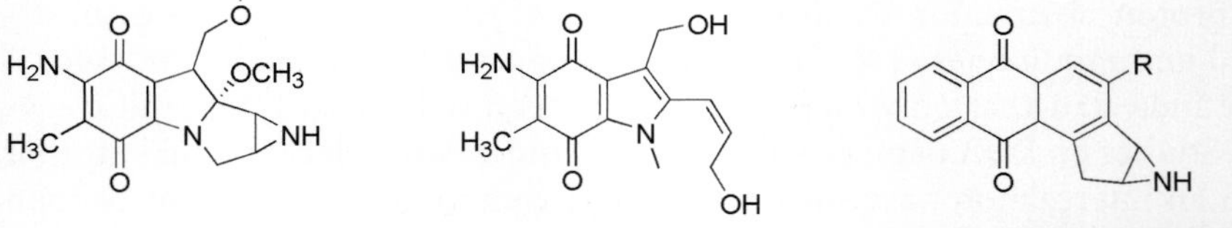

**16** **17**

**18a**: R = $O(CH_2)_2N(CH_2CH_2Cl)_2$
**18b**: R = $CH_2OCONHCH_3$

**Scheme 14**

hibits inhibitory activity against leukemic and solid tumor cell lines. A DNA-unwinding assay has indicated that **18b** is able to intercalate into dsDNA; and it also exhibits topoisomerase II inhibitor activies [242].

The propensity of anthraquinone derivatives to intercalate and to generate radical species upon irradiation in biological systems has been studied intensively [243]. Two general mechanisms for the photoinduced DNA damage are possible: (a) photoinduced electron transfer between electron-accepting anthraquinones and the nucleic bases to give the radical cations of the latter (with guanine as the most easily oxidized base), and (b) hydrogen abstraction from the sugar residues by groove-bound or non-associated anthraquinone molecules.

The anthraquinone derivatives may oxidize DNA when they are randomly bound to the DNA or attached to it covalently at particular locations. Radical cations introduced in the DNA by the excited anthraquinone cause damage both within the binding pocket and locally separated from the anthraquinone binding site. A mechanism has been proposed for long-range charge transport in DNA which depends on the spontaneous distortion of the DNA structure and which is called "phonon-assisted polaron hopping" [243–245].

Spectroscopic and thermodynamic studies of anthraquinone **19a** strongly suggest its binding to duplex DNA by intercalation with little or no base sequence selectivity [81]. The interactions of derivatives **19a** and **19b** with DNA are essentially identical. Irradiation (350 nm) of intercalated **19a** leads to base oxidation of DNA, as shown by the detection of single-strand breaks, which only appear after piperidine treatment of the photolysate. The oxidation occurs preferentially at the 5′-G part of GG steps [246]. The one-electron oxidation of DNA is thermodynamically favored [247, 248], because guanine has the lowest oxidation potential of the four common DNA bases [208, 209]. Kinetic factors resulting from sequence-dependent activation energies must also be considered. More evidence for an electron transfer as the initial step in DNA damage was provided by experiments with anthraquinone **19c**, which also intercalates in DNA [249]. The structures of **19a** and **19c** differ in the "orientation" of the amide that connects the anthraquinone and alkylammonium groups. In derivative **19a**, the anthraquinone is attached to the amide carbonyl group, whereas in **19c**, it is connected to the amide nitrogen atom. This structural difference leads to a $\pi\pi^*$ configuration of **19c** in its lowest

O
R
O
**19a–c**

**19a**: R = $C(O)NH(CH_2)_2NH_3^+$
**19b**: R = $SO_2N[(CH_2)_3NH_3]_2^{2+}$
**19c**: R = $NHC(O)(CH_2)_2HNC(CH_2CH_3)$

**Scheme 15**

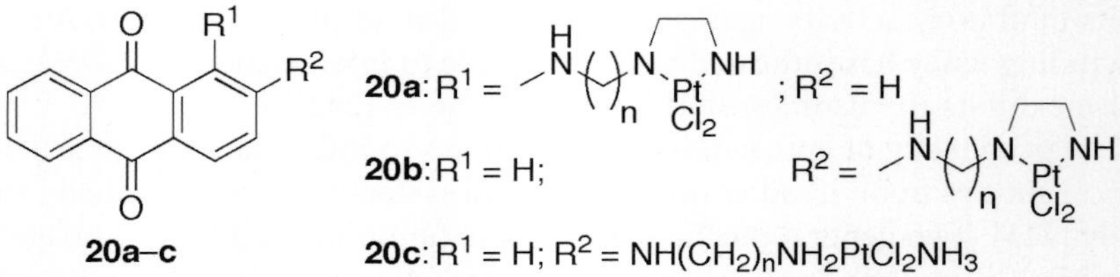

**20a–c**

**20a**: $R^1$ = NH(CH$_2$)$_n$-N(CH$_2$CH$_2$)NH-PtCl$_2$ ; $R^2$ = H

**20b**: $R^1$ = H; $R^2$ = NH(CH$_2$)$_n$-N(CH$_2$CH$_2$)NH-PtCl$_2$

**20c**: $R^1$ = H; $R^2$ = $NH(CH_2)_nNH_2PtCl_2NH_3$

**Scheme 16**

excited state, which is still capable of being involved in electron-transfer reactions, but is no longer able to abstract hydrogen atoms. In contrast, the lowest excited state of **19a** has an $n\pi^*$ configuration, from which hydrogen abstraction is possible. Nevertheless, the irradiation of intercalated anthraquinone **19c** leads to DNA damage with the same selectivity and same efficiency as **19a**. These results indicate that both derivatives **19a** and **19c** react with the nucleic bases by the same mechanism, i.e., photoinduced electron transfer.

Anthraquinones have been connected with transition metal complexes. The resulting conjugates have been studied with respect to their intercalation ability in dsDNA. A few combinations of anthraquinone with platinum complexes of the type *cis*-$PtLL'Cl_2$ [L=AQ-$NH(CH_2)_n$-$NH_2$ and $L' = NH_3$ or $L' = L$] **20** in which the anthraquinone is linked to the diaminedichloroplatinum(II) moiety via positions 1 and 2 have been prepared [250]. Preliminary in-vitro screening for antileukemic activity revealed an interesting structure–activity relationship of these compounds and led to further detailed studies to elucidate the DNA binding properties. The complexes **20** bind covalently to ct DNA due to the reaction with the platinum moiety. Moreover, the anthraquinones with a substituent in position 2, i.e., **20b** and **20c** ($K = 10^4\ M^{-1}$, ct DNA) showed a significantly lower binding affinity towards DNA than anthraquinones substituted in position 1, i.e., **20a** ($K = 10^6\ M^{-1}$) [251]. The antitumor activity, however, is more or less insensitive to these parameters and depends only on the length of the linker between the anthraquinone and the metal complex. The compounds with the shorter linker chains were as active as the known *cis*-platin-complex [250, 252–258], while those with longer chains were essentially inactive.

Furthermore 1- and 2-substituted copper–cyclam–anthraquinone conjugates **21** have been synthesized [259]. It was shown that the cyclam–anthraquinone conjugates intercalate into plasmid DNA with a higher affinity than anthraquinones without the macrocyclic ligand. Moreover, the equilibrium constant for the binding of anthraquinone **21a** to salmon sperm DNA with complexed copper is $4.7 \times 10^3\ M^{-1}$, whereas the uncomplexed anthraquinone–cyclam conjugate has a binding constant of $6.2 \times 10^3\ M^{-1}$. Also, the substances with longer side chains cause substantially more unwinding of the plasmid.

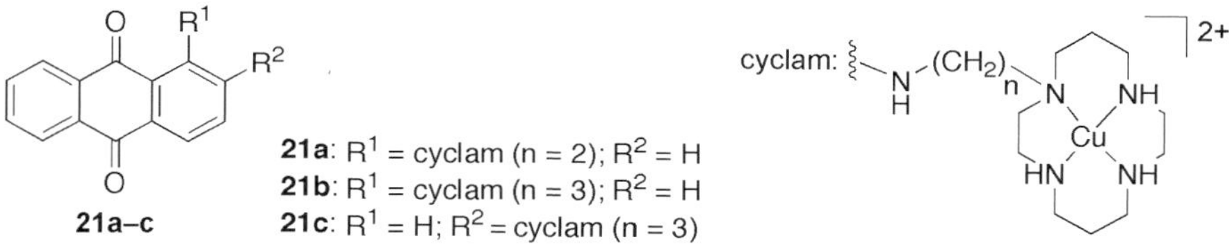

21a–c

**21a**: $R^1$ = cyclam (n = 2); $R^2$ = H
**21b**: $R^1$ = cyclam (n = 3); $R^2$ = H
**21c**: $R^1$ = H; $R^2$ = cyclam (n = 3)

cyclam: $-NH-(CH_2)_n-$[Cu cyclam]$^{2+}$

**Scheme 17**

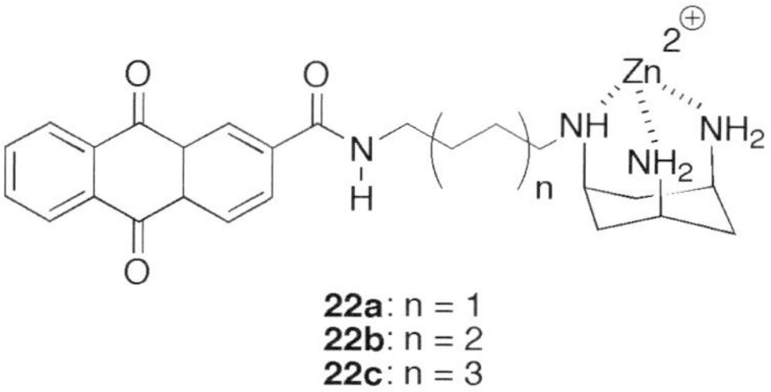

**22a**: n = 1
**22b**: n = 2
**22c**: n = 3

**Scheme 18**

Recently, the synthesis of a series of zinc complex–intercalator conjugates **22** has been reported [260], whose structure includes a chelating *cis-cis*-1,3,5-triaminocyclohexane subunit linked to an anthraquinone moiety via alkyl spacers of different lengths. These compounds have been investigated as hydrolytic cleaving agents for plasmid DNA. The conjugation of the metal complex with the anthraquinone group leads to a 15-fold increase in the cleavage efficiency compared with the anthraquinone not connected to a zinc-triaminocyclohexane complex. Comparison of the reactivities of the different complexes reveals a remarkable increase in DNA cleavage with the length of the spacer. A significantly shorter spacer may decrease or even cancel the advantages owed to the increased DNA affinity. The spacer must be long enough to fold unhindered into the groove to position the metal complex close to a phosphate group.

## 3.3 Cyanine Dyes

The cyanine derivatives of the general structure **23** [261] and the corresponding bis-cyanines **24** are particularly useful as fluorescent dyes. They exhibit remarkably high affinity for association with nucleic acids along with a significant change of photophysical properties upon DNA binding, which is used for DNA detection and quantification in a variety of techniques such as the polymerase chain reaction [262], DNA fragment sizing [263, 264], DNA staining [265], DNA damage detection [266, 267], flow cytometry [268], and evaluation of biological activity [269, 270]. The asymmetric cyanine dyes oxazole yellow **23a** (commonly named YO or YO-PRO-1) and the bis-oxazole

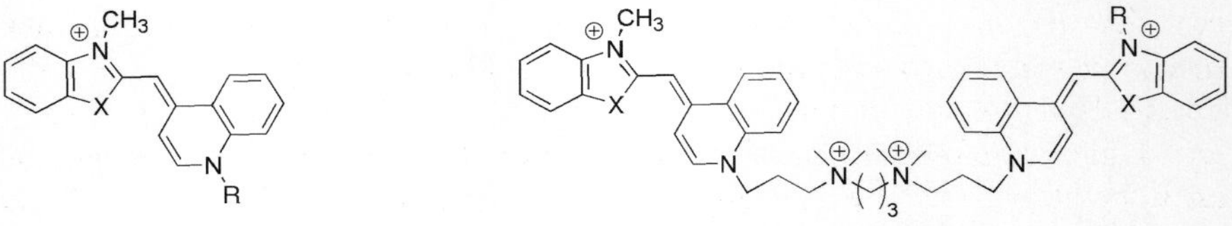

23a: X = O; R = $(CH_2)_3N^+(CH_3)_3$
23b: X = S; R = $CH_3$
23c: X = S; R = $(CH_2)_3N^+(CH_3)_3$

24a: X = O; R = $CH_3$
24b: X = S; R = $CH_3$
24c: X = S; R = $CH_2Ph$
24d: X = S; R = $CH_2CH_3$

**Scheme 19**

**24a** (commonly named YOYO or YOYO1) as well as its congeners thiazole orange (**23b**) (commonly named TO), its derivative **23c** (commonly named TO-PRO1) and the corresponding bis-thiazole **24b** (often named TOTO or TOTO1) have ideal properties for different fluorescence applications. The unbound dyes show no fluorescence in solution but high emission quantum yields when bound to DNA [271], presumably owing to the limited opportunity for photochemical deactivation by, e.g., *cis–trans* isomerization within the interior of the DNA [272]. The high binding constants and the large molar extinction ceffiicients [273] result in a significant difference in fluorescence intensity between the stained DNA molecules and the background.

The predominant binding mode for YO (**23a**) with DNA is monointercalation, whereas YOYO (**24a**) is a typical bisintercalator with rather large binding constants ($K = 6.0 \times 10^8\ M^{-1}$ phosphate buffer, pH 7.0, [NaCl] = 100 mM, ct DNA [274]). Another interaction from **23a** and **24a** with DNA is the external binding to the DNA helix that occurs, especially with YOYO (**24a**) at higher dye-to-DNA ratios [275, 276]. The extent and mode of association of YO and YOYO to DNA is not only dependent on the concentration of dye and DNA, but is also influenced by the counter cations present in the solution. In buffer solutions with tetraalkylammonium cations, a stronger interaction has been detected than with sodium cations. This observation is important for electrophoresis applications in which the stabilization of the dye–DNA complex influences the separation efficiency. In addition, studies with asymmetric monomethine cyanine dyes with a varying number of cationic functionalities have demonstrated the significant influence of the positive charge on the binding properties of cyanine dyes [277, 278].

Another feature of these cyanine dyes is their tendency to photocleave DNA. The irradiation of DNA in the presence of YO (**23a**) or YOYO (**24a**) causes single-strand breaks as determined by a relaxation assay with supercoiled plasmid DNA [279]. In both cases, single-strand breaks are observed to such an extent that eventually double-strand cleavage takes place. With the bisintercalated YOYO (**24a**), the yield of double-strand cleavage is 5 times higher than for YO (**23a**) owing to accumulation effects and because of the very slow dissociation of the bisintercalated dimer, which results in several cleavage sites close to the binding pockets. Moreover the photocleavage mechanism changes with the binding mode. Photocleavage by an externally bound dye is oxygen-dependent, owing to photosensitized formation of singlet oxygen, which subsequently oxidizes DNA. In contrast, photocleavage by an intercalated dye is essentially oxygen-independent and leads to a direct cleavage of the phosphate backbone. In the presence of DNA, light-induced photobleaching occurs by the reaction of the bound and free dyes with reactive oxygen species generated, for example, by Fenton-generated hydroxyl radicals (**23a**) or singlet oxygen (**24a**) [280].

Analogous to YO (**23a**) and YOYO (**24a**), TO (**23b**) monointercalates and TOTO (**24b**) bisintercalates into DNA (**24b:** $K = 1.1 \times 10^9\ M^{-1}$ phosphate buffer, pH 7.0, [NaCl] = 100 mM, ct DNA [274]). The structure of the bisintercalated complex with the self-complementary oligonucleotide d(CGCTAGCG)$_2$ has been established by NMR studies (Fig. 5) [281]. Bisintercalation of **24b** occurs with the benzothiazole ring system stacked between the nucleic base pairs. The *N*-methyl group of the benzothiazole is centered in the major groove, and the linker between the two chromophores is posi-

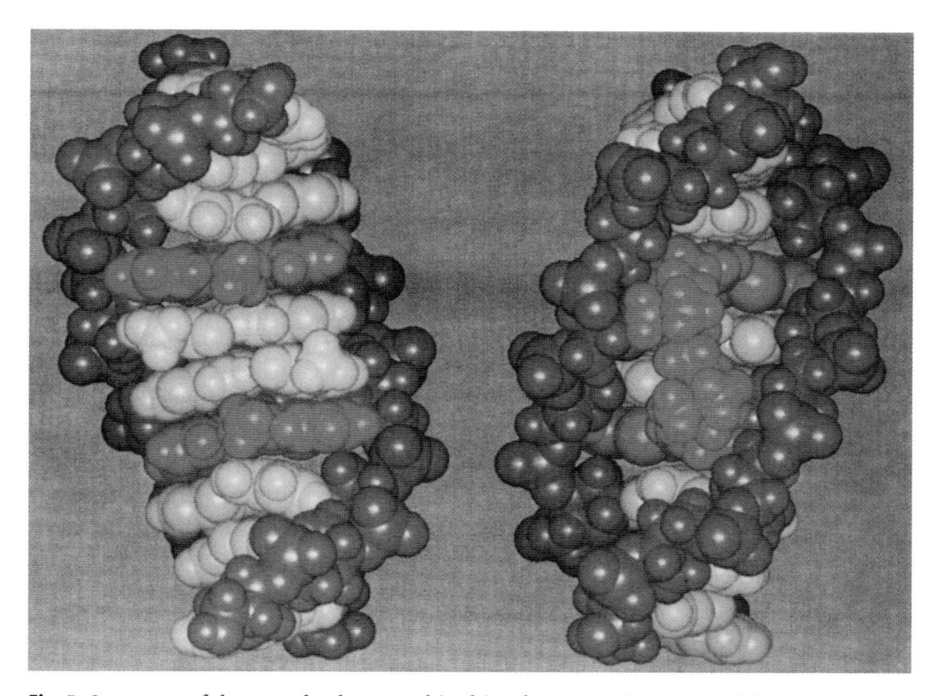

**Fig. 5** Structure of the complex between bis-thiazole orange (TOTO) and d(CGCTAGCG)$_2$. *Left* view into the minor groove, *right* view into the major groove. The TOTO molecule is shown in *red*, the nucleic bases are in *light blue*, and the sugar-phosphate backbones are in *dark blue* [Reprinted with permission from Spielmann HP, Wemmer DE, Jacobsen JP (1995) Biochemistry 34:8542. Copyright (1995) American Chemical Society]

tioned in the minor groove crossing from one side of the groove to the other. Upon binding to dsDNA these dyes exhibit a large enhancement of fluorescence intensity, which leads to a fluorescence quantum yield which is about 3,000 times higher than that of the free dye in solution [281]. Thiazole orange **23b** has been proposed to bind with moderate affinity single-stranded poly(dA) and poly(dG) [282]; however, fluorescence quantum yield measurements suggest that **23b** as well as **23a** bind preferentially to GC-rich regions in ct DNA [283]. Moreoever, similar binding constants are reported for binding of TO (**23b**) and YO (**23a**) to several types of DNA with varying GC-to-AT ratios [284], which indicates that **23a** and **23b** exhibit DNA sequence selectivity when binding to DNA. Recent results have shown that the bisintercalator TOTO (**24a**) binds about 100 times more strongly to the d(5-C$^{me}$CIG-3′)$_2$ site ($^{me}$C is 5-methylcytosine, U is uracil and I is inosine) than to other binding sites, which is in contrast to the proposed preference for (5′-CTAG-3′) [281]. From detailed NMR studies, it has been also concluded that the order of preference is $^{me}$CG > CG > CI > TA for the flanking base pair and $^{me}$CI > CI > TA > CG > UA for the central base pair ($^{me}$C is 5-methylcytosine,U is uracil and I is inosine) [285]. Most likely, the major contribution to the preferential binding in a particular binding site comes from the associative interaction of the thiazole orange chromophore with the DNA bases [285], because the linker does not influence the sequence-selective binding despite its resemblance to spermidine or spermine [281, 286–289]. The intercalation of several analogues of **24b** with different linkers has been studied by NMR spectroscopy, which revealed a significant influence of the minor-groove-bound linker chain [290–292]. The site selectivity is not influenced by linker length, but with an enhanced linker length the dyes can bisintercalate into two (5′-CpT-3′):(5′-ApG-3′) sites that are separated by one or two base pairs [290–292].

The binding mode of TOTO (**24b**) and its derivatives TOTOBzl (**24c**) and TOTOEt (**24d**) is similar to the parent molecule **24b**, i.e., the chromophore is sandwiched between two base pairs in a (5′-CpT-3′):(5′-ApG-3′) site. As in complexes of TOTO with DNA, the linker spans over two base pairs in the minor groove, whereas the benzyl and the ethyl group are pointing out of the major groove [290, 293].

The intercalating cyanine dyes **25a** (commonly named BO) and **25b** (commonly named PO) are structurally related to YO (**23a**) and TO (**23b**) and also exhibit enhanced fluorescence intensity upon DNA binding; but they show no sequence selectivity [294, 295]. In contrast, picogreen [296–298], a derivative of YO (**23a**), binds preferentially to dsDNA at alternating GC base pairs. This site-selective intercalation has been explored by time-resolved fluorescence spectroscopy, transient absorption spectroscopy and CD spectroscopy [299, 300].

Cyan 2 (**26**) is another representative cyanine dye, which intercalates into dsDNA with high binding affinity for GC sequences. The DNA complexes with

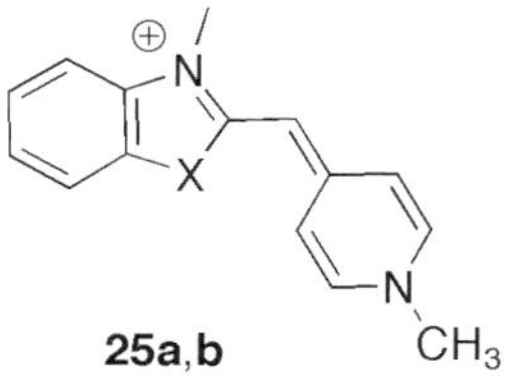

**25a**: X = S
**25b**: X = O

**Scheme 20**

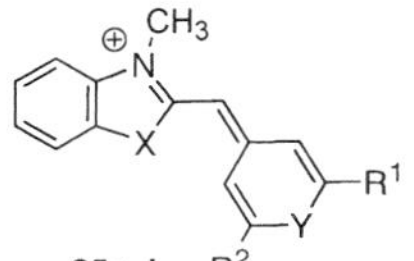

**25c** : X = O; Y = O; $R^1 = R^2 = CH_3$
**25d** : X = O; Y = $NCH_3$; $R^1 = R^2 = CH_3$
**25e** : X = S; Y = O; $R^1 = R^2 = CH_3$
**25f** : X = S; Y = O; $R^1 = CH_3$; $R^2$ = Ph
**25g** : X = S; Y = O; $R^1 = R^2$ = Ph
**25h** : X = S; Y = $NCH_3$; $R^1 = R^2 = CH_3$
**25i** : X = S; Y = $NCH_3$; $R^1 = R^2$ = Ph
**25j** : X = S; Y = $NCH_3$; $R^1 = CH_3$; $R^2$ = Ph

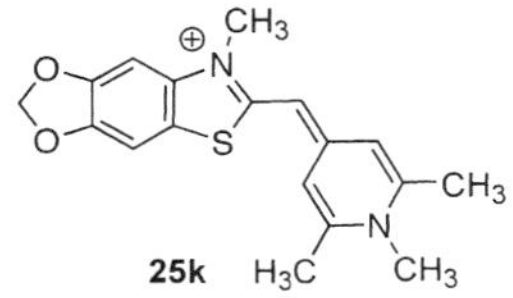

**Scheme 21**

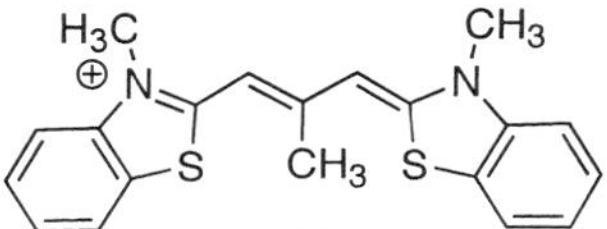

**Scheme 22**

this dye are stable even in solutions with high ionic strength. LD spectroscopic studies have shown that **26** is intercalated almost perpendicularly to the DNA helix axis [301].

Along with the "classical" theory of intercalation, another model for the intercalation of cyanine dyes to dsDNA has been proposed, namely the "half-intercalation model". According to this model, the heterocycle with higher positive charge is placed in the DNA groove, attracted by the negative charge of the phosphate backbone, and the less positively charged part of the dye is intercalated. For a systematic study of this model, the monomethine pyrylium and pyridinium cyanine derivatives **25c–j** have been synthesized along with corresponding derivatives with a 5,6-methylenedioxy-[*d*] benzothiazole residue, e.g. **25k** [302]. It has been shown that the fluorescence properties of the DNA-bound dyes are influenced by sterically demanding substituents at the peripheral sides of the dyes. Furthermore, a considerable increase in the fluorescence intensity in the presence of DNA has been observed for dyes with bulky phenyl groups, although these dyes do not fully intercalate between the DNA bases.

The change of the Stokes shift upon DNA binding of cyanine dyes **25c–j** has been explained by the half-intercalation of these dyes. As the intercalation pocket and the DNA groove represent two environments with signifantcly different properties, the intercalated part of the dye is in an environment

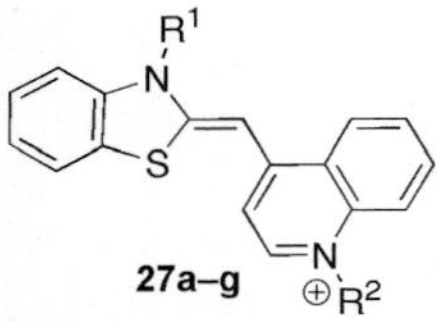

**27a–g**

**27a**: $R^1 = CH_3$; $R^2 = (CH_2)_{10}C(O)$-Lys-Ala-Gln-Ser-Trp-Gly-Lys-Ser-Ala-Lys
**27b**: $R^1 = CH_3$; $R^2 = (CH_2)_{10}C(O)$-Gly-Ala-Gln-Val-Trp-Gly-Trp-Ser-Ala-Lys
**27c**: $R^1 = (CH_2)_{10}C(O)$-Lys-Ala-Gln-Ser-Trp-Gly-Lys-Ser-Ala-Lys; $R^2 = CH_3$
**27d**: $R^1 = (CH_2)_{10}C(O)$-Gly-Ala-Gln-Val-Trp-Gly-Trp-Ser-Ala-Lys; $R^2 = CH_3$
**27e**: $R^1 = CH_3$; $R^2$ = Gly-Lys
**27f** : $R^1 = CH_3$; $R^2$ = Tyr-Lys
**27g**: $R^1 = CH_3$; $R^2$ = Trp-Lys

**Scheme 23**

with different polarity to the groove-bound moiety. This positioning of the two dye parts results in an overall electron distribution which differs from that in aqueous solution. This distortion of the electron density upon half-intercalation has been suggested to be the cause for the Stokes shift [303]. In conclusion, according to the half-intercalation model, partial intercalation of cyanine dyes is mainly controlled by steric interactions and by the electron density distribution of the dye molecule.

Selected peptidyl-intercalator conjugates, e.g. **27a–g**, have been synthesized with the intention of using cyanine dyes as intercalating functionalities, which deliver a peptide to a particular position of a nucleic acid [304]. These conjugates are expected to exhibit photonuclease activity and to provide model systems for amino-acid-promoted DNA damage. The peptide moiety was connected to the dye at the quinoline nitrogen (**27a,b**) or at the benzothiazole nitrogen atom (**27c,d**). All of these conjugates display fluorescence enhancement upon addition of DNA. Most interestingly, those derivatives substituted at the benzothiazole nitrogen (**27c**, **d**) have higher Stern–Volmer quenching rates, $k_q$, when the DNA–cyanine complex is titrated with a quencher such as $Ru(NH_3)_6^{3+}$, which indicates that in these cases the fluorophore is more accessible to external molecules when the peptide moiety is bound to the benzothiazole nitrogen. This is also consistent with the lower quantum yield values of the derivatives **27a** and **27b** which suggest that the binding mode thereof involves a conformation which allows less stacking and less protection than observed for the conjugates **27c** and **27d** [304].

The photo-induced damage of supercoiled plasmid DNA with peptide–intercalator conjugates has been investigated [305]. Upon irradiation of DNA in the presence of **27f** and **27g** under aerobic conditions, frank strand breaks are induced, while conjugates such as **27e**, which do not carry an aromatic amino acid, do not lead to significant DNA cleavage. Although the cyanine dyes are known to photosensitize the formation of singlet oxygen, the latter intermediate is not a major factor in DNA damage, since it mainly leads to base oxidations [306]. Thus, it has been proposed that the amino acid residues react directly with singlet oxygen to produce the corresponding peroxides [307]. Subsequent fragmentations of the peroxide lead to radical intermediates, which finally induce the strand breaks.

A conjugate between thiazole orange and the zinc-finger moiety of glucocorticoid receptor DNA-binding domain (GR-DBD) has been synthesized [308] that exhibits a remarkably large binding constant ($K \approx 10^8$ $M^{-1}$). Notably, the conjugate retains the sequence-selective binding properties of the GR-DBD. Upon irradiation of the complexed peptide–dye conjugate, the thiazole moiety has been shown to induce DNA damage. In a similar approach, peptide–cyanine dye conjugates have been synthesized which combine the sequence-selective binding properties of helix-turn-helix (HTH) DNA binding motifs with intercalating cyanine dyes [309]. The HTH-motif gets its name from a region of high secondary structure similarity, consisting of a recognition helix, which spans the major groove and exhibits the majority of sequence selective contacts. A second helix stabilizes the folded structure by forming a hydrophobic pocket and often confers some additional binding stabilization. These motifs recognize a wide range of short DNA sequences [310]. The investigation of these conjugates showed that the fluorescence of these substances is 1000–5000 fold enhanced upon intercalation. Additional thermodynamic studies, on the effects of the sequence selectivity and dye-labeling on the change in heat capacity and number of ions displaced upon association of the peptide–cyanine conjugates, show that non-selective interactions are stabilized by the formation of the contact ion pairs between the phosphate groups of the DNA backbone and the basic amino acid side chains of the protein. The selective binding of these complexes seems to be related to hydrophobic effects, likely to the reduction of the water-accessible nonpolar surface area [311].

## 3.4 Phenanthridinium Ions

Ethidium bromide (**28a**) and its analogue propidium (**28b**) are probably the best-known DNA-binding dyes in the phenanthridinium class and are considered to be the "standard intercalators" [312, 313]. At large dye-to-DNA ratios, however, ethidium bromide (**28a**) also binds to secondary binding sites along the DNA backbone, presumably by electrostatic association [314, 315] or via hydrogen bonds [316]. It has been proposed that the latter binding takes place preferentially in minor grooves with high GC content [317]. Whereas

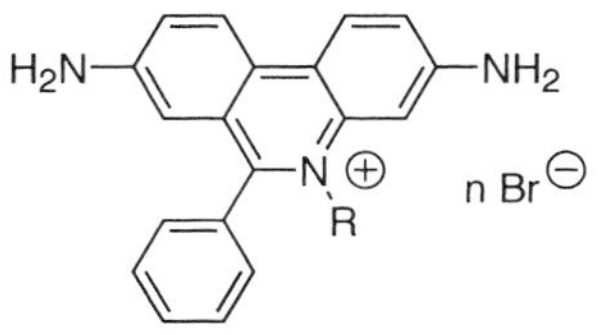

$H_2N$ $NH_2$ $N^{\oplus}$ R n $Br^{\ominus}$

**28a**: R = $CH_2CH_3$; n = 1
**28b**: R = $(CH_2)_3N^+CH_3(CH_2CH_3)_2$; n = 2

**Scheme 24**

the intercalation may be regarded as "non-competitive" binding mode, the backbone association is "competitive", since it competes for the binding sites with the metal cations, which are usually present in solution [318]. Thus, in any study on interactions of phenanthridinium ions (and other cationic dyes) these two binding modes with DNA need to be considered, unless the metal ion concentration is significantly larger than the dye concentration. In this particular case, the dye cannot compete for the backbone binding and intercalation takes place almost exclusively. There is some inconsistency concerning the binding constants for the complexes in the two binding sites. In early work, it was stated that the binding constant for the backbone association to ct DNA (mean value: $6 \times 10^5\ M^{-1}$) is larger than that of the intercalation complex with the same nucleic acid (mean value: $1 \times 10^4\ M^{-1}$) due to an unfavorable entropic term in the latter case [318]. Nevertheless, in more recent investigations it has been claimed that the fluorescent intercalation complex exhibits stronger binding (e.g. $8 \times 10^4\ M^{-1}$, [NaCl] = 0.05 M, DNA: *Escherichia coli*) than the backbone or groove-bound ethidium bromide (e.g. $3 \times 10^3\ M^{-1}$, [NaCl] = 0.05 M, DNA: *E. Coli*) [317, 319].

Detailed studies have been performed to elucidate the structure of the DNA–ethidium complex by X-ray diffraction analysis [320, 321] and NMR spectroscopy [322–324]. It has been shown by molecular modeling studies that the intercalation of **28a** takes place from the minor-groove site and that it changes the structure of B-DNA significantly, i.e., unwinding and subsequent lengthening the DNA (ca. 270 pm), displacement of the B-DNA helical axis (ca. 100 pm), as well as twisting (10°) and tilting (8°) of neighboring base pairs. Moreover, the neighbor-exclusion rule is followed, i.e., one binding site consists of four base pairs, which do not associate to neighboring binding pockets. Besides intercalation of ethidium bromide (**28a**) in dsDNA, it also binds strongly to G-quartets [325, 326], triplex DNA [327], and DNA hairpins [328].

The addition of DNA to a solution of ethidium bromide (**28a**) results in a bathochromic shift of the absorption maximum (ca. 40 nm) [106]. Ethidium bromide exhibits very low emission quantum yields in water or alcoholic solvents; however, in non-polar organic solvents the quantum yield increases slightly [329]. Most notably, when ethidium is intercalated into dsDNA, it is highly fluorescent [330]. It has been proposed that an excited-state proton transfer between ethidium and the solvent leads to a non-emissive deactivation of the excited state, which is suppressed within the hydrophobic interior of the DNA [329]. Moreover, an increase in the energy gap between the excited singlet and triplet states upon intercalation has been suggested to be the reason for emission enhancement [331]. This significant fluorescence enhancement upon DNA binding of the phenanthridinium ions **28a** and **28b** is the reason for their extensive use in analytical biochemistry, biology and medicine. Nevertheless, the use of emission enhancement as a quantitative tool needs to be approached with care. As the emission of intercalated ethidium bromide may be quenched by non-intercalated dye molecules [332], the amount of unbound ethidium bromide needs to be taken into consideration prior to quantitative analysis. Interestingly, ethidium bromide is essentially non-fluorescent when it is intercalated into binding sites that contain 7-deazaguanine as an artificial base [333]. Although a clear rationalization of this effect is still missing, significant changes in the electronic structure of ethidium bromide next to deazaguanine bases are indicated by absorption spectroscopy and have been suggested to be responsible for the low emission quantum yield. The fluorescence lifetime and intensity of intercalated ethidium bromide are significantly enhanced in $D_2O$ as a solvent [334].

DNA damage by ethidum bromide is widely used for analytical purposes [10]. Thus, most assays for DNA analysis by gel electrophoresis use ethidium bromide (**28a**) or propidium bromide (**28b**) to conveniently visualize the correponding DNA fragments in the gel [335]. When using this method, however, it should be noted that different binding constants to the separated DNA fragments may lead to different emission intensities. The hybridization and renaturation kinetics of complementary DNA strands may be monitored by a continuous fluorescence technique which is also based on the fluorescence enhancement in intercalated ethidium bromide [336]. The different emission lifetimes of free and intercalated ethidium bromide have been used to quantify cellular DNA and RNA by flow cytometry [337] or to detect free and intercalated ethidium bromide at low concentrations in living cells by time-resolved microspectrofluorimetry [338]. Also, the influence of external factors such as solvents, counter ions, pH etc. on the lifetimes of DNA-bound fluorophores is used in phase-sensitive flow cytometry, where the lifetime of intercalated ethidium bromide could be monitored in subpopulations of cells to distinguish apoptotic cells from non-apoptotic ones by their differences in emission lifetimes [339].

The relative binding affinity of non-chromophoric intercalators may also be determined with the help of ethidium bromide [340–342]. Thus, the ability of the intercalator under investigation to replace an ethidium bromide molecule in an intercalation site is used as a relative measure for its binding affinity, as the replacement of ethidium bromide in DNA is monitored by the decrease of emission intensity ($\lambda_{ex} = 546$ nm, $\lambda_{em} = 595$ nm). In an especially useful variation of this assay, hairpin deoxyoligonucleotides are used as host molecules. From the resulting complex with ethidium bromide the latter may be displaced by another intercalator as monitored by emission spectroscopy, which allows the determination of binding selectivity, stoichiometry, and binding site size of intercalators [116]. A similar method is used for the detection of DNA alkylation by fluorescence spectroscopy: methylated DNA has fewer binding sites for ethidium bromide than the unaltered nucleic acid, so that methylation of DNA, which is saturated with the intercalator, leads to a decrease in emission intensity [343]. Note that both assays only work if the ethidium bromide is the only absorbing species at the applied excitation

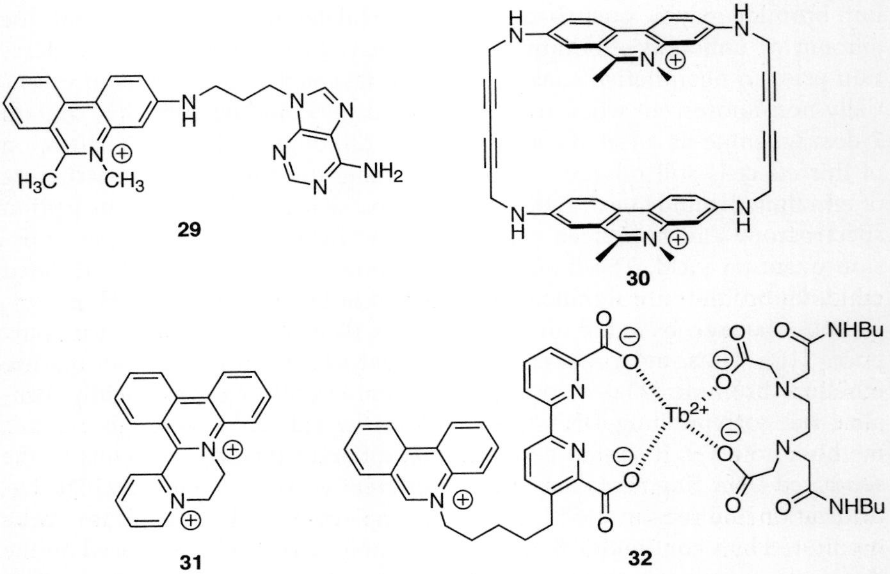

**Scheme 25**

wavelength. Recently, the kinetics of the RNA cleavage by deoxyribozymes were investigated with the aid of ethidium bromide fluorescence [344]. Thus, upon binding of the deoxyribozyme to the corresponding RNA strands, intercalation of ethidium bromide into the newly formed double strand takes place, along with an emission enhancement, whereas upon product formation the double strand collapses and the emission intensity decreases. This method, however, can only be used if ethidium bromide does not suppress the activity of the ribozyme.

Ethidium bromide (**28a**) and its derivatives also possess potential as drugs due to their interaction with DNA [345–347]. Therefore, several phenanthridinium derivatives have been synthesized and investigated with respect to their interactions with nucleic acids. Among these are phenanthridinium–nucleobase conjugates such as **29** [348, 349], phenanthridinium cyclobis-intercalands such as **30** [350], and bisphenanthridinium viologens such as **31** [351]. The phenanthridinium-linked terbium complexes **32** may be used as a DNA probe with a monitor wavelength at the terbium emission [352]. Whereas the unbound conjugate is essentially non-emissive due to intramolecular quenching, the intercalation of the phenanthridinium moiety into DNA suppresses this deactivation process, and terbium emission takes place.

Ethidium bromide has been included into synthetic oligonucleotide strands, e.g. in **33**, to serve as an electron-accepting functionality to study electron transfer processes within dsDNA [353, 354]. Nevertheless, in these

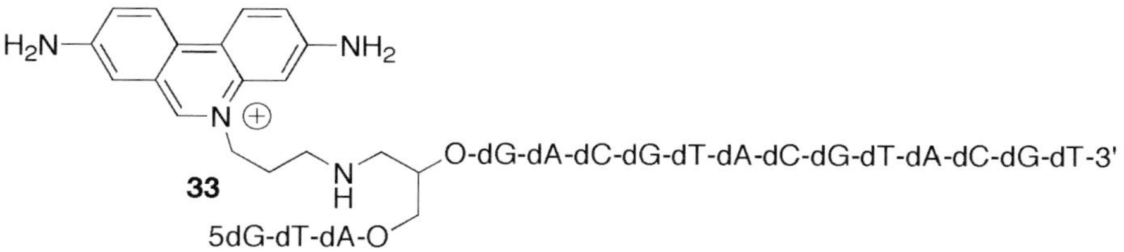

H2N NH2 N⊕ 33 N H O-dG-dA-dC-dG-dT-dA-dC-dG-dT-dA-dC-dG-dT-3' 5dG-dT-dA-O

**Scheme 26**

cases, the ethidium bromide molecule is positioned at the 5′-end of one single strand and does not act as an intercalator. In another approach, a modified DNA single strand has been synthesized with ethidium bromide in an internal position (**33**). Notably, when this oligonucleotide is annealed with a complementary DNA single strand, the ethidium bromide is intercalated in the double strand and serves as an artificial base [355, 356].

Meanwhile a compilation of review articles has been published (Topics in Current Chemistry: DNA Binders and Related Subjects, Volume Editors: Waring, M.J., Chaires, J.B., Vol. 253, 2005) wich also covers several aspects of the association of organic molecules with nucleic acids.

**Acknowledgements** Generous financial support from the Deutsche Forschungsgemeinschaft is gratefully acknowledged. We thank Mr. Anton Granzhan for his careful proofreading of the final manuscript.

## References

1. Dickerson RE (1992) Methods Enzymol 211:67
2. Ulyanov NB, James TL (1995) Methods Enzymol 261:90
3. Demeunynck M, Bailly C, Wilson WD (eds) (2002) DNA and RNA binders. Wiley-VCH, Weinheim
4. Thurston DE (1999) Br J Cancer 80:65
5. Hurley LH (2002) Natl Rev Cancer 2:188
6. Foye WO (ed) (1995) Cancer chemotherapeutic agents. ACS, Washington, DC
7. Neidle S, Thurston DE (1994) In: Kerr DJ, Workman P (eds) New targets for cancer chemotherapy. CRC Press, Boca Raton, FL, p 159
8. Propst CL, Perun TL (eds) (1992) Nucleic acid targeted drug design. Dekker, New York
9. Baguley BC (1991) Anti-Cancer Drug Des 6:1
10. Prento P (2001) Biotech Histochem 76:137
11. Blackburn GM, Gait MJ (eds) (1996) Nucleic acids in chemistry and biology. IRL, Oxford, UK, p 329
12. Blackburn GM, Gait MJ (eds) (1996) Nucleic acids in chemistry and biology. IRL, Oxford, UK, p 375
13. Gottesfeld JM, Neely L, Trauger JW, Baird EE, Dervan PB (1997) Nature 387:202
14. Takahashi T, Tanaka H, Matsuda A, Doi T, Yamada H, Matsumoto T, Sasaki D, Sugiura Y (1998) Bioorg Med Chem Lett 8:3303

15. Nicolaou KC, Smith BM, Pastor J, Watanabe Y, Weinstein DS (1997) Synlett:401
16. Depew KM, Zeman SM, Boyer SH, Denhardt DJ, Ikemoto N, Danishefsky SJ, Crothers DM (1996) Angew Chem Int Ed Engl 35:2797
17. Xiong Y, Ji YLN (1999) Coord Chem Rev 185/186:711
18. Erkkila KE, Odom DT, Barton JK (1999) Chem Rev 99:2777
19. Wagenknecht HA, Stemp EDA, Barton JK (2000) J Am Chem Soc 122:1
20. Zollinger H (2003) Color chemistry, 3rd edn. VCH, Weinheim, p 543
21. Benigni R, Giuliani A, Franke R, Gruska A (2000) Chem Rev 100:3697
22. Li Y, Dick WA, Tuovinen OH (2004) Biol Fert Soil 39:301
23. Wang YB, Yang JH, Wu X, Li L, Sun S, Su BY, Zhao ZS (2003) Anal Lett 36:2063
24. Wainwright M, Crossley KB (2002) J Chemotherapy 14:431
25. Nicklas JA, Buel E (2003) Anal Bioanal Chem 376:1160
26. Davis JT (2004) Angew Chem, Int Ed Engl 43:668
27. Hermann T (2000) Angew Chem, Int Ed Engl 39:1890
28. Chan PP, Glazer PM (1997) J Mol Med 75:267
29. Guelev VM, Hartig MT, Lokey RS, Iverson BL (1999) Chem Biol 7:1
30. Bailly C, Brana MF, Waring MJ (1996) Eur J Biochem 240:195
31. Liu Z, Hecker KH, Rill RL (1996) J Biomol Struct Dyn 14:331
32. Carrasco C, Joubert A, Tardy C, Maestre N, Cacho M, Brana MF, Bailly C (2003) Biochemistry 42:11751
33. Bailly C, Carrasco C, Joubert A, Bal C, Wattez N, Hildebrandt MP, Lansiaux A, Colson P, Houssier C, Cacho M, Ramos A, Brana MF (2003) Biochemistry 42:4136
34. Rogers JE, Weiss SJ, Kelly LA (2000) J Am Chem Soc 122:427
35. Gualev V, Sorey S, Hoffman, Iverson BL (2002) J Am Chem Soc 124:2864
36. Morrison H, Mohammad T, Kurukulasuriya R (1997) Photochem Photobiol 66:245
37. Amaral L, Kristiansen JE (2001) Int J Antimicrob Agents 18:411
38. Mazumder R, Ganguly K, Dastidar SG, Chakrabarty AN (2001) Int J Antimicrob Agents 18:403
39. Rohs R, Sklenar H (2004) J Biomol Struct Dyn A 21:699
40. Wagner SJ (2002) Transfusion Med Rev 16:61
41. Rohs R, Sklenar H (2001) Indian J Biochem Biophys 38:1
42. Tuite E, Kelly JM (1995) Biopolymers 35:419
43. Liu ZR, Rill RL (1996) Anal Biochem 236:139
44. Bevers S, Schutte S, McLaughlin LW (2000) J Am Chem Soc 122:5905
45. Balakin KV, Korshun VA, Mikhalev II, Malev GV, Malakhov AD, Prokhorenko IA, Berlin YA (1998) Biosens Bioelectron 7/8:771
46. Fedorova OS, Koval VV, Karnaukhova SL, Dobrikov MI, Vlossov VV, Knorre DG (2000) Mol Biol 34:814
47. Hurley LH, Wheelhouse RT, Sun D, Kerwin SM, Salazar M, Fedoroff OY, Han FX, Han HY, Izbicka E, Von Hoff DD (2000) Pharmacol Therapeut 85:141
48. Pilch DS, Yu C, Makhey D, LaVoie EJ, Srinivasan AR, Olson WK, Sauers RS, Breslauer KJ, Geacintov NE, Liu LF (1997) Biochemistry 36:12542
49. Persil O, Santai CT, Jain SS, Hud NV (2004) J Am Chem Soc 126:8644
50. Jain SS, Polak M, Hud NV (2003) Nucleic Acid Res 31:4608
51. Bejune SA, Shelton AH, McMillin DR (2003) Inorg Chem 42:8465
52. Lee S, Lee YA, Lee HM, Lee JY, Kim DH, Kim SK (2002) Biophys J 83:371
53. Wall RK, Shelton AH, Bonaccorsi LC, Bejune SA, Dubé, McMillin DR (2001) J Am Chem Soc 123:11480
54. Guliaev A, Leontis NB (1999) Biochemistry 38:15425

55. Kruk NN, Shishporenok SI, Korotky AA, Galievsky VA, Chirvony VS, Turpin PY (1998) J Photochem Photobiol B: Biol 45:67
56. Hannah KC, Armitage B (2004) Acc Chem Res 37:845
57. Pal SK, Zeweil AH (2004) Chem Rev 104:2099
58. McMillin DR, McNett KM (1998) Chem Rev 98:1201
59. Wall RK, Shelton AH, Bonaccorsi LC, Bejune CA, Dubé D, David R, McMillin D (2001) J Am Chem Soc 123:11480
60. Wemmer DE (2001) Biopolymers 52:197
61. Bailly C, Chaires JB (1998) Bioconjugate Chem 9:513
62. Pommier Y, Kohlhagen G, Bailly C, Waring M, Mazumder A, Kohn KW (1996) Biochemistry 35:13303
63. Liu C, Chen FM (1994) Biochemistry 31:1419
64. Pjura P, Grzeskowiak K, Dickerson ER (1987) J Mol Biol 197:257
65. Tidwell RR, Boykin WD (2002) In: Demeunynck M, Bailly C, Wilson WD (eds) DNA and RNA binders. Wiley-VCH, Weinheim, p 414
66. Baraldi PG, Bovero A, Fruttarolo F, Preti D, Aghazadeh T, Pavani MG, Romagnoli (2004) Med Res Rev 24:475
67. Lerman LS (1963) Proc Natl Acad Sci USA 49:94
68. Medhi C, Mitchell JBO, Price SL, Tabor AB (1999) Biopolymers 52:84
69. Gago F (1998) Methods 14:277
70. Ren J, Jenkins T, Chaires JB (2000) Biochemistry 39:8439
71. Reha D, Kabelac M, Ryjacek F, Sponer J, Sponer JE, Elstner M, Suhai S, Hobza P (2002) J Am Chem Soc 124:3366
72. Sartorius J, Schneider HJ (1997) J Chem Soc Perkin Trans 2:2319
73. For a collection of data on DNA and DNA complexes see: http://ndbserver.rutgers.e
74. Herzyk P, Neidle S, Goodfellow JM (1992) J Biomol Struct Dyn A 10:97
75. Friedman RA, Manning GS (1984) Biopolymers 23:2671
76. Rao S, Kollman P (1987) Proc Natl Acad Sci USA 84:5735
77. Frederick CA, Williams LD, Ughetto G, Van der Marel GA, Van Boom JH, Rich A, Wang AHJ (1990) Biochemistry 29:2538
78. Adams A, Guss JM, Collyer CA, Denny WA, Wakelin LPG (1999) Biochemistry 38:9221
79. Rohs R, Sklenar H, Lavery R, Roder B (2000) J Am Chem Soc 122:2860
80. Lisgarten JN, Coll M, Portugal J, Wright CW, Aymami J (2002) Nat Struct Biol 9:57
81. Armitage BA, Yu C, Devadoss C, Schuster GB (1994) J Am Chem Soc 116:9847
82. Le Pecq JB, Le Bret M, Barbet J, Roques B (1975) Proc Natl Acad Sci USA 72:2915
83. Denny WA, Baguley BC, Cain BF, Waring MJ (1983) In: Neidle S, Waring MJ (eds) Antitumour acridines. Molecular aspects of anti-cancer drug action. Verlag Chemie, England, p 1
84. Wakelin LPG, Waring MJ (1978) Biochemistry 17:5057
85. King HD, Wilson WD, Gabbay EJ (1982) Biochemistry 21:4982
86. Wakelin LPG (1986) Med Res Rev 6:275
87. Spicer JA, Gamage SA, Rewcastle GW, Finlay GJ, Bridewell, DJ, Baguley BC, Denny WA (2000) J Med Chem 43:1350
88. Deady LW, Desneves J, Kaye AJ, Finlay GJ, Baguley BC, Denny WA (2000) Bioorg Med Chem 8:977
89. Gamage SA, Spicer JA, Finlay GJ, Stewart AJ, Charlton P, Baguley BC, Denny WA (2001) J Med Chem 44:1407
90. Mekapati SB, Denny WA, Kurup A, Hansch C (2001) Bioorg Med Chem 9:2757
91. Spicer JA, Gamage SA, Finlay GJ, Denny WA (2002) Bioorg Med Chem 10:19

92. Cherney RJ, Swartz, SG, Patten AD, Akamike E, Sun JH, Kaltenbach RF III, Seit SP, Behrens CH, Getahun Z, Trainor GL, Vavala M, Kirshenbaum MR, Papp LM, Stafford MP, Czerniak PM, Diamond RJ, McRipley RJ, Page RJ, Gross JL (1997) Bioorg Med Chem Lett 7:163
93. O'Reilly S, Baker SD, Sartorius S, Rowinsky EK, Finizio M, Lubiniecki GM, Grochow LB, Gray JE, Pieniaszek HJ Jr, Donehower RC (1998) Ann Oncol 9:101
94. Thompson J, Pratt CB, Stewart CF, Avery L, Bowman L, Zamboni WC, Pappo A (1998) Invest New Drugs 16:45
95. Bousquet PF, Brana MF, Conlon D, Fitzgerald KM, Perron D, Cocchiaro C, Miller R, Moran M, George J, Qian XD, Keilhauer G, Romerdahl CA (1995) Cancer Res 55:1176
96. Villalona-Calero MA, Eder JP, Toppmeyer DL, Allen LF, Fram R, Velagapudi R, Myers M, Amato A, Kagen-Hallet K, Razvillas B, Kufe DW, Von Hoff DD; Rowinsky EK (2001) J Clin Oncol 19:857
97. Far S, Kossanyi A, Verchère-Béaur C, Gresh N, Taillandier E, Perrée-Fauvet M (2004) Eur J Org Chem:1781
98. Fechter EJ, Olenyuk, Dervan PB (2004) Angew Chem Int Edn Engl 43:3591
99. Wakelin LPG, Romanos M, Chen TK, Glaubiger D, Canellakis ES, Waring MJ (1978) Biochemistry 17:5057
100. Assa-Munt N, Denny WA, Leupin W, Kearns DR (1985) 24:1441
101. von Hippel PH, McGhee JD (1972) Ann Rev Biochem 41:231
102. Leng F, Chaires JB, Waring MJ (2003) Nucleic Acids Res 31:6191
103. Haq I, Ladbury JE, Chowdhry BZ, Jenkins TC, Chaires JB (1997) J Mol Biol 271:244
104. Chaires J (1997) Biopolymers 44:201
105. Gabelica V, De Pauw E, Rosu F (1999) J Mass Spectrom 34:1328
106. Porumb H (1978) Prog Biophys Mol Biol 34:175
107. Warner IM, Soper SA, McGown LB (1996) Anal Chem 68:73
108. Sartorius J, Schneider HJ (1995) FEBS Lett 374:387
109. Eriksson M, Nordén B (2001) Methods Enzymol 340:68
110. Nordén B, Kurucsev T (1994) J Mol Recognit 7:141
111. Horvath JJ, Gueguetchkeri M, Gupta A, Penumatchu D, Weetall HH (1995) In: Biosensors and chemical sensor technology. ACS Symp Ser vol. 614, p 44
112. Hyun KM, Choi SD, Lee S, Kim SK (1997) Biochim Biophys Acta 1334:312
113. Hardenbol P, Wang JC, Van Dyke MW (1997) Bioconjugate Chem 8:617
114. Bailly C, Waring MJ (1995) J Biomol Struct Dyn 12:869
115. Ren J, Chaires JB (1999) Biochemistry 38:16067
116. Tse WC, Boger DL (2004) Acc Chem Res 37:61
117. Suh D, Oh YK, Chaires JB (2001) Process Biochem 37:521
118. Long EC, Barton JK (1990) Acc Chem Res 23:273
119. Suh D, Chaires JB (1995) Bioorg Med Chem 3:723
120. Förster T (1948) Ann Phys 2:55
121. Haq I (2002) Arch Biochem Biophys 403:1
122. Han F, Chalikian TV (2003) J Am Chem Soc 125:7219
123. Hopkins HP Jr, Fumero J, Wilson WD (1990) Biopolymers 29:449
124. Taquet A, Labarbe R, Houssier C (1998) Biochemistry 37:9119
125. Moghaddam MS, Shimizu S, Chan HS (2005) J Am Chem Soc 127:303
126. Ajay, Murcko MA (1995) J Med Chem 38:4953
127. MacGregor RB Jr, Clegg RM, Jovin TM (1985) Biochemistry 26:4008
128. Finkelstein AJ, Janin J (1989) Protein Eng 3:1
129. Gilson, MK, Given JA, Bush BL (1997) Biophys J 72:1047
130. Holtzer A (1995) Biopolymers 35:595

131. Boresch S, Tettinger F, Leitgreb M, Karplus M (2003) J Phys Chem B 107:9535
132. Record MT Jr, Ha JH, Fisher MA (1991) Methods Enzymol 208:291
133. See also: Rick SW (2003) J Phys Chem B 107:9853
134. Chaires JB (1996) Anti-cancer Drug Des 11:569
135. For a recent compilation see: http://www.biophysics.org/education/resources.htm
136. Chiron J, Galy JP (2004) Synthesis 2004:313
137. Laursen JB, Nielsen J (2004) Chem Rev 1004:1663
138. Denny WA (2002) In: Demeunynck M, Bailly C, Wilson WD (eds) DNA and RNA binders. Wiley-VCH, Weinheim, p 482
139. Wirth M, Buchardt O, Koch T, Nielsen PE, Nordén B (1988) J Am Chem Soc 110:932
140. Adams A (2002) Curr Med Chem 9:1667
141. Tsann-Long S (2002) Curr Med Chem 9:1677
142. Antonini I (2002) Curr Med Chem 9:1701
143. Denny WA (2004) Med Chem Rev 1:257
144. Mannaberg J (1897) Arch Klin Med 59:185
145. Browning CG, Gilmour W (1913) J Pathol Bacteriol 24:127
146. Albert A (1966) The acridines, 2nd edn. Arnold, London, England
147. Zhu H, Clark SM, Benson SC, Rye AN, Mathies RA (1994) Anal Chem 66:1941
148. Rehn C, Pindur U (1996) Monatsh Chem 127:645
149. Gamage SA, Tepsiri N, Wilairat P, Wojcik SJ, Figgitt DP, Ralph RK, Denny WA (1994) J Med Chem 37:1486
150. McConnaughie AW, Jenkins TC (1995) J Med Chem 38:3488
151. Gamage SA, Figgitt DP, Wojcik SJ, Ralph RK, Ransijn A, Mauel J, Yardley V, Snowdon D, Croft SL, Denny WA (1997) J Med Chem 40:2634
152. Charmantray F, Demeunynck M, Carrez D, Croisy A, Lansiaux A, Bailly C, Colson P (2003) J Med Chem 46:967
153. Graves PR, Kwiek JJ, Fadden P, Ray R, Hardeman K, Coley AM, Foley M, Haystead TA (2002) Mol Pharmacol 62:1364
154. Girault S, Grellier P, Berecibar A, Maes L, Mouray E, Lemière P, Debreu MA, Davioud-Charvet E, Sergheraert C (2000) J Med Chem 43:2646
155. Sumner AT (1986) Histochemistry 84:566
156. Wainwright M (2001) J Antimicrob Chemother 47:1
157. Zwelling LA, Michaels S, Erickson LC, Ungerleider RS, Nichols M, Kohn KW (1981) Biochemistry 20:6553
158. Denny WA, Cain BF, Atwell GJ, Hansch C, Panthananickal AL (1982) J Med Chem 25:276
159. Cassileth PA, Gale RP (1986) Leukemia Res 10:1257
160. Sajewicz W, Dlugosz A (2000) J Appl Toxicol 20:305
161. Plymale DR, de la Iglesia FA (1999) J Appl Toxicol 19:31
162. Tuite E, Kim SK, Norden B, Takahashi M (1995) Biochemistry 34:16365
163. Hicks KO, Ohms SJ, Van Zijl PL, Denny WA, Hunter PJ, Wilsom WR (1997) Br J Cancer 76:894
164. Chen AY, Liu LF (1994) Annu Rev Pharmacol 34:191
165. Wang JC (1996) Annu Rev Biochem 65:935
166. Li TS, Liu LF (2001) Annu Rev Pharmacol 41:53
167. Pilch D, Liu HY, Li TK, Kerrigan JE, LaVoie EJ, Barbieri CM (2002) In: Demeunynck M, Bailly C, Wilson WD (eds) DNA and RNA binders. Wiley-VCH, Weinheim, p 576
168. Fechter EJ, Dervan PB (2003) J Am Chem Soc 125:8476
169. Bentin T, Nielsen PE (2003) J Am Chem Soc 125:6378

170. Fkyerat A, Demeunynck M, Constant JF, Michon P, Lhomme J (1993) J Am Chem Soc 115:9952
171. Alarcon K, Demeunynck M, Lhomme J, Carrez D, Croisy A (2001) Bioorg Med Chem 9:1901
172. Alarcon K, Demeunynck M, Lhomme J, Carrez D, Croisy A (2001) Bioorg Med Chem Lett 11:1855
173. Asseline U, Bonfils E, Dupret D, Thuong NT (1996) Bioconjugate Chem 7:369
174. Fukui K, Tanaka K (1996) Nucleic Acid Res 24:3962
175. Baruah H, Day CS, Wright MW, Bierbach U (2004) J Am Chem Soc 126:4492
176. Budiman ME, Alexander RW, Bierbach U (2004) J Am Chem Soc 126:8560
177. Temple MD, McFayden WD, Holmes RJ, Denny DA, Murray V (2000) Biochemistry 39:5593
178. Boldron C, Ross SA, Pitié M, Meunier B (2002) Bioconjugate Chem 13:1013
179. Wang L, Price HL, Juusola J, Kline M, Phanstiel IV O (2001) J Med Chem 44:3682
180. Joseph J, Eldho NV, Ramaiah D (2003) Chem Eur J 9:5926
181. Isobe H, Tomita N, Lee JW, Kim H-J, Kim K, Nakamura E (2000) Angew Chem Int Edn Engl 39:4257
182. Antonini I, Polucci P, Magnano A, Gatto B, Palumbo M, Menta E, Pescalli N, Martelli S (2003) J Med Chem 46:3109
183. Gamage SA, Spicer JA, Atwell GJ, Finlay GJ, Baguley BC, Denny WA (1999) J Med Chem 42:2383
184. Laugâa P, Markovits J, Delbarre A, LePecq JB, Roques BP (1985) Biochemistry 24:5567
185. Fechter EJ, Olenyuk B, Dervan PB (2004) Angew Chem Int Edn Engl 43:3591
186. Yang X, Robinson H, Gao Y-G, Wang A H-J (2000) Biochemistry 39:10950
187. Claude S, Lehn JM, Pérez de Vega MJ, Vigneron JP (1993) New J Chem 16:21
188. Jourdan M, Garcia J, Lhomme J, Teulade-Fichou MP, Vigneron JP, Lenh JM (1999) Biochemistry 39:14205
189. Veal JM, Li Y, Zimmerman SC, Lamberson CR, Cory M, Zon G, Wilson WD (1990) Biochemistry 29:10918
190. Lhomme J, Constant JF, Demeunynck M (1999) Biopolymers 52:65
191. Constant JF, Demeunynck M (2002) In: Demeunynck M, Bailly C, Wilson WD (eds) DNA and RNA binders. Wiley-VCH, Weinheim, p247
192. Lerman LS (1961) J Mol Biol 3:18
193. Armstrong RW, Kurucsev T, Strauss UP (1970) J Am Chem Soc 92:3174
194. Muller W, Crothers DM (1975) Eur J Biochem 54:267
195. Peacocke AR, Skerrett JNH (1956) J Chem Soc, Faraday Trans 52:261
196. Yang L, Weerasinghe S, Smith PE, Pettitt BM (1995) Biophys J 69:1519
197. Pack G, Wong L (1995) Chem Phys 204:279
198. Wang AHJ, Quigley GJ, Rich A (1979) Nucleic Acid Res 12:3879
199. Schelhorn T, Kretz S, Zimmermann HW (1992) Cell Mol Biol 38:345
200. Biver T, Secco F, Tinè MR, Venturini M (2003) Arch Biochem Biophys 418:63
201. Reddy BS, Seshardi TP, Sakore TD, Sobell HM (1979) J Mol Biol 135:787
202. Aggarwal A, Islam SA, Kuroda R, Neidle S (1984) Biopolymers 23:1025
203. Berman HM, Stalling W, Carrell HL, Glusker JP, Neidle S, Taylor G, Achari A (1979) Biopolymers 18:2405
204. Shieh HS, Berman HM, Dabrow M, Neidle S (1980) Nucleic Acid Res 8:85
205. Lyles MB, Cameron IL (2002) Cell Biol Int 26:145
206. Löber G, Achert G (1969) Biopolymers 8:595
207. Löber G, Kittler L (1978) Stud Biophys 73:25

208. Seidel AM, Schulz A, Sauer MHM (1996) J Phys Chem 100:5541
209. Steenken S, Jovanovic S (1997) J Am Chem Soc 119:617
210. Löber G (1965) Photochem Photobiol 4:607
211. Boyle RE, Nelson SS, Dollish FR, Olson MI (1962) Arch Biochem Biophys 96:47
212. Petit JM, Denis-Gay M, Ratinaud MH (1993) Biol Cell 78:1
213. Stormer U, Baumgartel H (1988) Acta Histochem 84:31
214. Darzynkiewics Z, Kapuscinski J (1990) In: Melamed MR, Lindmo T, Mendelsohn MI (eds) Flow cytometry and sorting, 2nd edn. Wiley-Liss, New York, p 291
215. Brake DG, Thaler R, Evenson DP (2004) J Agr Food Chem 52:2097
216. Brauns EB, Murphy CJ, Berg MA (1998) J Am Chem Soc 120:2449
217. Lown JW (1988) Anthracycline and anthracenedione-based anicancer agents. Elsevier, Amsterdam
218. Cheng CC, Zee-Cheng RKY (1983) Prog Med Chem 20:83
219. Gandolfi CA, Beggiolin G, Menta E, Palumbo M, Sissi C, Spinelli S, Johnson F (1995) J Med Chem 38:526
220. Wang S, Peng T, Yang CF (2003) Biophys J 104:239
221. Citarella RV, Wallace RE, Murdock KC, Angier RB, Durr FE, Forbes M (1982) Cancer Res 42:440
222. Foye WD, Vajragupata O, Sengupta SK (1982) J Pharm Sci 71:253
223. Kapuscinski J, Darzynkiewicz Z, Traganos F, Melamed MR (1981) Biochem Pharmacol 30:231
224. Zee-Cheng RKY, Cheng CC (1978) J Med Chem 21:291
225. Zee-Cheng RKY, Podrebarac EG, Menon CS, Cheng CC (1979) J Med Chem 22:501
226. Murdock KC, Child RG, Fabio PF, Angier RB, Wallace RE, Durr FE, Citarella RV (1979) J Med Chem 22:1024
227. Krapcho AP, Getahun Z, Avery KL, Vargas KJ, Hacker MP, Spinelli S, Pezzoni G, Manzotti C (1991) J Med Chem 34:2373
228. Gatto B, Zagotto G, Sissi C, Cera C, Uriate E, Palù G, Capranico G, Palumbo M (1996) J Med Chem 39:3114
229. Ijaz T, Tran P, Ruparelia KC, Teesdale-Spittle PH, Orr S, Patterson LH (2001) Bioorg Med Chem Lett 11:351
230. Abate C, Patel L, Rauscher FJ, Curran T (1990) Science 249:1157
231. Steullet V, Edwards-Benett S, Dixon DW (1999) Bioorg Med Chem 7:2531
232. David SS, William SD (1998) Chem Rev 98:1221
233. Sugjyama H, Fujiwara T, Ura A, Tashiro T, Yamamoto K, Kawanishi S, Saito I (1994) Chem Res Toxicol 7:673
234. Mc Hugh PJ, Knowland J (1995) Nucleic Acid Res 23:1664
235. McKnight RE, Zhang J, Dixon DW (2004) Bioorg Med Chem Lett 14:401
236. Koyama M, Kelly TR, Watanabe KA (1988) J Med Chem 31:283
237. Gourdie TA, Prakash AS, Wakelin LPG, Woodgate PD, Denny WA (1991) J Med Chem 34:240
238. Köhler B, Su TL, Chou TC, Jiang XJ, Watanabe KA (1993) J Org Chem 58:1680
239. Satorelli AC (1988) Cancer Res 48:775
240. Oostveen EA, Speckamp WN (1987) Tetrahedron 43:255
241. Carter SK, Crooke ST (1979) Mitomycin C: Current Status and New Developments. Academic Press, New York
242. Kim JY, Su T-S, Chou T-C, Koehler B, Scarborough A, Ouerfelli O, Watanabe KA (1996) J Med Chem 39:2812
243. Schuster GB (2000) Acc Chem Res 33:253
244. Armitage B (1998) Chem Rev 98:1171

245. Schuster GB, Landman U (2004) Top Curr Chem 236:139
246. Ly D, Kan Y, Armitage B, Schuster GB (1996) J Am Chem Soc 118:8747
247. Saito I, Nakamura T, Nakatani K, Yoshioka Y, Yamaguchi K, Sugiyama H (1998) J Am Chem Soc 120:12686
248. Prat F, Houk KN, Foote CS (1998) J Am Chem Soc 120:845
249. Breslin DT, Schuster GB (1996) J Am Chem Soc 118:2311
250. Gibson D, Mansur N, Gean FK (1995) J Inorg Biochem 58:79
251. Barcelo F, Minchner DJ, Crampton MR, Brown JR, Shaw G (1991) Anti-cancer Drug Des 6:37
252. Pinto AL, Lippard SJ (1984) Biochem Biophys Acta 780:167
253. Sherman SE, Lippard SJ (1987) Chem Rev 87:1153
254. Reedijk J, Fichtinger-Shepman AMJ, Van Oosteran AT, Van de Putte O (1987) Struct Bonding 67:53
255. Johnson NP, Lapatoule P, Razaka H (1986) In: McBrien DC, Slatter T (eds) Biochemical mechanisms of platinum antitumor drugs. IRL, Washington DC, p 1
256. Roberts JJ, Knox RJ, Frielos F, Lyndall DA (1986) In: McBrien DC, Slatter T (eds) Biochemical mechanisms of platinum antitumor drugs. IRL, Washington DC, p 29
257. Sherman SE, Gibson D, Wang HJ, Lippard SJ (1985) Science 230:412
258. Sherman SE, Gibson D, Wang HJ, Lippard SJ (1988) J Am Chem Soc 110:7368
259. Ellis LT, Perkins DF, Turner P, Humbley TW (2003) J Chem Soc Dalton Trans 13:2728
260. Boseggia E, Gatos M, Lucatello L, Mancin F, Moro S, Palumbo M, Sissi C, Tecilla P, Tonatello U, Zagotto G (2004) J Am Chem Soc 126:4543
261. Mishra A, Behera RB, Behera PK, Mishra BK, Berhera GB (2000) Chem Rev 100:1973
262. Schwartz HE, Ulfelder KJ (1992) Anal Chem 64:1737
263. Figeys D, Arriaga E, Renborg A, Dovichi NJ (1994) J Chromatogr 669:205
264. Yan XM, Grace WK, Yoshida TM, Habbersett RC, Velappan N, Jett JH, Keller RA, Marrone BL (1999) Anal Chem 71:5470
265. Kricka LJ (2002) Ann Clin Biochem 39:114
266. Rogers KR, Apostol A, Madsen SJ, Spencer CW (1999) Anal Chem 71:4423
267. Elmendorff Dreikorn K, Chauvin C, Slor H, Kutzner J, Batel R, Muller WE, Schroder HC (1999) Cell Mol Biol 45:211
268. Hirons GT, Fawcett JJ, Crissman HA (1994) Cytometrie 15:129
269. Blaheta RA, Kronenberger B, Woitaschek D, Weber S, Scholz M, Schuldes H, Encke A, Markus BH (1998) J Immunol Methods 211:159
270. Choi SJ, Szoka FC (2000) Anal Biochem 218:95
271. Rye HS, Yue S, Wemmer DE, Quesada MA, Haugland RP, Mathies RA, Glazer AN (1992) Nucleic Acid Res 20:2803
272. Carlsson C, Larsson A, Jonsson M, Albinsson B, Nordén B (1994) J Phys Chem 98:10313
273. Glazer AN, Rye HS (1992) Nature 359:859
274. Gurrieri S, Sam Wells K, Johnson ID, Bustamante C (1997) Anal Biochem 249:44
275. Larsson A, Carlsson C, Jonsson M, Albinsson N (1994) J Am Chem Soc 116:8459
276. Larsson A, Carlsson C, Jonsson M (1995) Bioploymers 36:153
277. Timtehava I, Maximova V, Deligeorgiev T, Zaneva D, Ivanov I (2000) Photochem Photobiol 130:7
278. Timcheva II, Maximova VA, Deligeorgiev TG, Gadjev NI, Sabnis RW, Ivanov IG (1997) FEBS Lett 405:141
279. Akerman B, Tuite E (1996) Nucleic Acid Res 24:1080
280. Kanony C, Akerman B, Tuite E (2001) J Am Chem Soc 123:7985
281. Spielmann HP, Wemmer DE, Jacobsen JP (1995) Biochemistry 34:8542

282. Nygren J, Svanvik N, Kubista M (1998) Biopolymers 46:39
283. Netzel TL, Nafisi K, Zhao M, Lenhardt JR, Johnson I (1995) J Phys Chem 99:17936
284. Petty JT, Bordelon JA, Robertson ME (2000) J Phys Chem 104:7221
285. Bunkenborg J, Munch Stidsen M, Jacobsen JJ (1999) Bioconjugate Chem 10:824
286. Braunlin WH, Strick TJ, Record MT (1986) Biopolymers 21:1301
287. Podmanabhan S, Rickey B, Anderson CF, Record MT (1988) Biochemistry 27:4367
288. Podmanabhan S, Brushaber VM, Anderson CF, Record MT (1991) Biochemistry 30:7550
289. Wemmer DE (1985) J Mol Biol 185:457
290. Stærk D, Hamed AA, Pedersen EB, Jacobsen JP (1997) Bioconjugate Chem 8:869
291. Bondesgaard K, Jacobsen JP (1999) Bioconjugate Chem 10:824
292. Bunkenborg J, Gadjev NI, Deliegeorgiev T, Jacobsen JP (2000) Bioconjugate Chem 11:861
293. Petersen M, Hamed AA, Pedersen EB, Jacobsen JJ (1999) Bioconjugate Chem 10:66
294. Isacsson J, Westmann G (2001) Tetrahedron Lett 42:3207
295. Haugland RP (1996) In: Haugland RP (ed) Handbook of fluorescent probes and research chemicals, 6th edn. Molecular Probes, Eugene, p 144
296. Haugland RP, Yue ST, Millard PJ, Roth BL (1995) US Patent 5436134
297. Yue ST, Singer VL, Roth BL,Mozer TJ, Millard PJ, Jones LJ, Xiaokui J, Haugland RP (1996) WO Patent 96/13552
298. Singer VL, Jones LJ, Yue ST, Haugland RP (1997) Anal Biochem 249:228
299. Cosa G, Focsaneanu KS, McLean JRN, Scaiano JC (2000) J Chem Soc Chem Commun 689
300. Schweitzer C, Scaiano JC (2003) Phys Chem Chem Phys 5:4911
301. Yarmoluk SM, Lukashov SS, Losytskyy MY, Åkerman B, Kornyushyna OS (2002) Spectrochim Acta Part A 58:3223
302. Yarmoluk SM, Lukashov SS, Ogul'chansky TY, Losytskyy MY, Kornyushyna OS (2001) Biopolymers 62:219
303. Ishchenko AA (1994) The structure and the spectroscopic properties of polymethine dyes. Nahova Dumka, Kiev
304. Carreon JR, Mahon KP, Kelley SO (2004) Org Lett 6:517
305. Mahon KP, Otriz-Meoz RF, Prestwich EG, Kelley SO (2003) J Chem Soc, Chem Commun: 1956
306. Blazek ER, Peak JG, Peak MJ (1989) It may be noted that even singlet oxygen can induce frank strand breaks in plasmid DNA. Photochem Photobiol 49:607
307. Wright A, Bubb WA, Hawkins CL, Davies MJ (2002) Photochem Photobiol 76:35
308. Thomson M, Woodbury NW (2000) Biochemistry 39:4327
309. Thompson M, Woodbury NW (2001) Biophys J 81:1793
310. Bruist MF, Horvath SJ, Hood LE, Steits TA, Simon MI (1987) Science 235:777
311. Livingstone JR, Spolar RS, Record MT (1991) Biochemistry 30:4237
312. Crothers DM (1968) Biopolymers 6:575
313. Waring MJ (1965) J Mol Biol 13:269
314. Normeier E (1992) J Phys Chem 96:6045
315. Byrne CD, de Mello AJ (1998) Biophys Chem 70:173
316. Monaco RR, Hausheer FH (1993) J Biomol Struct 10:675
317. Borisova OF, Shchyolkina AK, Karapetyan AT, Surovaya AN (1998) Mol Biol 32:718
318. Pauluhn J, Zimmermann HW (1978) Ber Bunsenges Phys Chem 82:1265
319. Talavera EM, Guerrero P, Ocana F, Alvarez-Pez JM (2002) Appl Spectrosc 56:362
320. Lippard SJ, Bond PJ, Wu KC, Bauer WR (1976) Science 194:726
321. Tsai CC, Jain SC, Sobell HM (1977) J Mol Biol 114:301

322. Chandrasekaran S, Jones RL, Wilson WD (1985) Biopolymers 24:1963
323. Davies DB, Veselkov A, Veselkov AN (1999) Mol Phys 97:439
324. Davies DB, Veselkov AN (1996) J Chem Soc, Faraday Trans 92:3545
325. Guo Q, Lu M, Marky LA, Kallenbach NR (1992) Biochemistry 31:2451
326. Koeppel F, Riou JF, Laoui A, Mailliet P, Arimondo PB, Labit D, Petitgenet O, Hélène C, Mergny JL (2001) Nucleic Acids Res 29:1087
327. Tuite E, Nordén B (1995) Bioorg Med Chem 3:701
328. Rentzeperis D, Medero M, Marky LA (1995) Bioorg Med Chem 6:751
329. Olmstedt J III, Kearns DR (1977) Biochemistry 16:3647
330. LePecq JB, Paoletti C (1967) J Mol Biol 27:87
331. Hudson B, Jacobs R (1975) Biopolymers 14:1309
332. Heller DP, Greenstock CL (1994) Biophys Chem 50:305
333. Latimer LJP, Lee JS (1991) J Biol Chem 266:13849
334. Sailer BL, Nastasi AJ, Valdez JG, Steinkamp JA, Crissman HA (1997) J Histochem Cytochem 45:165
335. Sharp PA, Sugden B, Sambrook J (1973) Biochemistry 12:3055
336. Yguerabide J, Ceballos A (1995) Anal Biochem 228:208
337. Cui HH, Valdez JG, Steinkamp JA, Crissman HA (2003) Cytometry Part A 52A:46
338. Tramier M, Kemnitz K, Durieux C, Coppey-Moisan M (2004) J Microsc 213:110
339. Sailer BL, Nastasi AJ, Valdez JG, Steinkamp JA, Crissman HA (1996) Cytometry 25:164
340. Cai J, Soloway AH, Barth RF, Adams DM, Hariharan JR, Wyzlic IM, Radcliffe K (1997) J Med Chem 40:3887
341. Edwards ML, Snyder RD, Stemerick DM (1991) J Med Chem 34:2414
342. Stewart KD (1988) Biochem Biophys Res Commun 152:1441
343. Lown JW (1982) Acc Chem Res 15:381
344. Ferrari D, Peracchi A (2002) Nucleic Acids Res 30:112
345. Tettey JNA, Skellern GG, Midgley JM, Grant MH, Pitt AR (1999) Chem Biol 123:105
346. Luedtke NW, Liu Q, Tor Y (2003) Bioorg Med Chem 11:5235
347. Sawyer JR (1995) Hum Genet 95:49
348. Juranovic I, Meic Z, Piantanida I, Tumir LM, Zinic M (2002) J Chem Soc Chem Commun:1432
349. Tumir, LM, Piantanida I, Novak P, Zinic M (2002) J Phys Org Chem 15:599
350. Piantanida I, Palm BS, Cudic P, Zinic M, Schneider HJ (2001) Tetrahedron Lett 42:6779
351. Colmenarejo G, Bácena M, Guitiérrez-Alnso MC, Montero F, Orellana G (1995) FEBS Lett 374:426
352. Sammes PG, Shek L, Watmore D (2001) J Chem Soc Chem Commun:1625
353. Hall HB, Kelley SO, Barton JK (1998) Biochemistry 37:15933
354. Kelley SO, Holmlin RE, Stemp EDA, Barton JK (1997) J Am Chem Soc 119:9861
355. Amann N, Huber R, Wagenknecht HA (2004) Angew Chem Int Ed Engl 43:1845
356. Huber R, Amann N, Wagenknecht HA (2004) J Org Chem 69:744

Top Curr Chem (2005) 258: 205–255
DOI 10.1007/b136670

Published online: 8 August 2005

# Two-Dimensional Dye Assemblies on Surfaces Studied by Scanning Tunneling Microscopy

Steven De Feyter (✉) · Frans De Schryver

Department of Chemistry, Katholieke Universiteit Leuven, Celestijnenlaan 200 F, 3001 Leuven, Belgium
*steven.defeyter@chem.kuleuven.be, frans.deschryver@chem.kuleuven.ac.be*

**Abstract** This chapter gives an overview of the ordering of dyes on atomically flat conductive surfaces as revealed by scanning tunneling microscopy. Scanning tunneling microscopy provides detailed insight into the arrangement of dyes on various surfaces and unravels novel properties which arise as a result of the interaction of the dyes with the substrate. These high-resolution studies have been motivated by the importance of interfacial layers in view of material and device properties. Typical examples are provided of popular dye systems, including conjugated oligomers and polymers. In addition, the supramolecular ordering of mixtures of dyes and two-dimensional chirality is discussed. Scanning tunneling microscopy goes beyond imaging and its ability to manipulate dye molecules, both mechanically and chemically, is discussed. Finally, the use of scanning tunneling microscopy as a spectroscopy tool for probing the electronic properties of dye layers on surfaces is highlighted.

**Keywords** Dye · Monolayer · Supramolecular chemistry · Scanning tunneling microscopy

**Abbreviations**

C18I 1-Iodooctadecane
DC Decacyclene
EC-STM Electrochemical scanning tunneling microscopy
HBC Hexabenzocoronene
HOMO Highest occupied molecular orbital
HOPG Highly oriented pyrolytic graphite
*I–V* Current–voltage
LDOS Local density of states
LUMO Lowest unoccupied molecular orbital
NTCDI Naphthalene tetracarboxylic diimide
PAH Polyaromatic hydrocarbon
PTCDA 3,4,9,10-Perylene tetracarboxylic dianhydride
PTCDI 3,4,9,10-Perylene tetracarboxylic diimide
SAM Self-assembled monolayer
SCE Saturated calomel electrode
STM Scanning tunneling microscopy
STS Scanning tunneling spectroscopy
UHV Ultrahigh vacuum

# 1 Introduction

## 1.1 Organic Dyes and Surfaces

Organic dyes and molecules in general are rich in properties and are promising candidates to be components in devices such as light-emitting diodes and field-effect transistors. "Dyes" is used here in a broad sense and includes typical compounds such as perylene-type dyes, porphyrins and phthalocyanines, as well as conjugated oligomers and polymers. In many cases, well-structured surfaces are required. Interfacial structures and their properties might differ from those in the bulk and these interfacial structures affect device properties. For instance, the interfacial layer may control the charge transport properties between the layer and the substrate. It is therefore important to understand the ordering of the molecules at the surface to control their properties. Self-assembly, which is the spontaneous formation of highly organized functional supramolecular architectures, will be of great value to achieve ordered monolayer formation at interfaces both under ultrahigh vacuum (UHV) conditions and ambient conditions, as well as at the liquid–solid interface.

## 1.2 Scanning Tunneling Microscopy

The invention of the scanning tunneling microscope in the early 1980s [1] opened new ways to investigate surface phenomena on a truly atomic scale thanks to the very localized nature of the probing. In scanning tunneling microscopy (STM), a metallic tip is brought very close to a conductive substrate and by applying a voltage between both conductive media, a tunneling current through a classically impenetrable barrier may result between the two electrodes. The direction of the tunneling depends on the bias polarity. The exponential distance dependence of the tunneling current provides excellent means to control the distance between the probe and the surface and very high resolution (atomic) on atomically flat conductive substrates can be achieved. For imaging purposes, the tip and substrate are scanned precisely relative to one another and the current is accurately monitored as a function of the lateral position. The contrast in STM images reflects both topography and electronic effects. There are mainly two modes of operation. In the constant-height mode, the absolute vertical position of the probe remains constant during raster-scanning. The tunneling current is plotted as a function of the lateral position. In the constant-current mode, the signal of the probe is kept constant through readjusting the vertical position of the probe (or the sample). The vertical position of the probe is plotted as a function of the lateral position.

The initial applications of STM dealt with imaging of semiconductor, inorganic and metal surfaces and later the first experiments on molecules adsorbed on surfaces were reported. The method requires atomically flat conductive substrates and immobilization of the molecules on the time scale of recording an image, and, owing to the distance dependence of the tunneling process, only very thin layers—typically monomolecular—are probed.

As it turned out, STM is a very versatile technique. The first experiments on molecular adlayers were carried out under UHV conditions on atomically flat metal terraces [2]. Organic molecular beam epitaxy is the method of choice to deposit molecules under UHV conditions [3]. Provided that the molecule–substrate interaction is strong enough individual isolated molecules can be immobilized and visualized. To overcome the problem of a too high molecular mobility on a given substrate two approaches have been followed: (1) controlling the temperature under UHV conditions and (2) increasing the monolayer coverage. Successful STM imaging requires balanced adsorbate–substrate interactions: a too strong interaction immobilizes the molecules and may impede self-assembly into ordered 2D layers. A too weak adsorbate–substrate interaction leads to a too high mobility and STM imaging often becomes impossible.

The majority of studies on the ordering and properties of thin films of dyes (monolayers) by STM have been performed under UHV conditions. Under UHV conditions, both individual atoms as well as relatively large molecules can be investigated with very high spatial resolution, under well-defined conditions, on a large variety of substrates, over a wide temperature range, with control of the surface coverage. However, not all species can be adapted to UHV, such as those with relatively low thermal stability and/or high molecular weight (i.e., macromolecules).

UHV conditions are indeed not necessary to achieve high-quality STM imaging. Under ambient conditions, liquid-crystalline compounds have been visualized as well and a typical substrate is highly oriented pyrolytic graphite (HOPG) [4, 5]: it is electrically conductive, atomically flat, inert and is easy to clean. Also liquidlike organic compounds or isotropic organic solutions allow STM imaging [6, 7]. Self-assembled monolayers (SAMs) form spontaneously by physisorption at the liquid–solid interface and stable imaging is achieved if the molecules are laterally immobilized by adsorbate–adsorbate and adsorbate–substrate interactions. In isotropic organic solutions, the typical organic solvents used have a low vapor pressure (experiments are typically not carried out in a closed container), are (electro)chemically inert and show no or less affinity to form ordered monolayers on the surface than the molecules of interest.

Spin-coating or drop-casting of diluted solutions followed by solvent evaporation allows imaging at the solid–air interface. Other approaches include transfer of films at the water–air interface to the substrate by means of the Langmuir–Blodgett technique (vertical dipping) or the horizontal lifting method. In contrast to the situation at the solid–air interface, at the solid–liquid interface the adsorbed molecules are in equilibrium with those dissolved in the solution, leading to adsorption–desorption dynamics.

The formation of monolayers at the solid–liquid interface can also be induced under potential control in aqueous solutions [8]. The adsorption of organic molecules at the electrode–electrolyte interface can be considered as a very promising approach for the preparation of ordered adlayers [9]. Under electrochemical conditions, adsorbate–substrate interactions can be modulated by the surface charge density. Electrochemical environments offer therefore additional possibilities to control surface dynamics and monolayer structure via the surface charge [10, 11]. Typically, in situ electrochemical STM (EC-STM) experiments are carried out in aqueous solutions and water-soluble compounds are intuitively the first choice when it comes to inducing ordering at the electrified surfaces. For a long time it has been difficult to investigate large aromatic molecules. However, as recently shown (vide infra), this technique is not limited to the use of water-soluble compounds. Also compounds with extremely low solubility can be investigated at electrified surfaces. A key step is preadsorbing the molecules from an organic solution.

Control of the mobility of molecules at room temperature really allows advantage to be taken of the versatility of STM to operate at the interface between two media, one being an atomically flat conducting solid and the other being a gas (air) or a liquid or a liquid crystalline material or a gel. For the imaging process, the thickness of the film is, in principle, not that critical since the tip penetrates through the excess insulating organic material. Typically, monolayers are imaged as the tip is very close to the substrate under standard measuring conditions (tunneling bias approximately 1 V; tunneling current less than 1 nA). However, for highly conductive systems, multilayers have also been observed (vide infra). In some cases it is even possible to selectively image the first or second layer, by changing the tip height which is controlled by the tunneling current (vide infra).

Whatever the measuring conditions, the balance of molecule–molecule and molecule–substrate interactions determines the result; therefore, it is clear that the choice of the substrate plays an important role. In addition to the substrate, one should also consider the importance of the solvent used (at the liquid–solid interface). Though its role has not been investigated systematically, it plays an important role by, for instance, favoring one of the possible 2D polymorphous structures.

For sure, the thermodynamically stablest structures are not always formed and imaged on the surface. Kinetics plays an important role. Under UHV conditions, often annealing (increasing the temperature) is necessary before the deposited molecules "find" each other and order in specific patterns. At the liquid–solid interface, it is sometimes observed that during a measuring session the degree of ordering increases. A well-documented phenomenon at the liquid–solid interface is the observation that large domains grow at the expense of smaller domains (Ostwald ripening). In mixtures of different compounds, one of them often adsorbs faster, and equilibrium sets in only after a while.

STM provides in the first place an insight into the molecular ordering, typically within monolayers and for a few cases also in thin multilayers at conductive and atomically flat surfaces. However, making beautiful pictures in real space of 2D patterns is not the only characteristic of STM. To a good approximation, STM probes the local density of states (LDOS) and is therefore sensitive to the electronic structure of the substrate, the tip and the molecules adsorbed on the substrate, which we will call adsorbates. By using STM as a local spectroscopy tool by sweeping the bias voltage for a given current, giving rise to the so-called current–voltage or $I$–$V$ curves, detailed information can be obtained on the states involved in the tunneling process. In addition, STM can also be used to manipulate matter on the nanoscale, by mechanically moving molecules around or by using the tip as an electrochemically active probe.

## 2
## STM Imaging of Organic Dyes on Surfaces

A huge amount of research has been carried out on the self-assembly and properties of dyes at surfaces, investigated by STM. This review will necessarily only cover a selection of highlights on a variety of dyes. Smaller aromatic compounds such as benzene, naphthalene and so on will not be covered [12]. A number of the systems presented here are covered in detail in excellent reviews [13, 14]. The focus is on ordering and imaging, as well as on spectroscopy and manipulation.

### 2.1
### 3,4,9,10-Perylene tetracarboxylic dianhydride and 3,4,9,10-Perylene tetracarboxylic diimide

Organic epitaxy—that is the growth of molecular films with a commensurate relationship to their crystalline substrates—relies on successful recognition of preferred epitaxial sites. 3,4,9,10-Perylene tetracarboxylic dianhydride (PTCDA) (Fig. 1, **1**) is one of the most studied dyes on surfaces. It is a commercially available dye which shows sufficient thermal stability to allow purification by sublimation techniques and vapor deposition by molecular beam epitaxy. PTCDA crystals are organic semiconductors with potential application in molecular electronics. STM studies on various substrates such as Au(111), Au(100), Ag(110), $MoS_2$ and HOPG revealed that PTCDA forms well-defined monolayers [15–19]. In most cases, the layers of PTCDA grow in the herringbone pattern that corresponds to the (102) plane of the bulk

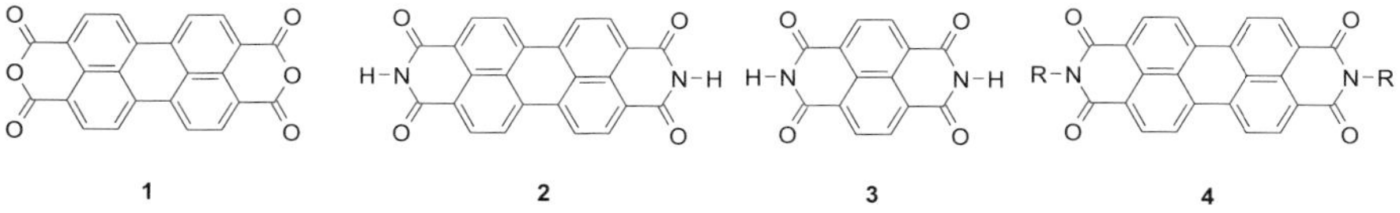

1 2 3 4

**Fig. 1** Perylene and naphthalene derivatives

crystal. This arrangement results from the minimization of the electrostatic interaction energy originating in the quadrupolar field associated with the molecule. Ordered monolayers of PTCDA different from those for the (102) bulk plane have been observed for epitaxial growth on Ag(110) [19, 20], forming almost a square unit cell, and Au(100) [21], where the (010)-bulk plane of PTCDA is found to interface the substrate. PTCDA has also been grown on top of a decanethiol monolayer by organic molecular beam deposition. Not only is the growth of PTCDA influenced by the underlying decanethiol SAM, but the deposition of PTCDA can also modify the decanethiol SAM [22]. One of the most prominent examples of organic epitaxy is PTCDA on Ag(111). It is an example of true commensurate epitaxy in contrast with most polyacenes and their derivatives, which show so-called quasiepitaxy. Normally, molecules which form layered van der Waals bonded crystals in their bulk states allow quasiepitaxy on inorganic substrates owing to a very shallow minimum of the interaction potential between the molecular layer and the substrate. PTCDA on Ag(111) is one of the exceptions as it forms almost perfectly ordered commensurate overlayers (Fig. 2). This form of site recognition is attributed to the existence of a local molecular reaction center in the extended $\pi$-electron system of the molecule. When deposition occurs at low temperature on Ag(111) no ordering is observed: the site-recognition reaction is blocked by a lack of

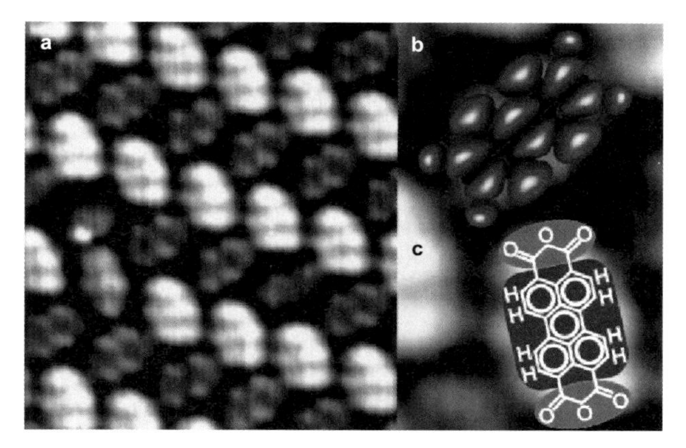


**Fig. 2** Scanning tunneling microscopy (*STM*) image (*left*) and models (*right*) of **1** on Ag(111). (Reproduced with permission from Ref. [23])

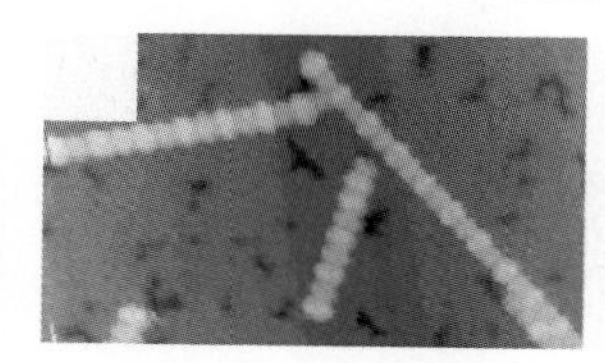
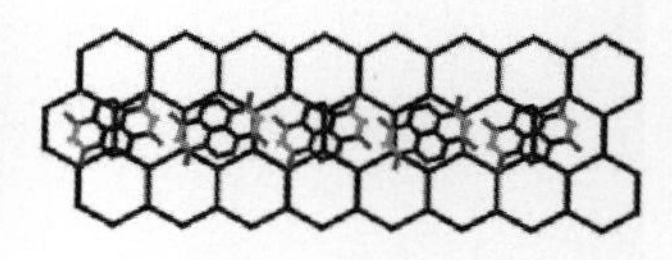

**Fig. 3** STM image (*left*) and model (*right*) of single rows of **3** on Ag/Si(111)-$\sqrt{3} \times \sqrt{3}R30°$ reconstruction recorded under ultrahigh vacuum (*UHV*) conditions. (Reproduced with permission from Ref. [25])

activation energy and the molecule is caught in a metastable chemisorption precursor. In this state PTCDA floats on the metallic electron density instead of reacting with it [23].

In contrast to PTCDA, 3,4,9,10-perylene tetracarboxylic diimide (PTCDI) (Fig. 1, **2**) forms 2D domains where the molecules are arranged in a row-like structure, as well as on HOPG $MoS_2$ [18] and on hydrogen-terminated Si(111) [24] caused by the formation of hydrogen bonds. A related molecule, naphthalene tetracarboxylic diimide (NTCDI) (Fig. 1, **3**), was shown to form 1D *single* rows on a Ag/Si(111)-$\sqrt{3} \times \sqrt{3}R30°$ reconstruction (Fig. 3) as the substrate which was chosen because it allows the molecules to diffuse freely (Fig. 2) [25]. The balance between intermolecular and molecule–substrate interactions is believed to determine whether 1D or 2D growth occurs.

Alkylation of the N site makes hydrogen bonding impossible. Alkylated PTCDIs (Fig. 1, **4**) have been investigated at the liquid–solid (HOPG, $MoS_2$) interface. The length of the hydrocarbon tail on the molecules was varied to observe the transition between surface structures driven by intermolecular core interactions and interactions between the hydrocarbon tails and the substrate. Both rectangular, or herringbone-like structures, and row structures were observed. The lattice constants, and thus the area per molecule of the rectangular structures, did not increase as expected when the alkyl chain length was increased, indicating protrusion of the alkyl tails into the solvent above the 2D layer. In the row structures, the alkyl chains were adsorbed [26].

## 2.2 Phthalocyanines and Porphyrins

Metal-substituted phthalocyanines (Fig. 4, **5**) were among the first molecules ever imaged under UHV conditions [27]. On Cu(100), the molecules are adsorbed in two different rotational orientations. The observed internal structures compare well with the calculated highest occupied molecular orbital (HOMO) of the free molecule. This type of dye molecule has been investigated on many other different substrates too and it typically appears as a cloverleaf with fourfold symmetry, reflecting the shape of the molecule. In the case of a planar metal-substituted phthalocyanine, the apparent height of

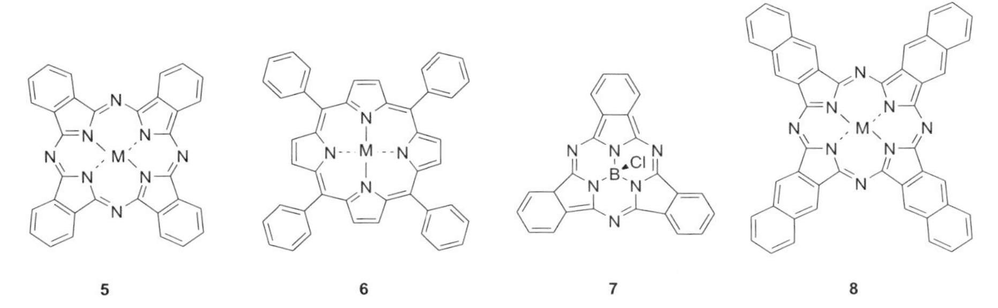


**Fig. 4** Phthalocyanine- and porphyrin-type dyes

the molecular center in the STM profile is crucially dependent on the chemical nature of the central metal atom. For instance both CuPc/CoPc [28, 29], where PC represents phthalocyanine, and NiPc/FePc [30] can be distinguished by their internal structures when adsorbed on a Au(111) substrate, owing to differences in the electronic structure of the central metal atom substituent. Chemical sensitivity is attained for selected mixed molecular layers of these molecules. Furthermore, nonplanar metal-substituted phthalocyanines such as PbPc [31] and SnPc [32] exhibit two different adsorption geometries (metal up or metal down). A similar metal-atom sensitivity was observed for metal-coordinated tetraphenyl porphyrins (Fig. 4, **6**) [33, 34].

High-quality monolayer formation and high-resolution imaging of phthalocyanines (**5**) and tetraphenyl porphyrins (**6**) was also achieved under potential control on Au(100) and Au(111) surfaces in 0.1 M $HClO_4$ by in situ imaging with EC-STM (Fig. 5) [35–37].

Other systems such as subphthalocyanines (Fig. 4, **7**) and naphthalocyanines (Fig. 4, **8**) have also been investigated. Subphthalocyanines consist of a conelike structure with threefold symmetry with a central B atom and an axial Cl head, missing one isoindoline unit of regular Pc (**7**) [38–43]. UHV imaging of free-base naphthalocyanines (**8**) on graphite also revealed a four-

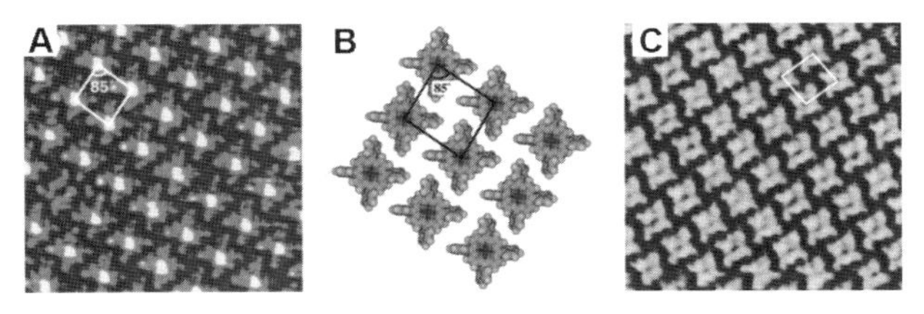


**Fig. 5** **a** STM image of Co**6** adlayer on the Au(111) surface in 0.1 M $HClO_4$ ($9 \times 9\,nm^2$). **b** Structural model for the Co**6** adlayer formed on the Au(111) surface. **c** STM image of the Cu**6** adlayer on the Au(111) surface in 0.1 M $HClO_4$ ($10 \times 10\,nm^2$). For Co**6**, the center of the molecules appears as a protrusion (*bright*), while for Cu**6**, the center of the molecule appears as a depression (*dark*). (Reproduced with permission from Ref. [35])

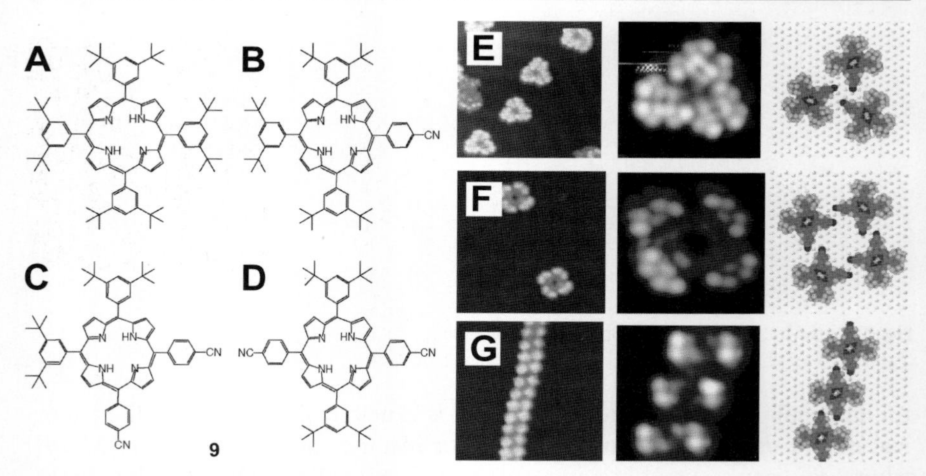


**Fig. 6** STM image at 63 K (UHV) of supramolecular architectures induced by the cyano groups of the porphyrin derivatives **9** on Au(111). **a–d** Chemical structures. **e–g** Respective STM images of the structures in **b–d**. (Reproduced with permission from Ref. [45])

leaf-clover shape of which the internal structure depends on the bias voltage. Occupied molecular states exhibit a decrease in the tunneling current at the molecular center, whereas unoccupied states show an increase in tunneling current at the center. The differences are consistent with the frontier molecular orbitals calculated for gas-phase molecules (see Sect. 8) [44].

By controlling selective intermolecular interactions surface-supported supramolecular structures are created whose size and aggregation pattern are rationally controlled by tuning the noncovalent dipole–dipole interactions between individual cyano-substituted porphyrins (Fig. 6, **9**) [45]. Using low-temperature STM, these molecules adsorbed on a gold surface form monomers, trimers, tetramers or extended 1D structures. Porphyrins with one cyano group self-assemble into well-defined trimers. Introduction of another cyano substituent at 90° leads to the formation of tetramers. If, on the other hand, both cyano groups are at an angle of 180°, "infinite" 1D structures (up to 100-nm long) are formed.

Also by hydrogen bonding, carboxylic acid functionalized porphyrins (Fig. 7, **10**) and phthalocyanines (Fig. 7, **11**) can self-assemble under ambient conditions, forming a network connected by hydrogen bonds [46]. A porphyrin derivative and stearic acid were codeposited on HOPG. It was observed that adsorption of the porphyrin derivative alone on the surface of HOPG did not yield observable molecular images, while in the presence of stearic acid, 2D islands of the porphyrin derivative were absorbed, surrounded by lamellae of stearic acid. In the porphyrin domains, the molecules are arranged with fourfold symmetry. The 2D ordering is a compromise

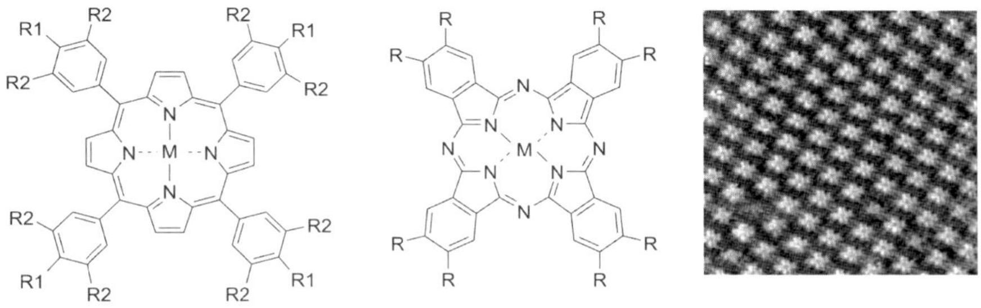

**10** R1 = COOH; R2 = H

**13** R1 = $OC_{14}H_{29}$; R2 = H

**14** R1 = H; R2 = *tert*-butyl

**11** R = COOH

**12** R = $OC_8H_{17}$

**Fig. 7** Substituted phthalocyanines and porphyrins. STM image of **12** on highly oriented pyrolytic graphite (*HOPG*) ($25 \times 25\,nm^2$) under ambient conditions. (Reproduced with permission from Ref. [47])

between the intermolecular interactions (hydrogen bonding) and the minimizations of the surface free energy: in the hypothetical closest packing, no hydrogen bonding is possible; those configurations with optimal hydrogen bonding would lead to large voids; the observed packing is a compromise between both.

On HOPG, under ambient conditions at the air–liquid interface, alkylated phthalocyanines and porphyrins can be observed. STM images reveal that **12** (Fig. 7), a copper phthalocyanine carrying eight octyloxy groups, forms uniform regions of molecular arrays with sizes ranging form tens to hundreds of nanometers [47] (Fig. 7). The phthalocyanine molecules, appearing as bright spots with fourfold symmetry, form ordered 2D arrays. The shaded zigzag lines interconnecting the bright regions correspond to the long alkyl chains. The phthalocyanine cores dominate the image contrast, indicating that tunneling through the phthalocyanine cores is more effective than through the alkyl chains. Through adjustment of the length of the attached alkane chains, one might be able to tailor the intermolecular spacing. In analogy, a tetraphenyl porphyrin carrying four tetradecyloxy groups (Fig. 7, **13**) forms well-ordered monolayers at the liquid–solid or air–solid interface [47]. Large uniform domains are formed. The porphyrins are aligned side by side, separated by the alkane lamellae, which is a direct indication of the effect of 2D crystallization of alkanes. The self-assembly is directly affected by the alkane substituents.

By increasing the size of the substituents, i.e., di-*tert*-butylphenyl groups, the porphyrin core can be separated and electronically decoupled from the substrate (Fig. 7, **14**). These spacer groups effectively act as "legs" which physically separate the porphyrin central ring from the substrate. The rotation angle around each of the four phenyl–porphyrin bonds is the predominant conformational factor that ultimately determines the shape of **14** when adsorbed on a surface. STM images of **14** on a metal surface consist typically of four bright lobes related to tunneling through the four di-*tert*-butylphenyl substituents which are in direct contact with the substrate. The central porphyrin part does not contribute noticeably to the image. Depending on the type of substrate different patterns are observed which are directly related to the conformation of the substituents, ranging from 90° on Cu(001) to 45° on Ag(110) to flat on Cu(111). Two different rotations were observed on Au(110) [48–50]. Nonplanar adsorption of the porphyrin moieties was observed on Au(111) [51].

The self-assembly of water-soluble porphyrins has been investigated under potential control by EC-STM. Tuning the substrate potential provides control of the balance between adsorbate–adsorbate and adsorbate–substrate interactions. Both a too high diffusivity and too strong molecule–substrate interactions might hinder the formation of stable SAMs. For instance, adsorption of **15** (Fig. 8) at high electrode potential (more than 0.5 V vs the saturated calomel electrode, SCE) on Au(111) in 0.1 M $H_2SO_4$ typically results in the formation of a disordered phase [11]. While isolated species were observed at + 0.5 V, the molecules can no longer be observed on the surface at – 0.3 V, which was attributed to increased dynamics. The adsorbate-adsorbate interaction is potential dependent. At + 0.5 V versus the SCE, the adsorbate–substrate interaction is strong and molecules are immobilized on the surface hindering the formation of an organized supramolecular structure. At – 0.3 V versus the SCE, the adsorbate–substrate interaction is too low to immobilize the molecules on the surface. The mobility of the molecules was controlled by tuning the electrode potential. At – 0.05 V versus the SCE, ordered structures were observed. A plausible explanation is that the electron donation from the $\pi$ orbitals of **15** to the substrate may be enhanced at high electrode poten-

**15** **16**

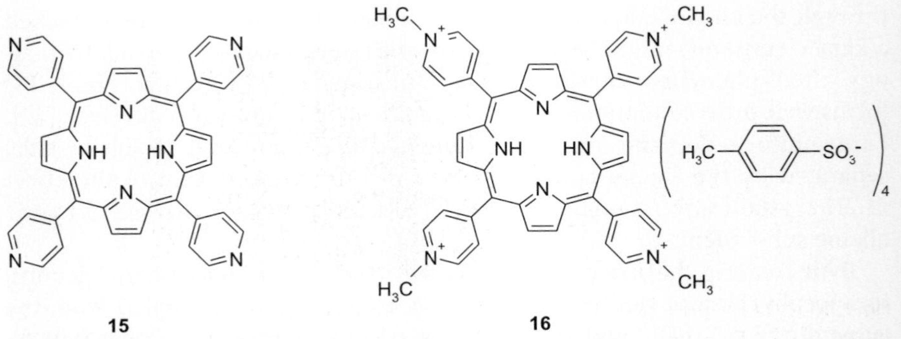

**Fig. 8** Water-soluble porphyrins

tial (positive surface charge density) and reduced at low electrode potential (negative surface charge density).

As is generally the case, the nature of the substrate also plays an important role as it determines the interaction strength with the molecules. Au(111) can be modified by an iodine monolayer, giving rise to the so-called iodine-modified gold (I-Au) [52]. Under potential control, **16** (Fig. 8) in 0.1 M $HClO_4$ gives rise to disordered structures on Au(111) (0.3 V vs the reversible hydrogen electrode) revealing irreversible adsorption and limited surface mobility. On I-Au(111) in 0.1 M $HClO_4$, highly ordered monolayers were formed [10]. Molecular rows consist of flat-lying molecules with two different rotational orientations on the surface. These different angles of rotation of the molecules in a given row can result from the fact that repulsive interactions between positively charged pyridinium units are minimized in the thermodynamically stable adlayer. The dynamics of monolayer formation could be observed.

## 2.3 Triphenylenes

Triphenylenes form another important class of $\pi$-conjugated materials. The self-assembly of alkoxy-substituted triphenylenes (Fig. 9, **17**) has been studied in detail on HOPG with a large variety in alkyl chain length ranging from five (**17**-5) to 20 (**17**-20) carbon atoms [53–55]. These studies have revealed that the molecular ordering depends very much on the length of the alkyl chains. The aromatic triphenylene cores appear bright (high tunneling current) and they are separated in space by the alkyl chains, which appear darker. For **17**-12, every seven molecules form an approximate hexagon with the seventh molecule in the center. For **17**-14, a characteristic dimerlike feature persists over a large area. The carbon chains of the molecule are divided into two oriented groups, which leads to the loss of original molecular symmetry. The triphenylene cores lie antiparallel to each other within the dimer. Upon increasing the length of the alkyl chains by two units (**17**-16), every two

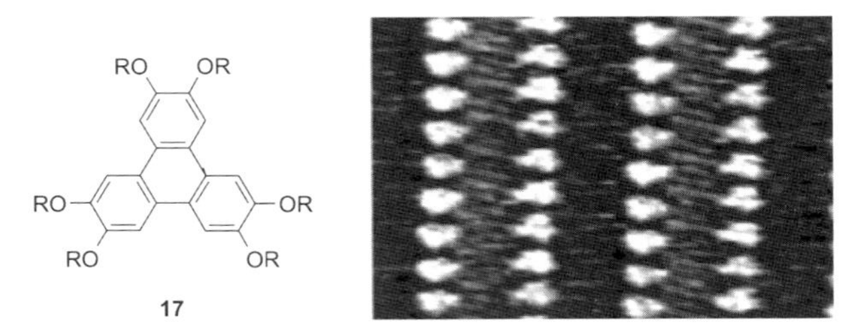


**Fig. 9** Model of triphenylene **17**. *R* is the alkyl chain. Its length is indicated as **17**-*X*. STM image of **17**-11 at the tetradecane–Au(111) interface. The image width corresponds to 17 nm. (Reproduced with permission from Ref. [56])

neighboring molecules orient in an antiparallel way to form dimers. When the number of methylene units reaches 18 (**17**-18) and 20 (**17**-20), the intermolecular spacing within the triphenylene arrays becomes less uniform than that in **17**-16. The aromatic cores are no longer uniformly spaced along the array, but rather display a sporadic spacing distribution which is due to the increased impact of the alkyl chains: the longer the alkyl chains, the more the packing is dominated by them and the steric hindrance of the aromatic cores becomes less important. Not only the length of the alkyl chains plays a role but also the substrate. **17**-11 has been investigated at the tetradecane–Au(111) interface [56]. **17**-11 molecules self-organize in domains which exhibit an ordered structure formed by double rows of aligned bright spots having an approximate diameter of 0.5 nm. The two preferential directions of growth are consequently parallel to the ⟨110⟩ and ⟨112⟩ directions of Au(111) and pairing is observed, which is a substrate-induced process as no **17**-11 dimers on graphite were observed under the same experimental conditions (Fig. 9).

## 2.4 Coronene and Graphite Models

Coronene (Fig. 10, **18**) is an interesting substance: the molecule is planar with a sixfold symmetry axis. It can be considered as the smallest possible flake of a graphite sheet saturated by hydrogen atoms. **18** typically forms a close-packed hexagonal arrangement of flat-lying molecules as observed under UHV conditions [57, 58]. Also, under potential control, a monolayer of **18** could be prepared, even though **18** is not water-soluble, and imaged on Au(111) in 0.1 M $HClO_4$ giving rise to the same arrangement [59, 60]. High-resolution images show that each molecule has the shape of a hexagon consisting of small spots (Fig. 10).

By synthetic means, it is possible to extend the size of coronene-type molecules and to introduce solubilizing alkyl groups. An early example is the study of alkyl-substituted peri-condensed hexabenzocoronenes (HBCs) (Fig. 11, **19b**, **19c**) [61, 62]. The molecular symmetry does not determine the

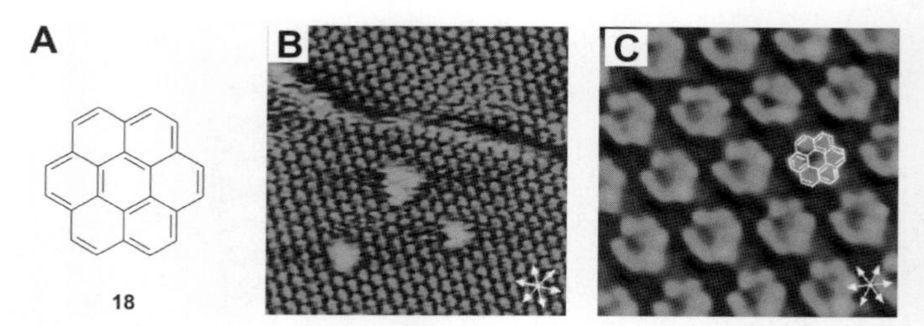


**Fig. 10** **a** Coronene **18**. **b**, **c** STM images of coronene on Au(111) in 0.1 M $HClO_4$: **b** $25 \times 25\ nm^2$; **c** $5 \times 5\ nm^2$. The *arrows* indicate the closed-packed direction of the substrate. (Reproduced with permission from Ref. [60])

packing pattern. **19b** exhibits the same two typical motifs as **19c**, namely a rhombic and a "dimer row" structure. Alkyl chains are often not observed, which is attributed to their mobility. Introducing a bromo function at the aromatic cores causes a completely different packing pattern. The bromine groups of different atoms seem to interact strongly and thereby induce a hexagonal superstructure of six trimers and a single molecule [62]. The nature of the alkyl chains also has a pronounced influence on the self-assembly of the molecules: in the case of optically active phenylene-alkyl side chains a so-called staircase architecture was formed (not all disks are at the same height with respect to the substrate), which has its origin in the interplay between intramolecular as well as intermolecular and interfacial interactions where a key role is played by the steric hindrance suffered by the side chains [63]. Tetra-*n*-alkyl and octa-*n*-alkyl derivatives of a flat **19a** could also be solubilized and the formation of monolayers from solution is possible even for these large disks. These studies provide insight into the electronic characterization of molecular graphite models. Three similarly directed strips with a distance of 0.4 nm were observed over each molecule, a value which is in agreement with the distance between benzene units of aromatic disks [64]. In line with these observations, a bishexa-peri-hexabenzocoronenyl (Fig. 11, **20**) also self-assembles at the liquid–solid interface. Submolecularly resolved STM in situ revealed a contrast, which reflects the structure of the aromatic parts of the molecule, and which shows that the aromatic molecules are oriented like a graphene layer on graphite [65].

**19a** is a polycyclic aromatic hydrocarbon (PAH) and (**21**) is an electron-poor PAH (Fig. 11) [66]. Films prepared by codeposition exhibit an oblique arrangement reflecting the formation of epitaxial composite layers of the electron donors and acceptors. Interestingly, on top of a layer of donor molecules,

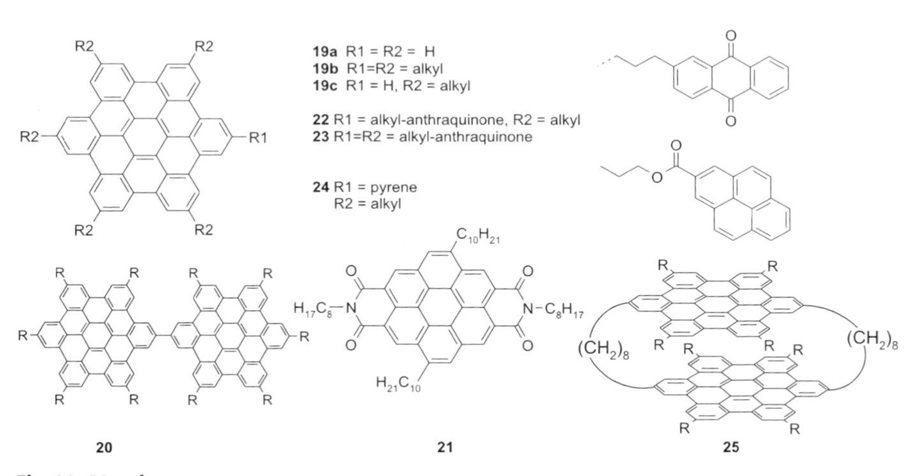


**Fig. 11** Hexabenzocoronenes

within the same layer between two physisorbed donor molecules, acceptor molecules are coadsorbed in a well-defined arrangement (Fig. 12). It was possible to image the first or the second layer selectively by changing the tunneling current. These studies were extended to covalently linked electron donor–acceptor systems. Soluble, alkylated hexa-peri-hexabenzocoronenes bearing tethered anthraquinones are shown by STM to self-assemble at the solution–graphite interface into either defect-rich polycrystalline monolayers or extended 2D crystalline domains, depending on the number of tethered anthraquinones [67]. For the dyads (Fig. 11, **22**), within the monolayers, the anthraquinones might intermolecularly or intramolecularly stack on top of the HBC cores or be partially solvated in the 3D supernatant solution as well as assemble on the HOPG substrate, defining the size of apparent voids. This behavior contrasts with that of the highly ordered monolayers formed from the hexaanthraquinone-substituted HBC (Fig. 11, **23**), where four anthraquinone units are adsorbed on graphite, with the other two most likely directed to the supernatant solution. On the other hand, a hexa-peri-hexabenzocoronene pyrene dyad (Fig. 11, **24**) self-assembled at the solution–solid interface shows uniform nanoscale segregation of the large (HBC) from the small (pyrene) $\pi$ systems, leading to a well-defined 2D crystalline monolayer. A higher affinity of pyrene for graphite compared with anthraquinone might explain the uniform nanoscale segregation. The difference in brightness—the brighter, the higher the tunneling current—between the HBC core and pyrene units can be related to the lower ionization potential of HBC compared with that of pyrene. The energy difference between the frontier orbitals of the adsorbed molecules and the Fermi level of the substrate (HOPG) is an important factor because it is a resonant tunneling process [68]. The ordering of hexa-peri-hexabenzocoronene cyclophanes (Fig. 11, **25**) is similar to the that of the

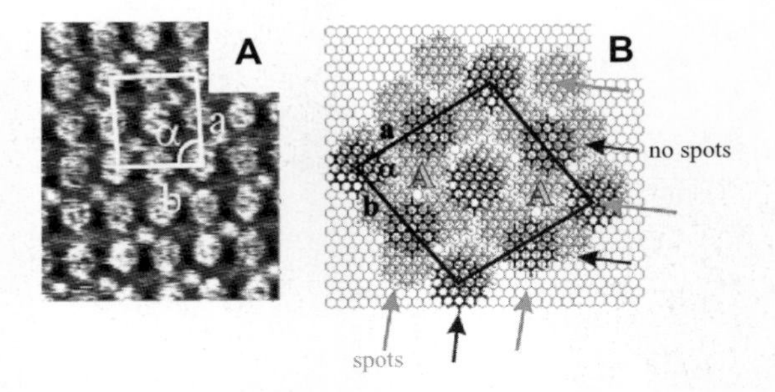


**Fig. 12** **a** STM image at the 1,2,4-trichlorobenzene–graphite interface of a mixture of **19a** and **21** (image size is $11 \times 13\ nm^2$). It is an image of a second epitaxial layer with a donor–acceptor stoichiometry of 2 : 1 on top of a first epitaxial layer of **19a**. The *large features* are **19a**, the *smaller and brighter features* are **21**. **21** packs only every second row probably because of its preferential adsorption on the electron donor disk of **19a**, which is exposed in the underlying first $C_{42}H_{18}$ layer (see model **b**). (Adapted with permission from Ref. [66])

corresponding **19** [69]. This family of hexa-peri-hexabenzocoronene derivatives is very well suited to provide insight into the electronic properties of the compounds which are revealed by scanning tunneling spectroscopy (STS).

## 2.5 Decacyclene

In many cases, the metal surface is considered as a static checkerboard that provides bonds and specific adsorption sites to the molecules. When the adsorbed molecules become large and complex, the complexity of the interaction between the substrate and the molecules may increase. Anchoring of large molecules and the subsequent self-assembly of molecular nanostructures on a metal surface can be associated with a local disruption of the surface layer underneath molecules [70]. With the adsorption of large molecules, such as hexa-*tert*-butyl decacyclene (Fig. 13, **26**) on a Cu(110) surface, metal atoms can be dug out of the surface, resulting in a "trench base" for anchoring of the molecules. In the case of the individual molecules, all six *tert*-butyl lobes appear to have nearly the same height. For molecules within a double row, the three lobes at the rim of the row are imaged much more brightly than the three lobes pointing toward the interior of the double row. The molecules in the double rows appear to be tilted. After the adsorbed molecules had been moved aside, images showed that 14 Cu atoms were expelled from the surface in two adjacent rows, creating a chiral hole in the surface [71]. The three more dimly imaged *tert*-butyl lobes of each molecule are located on top of the missing Cu atoms, increasing molecule–substrate in-

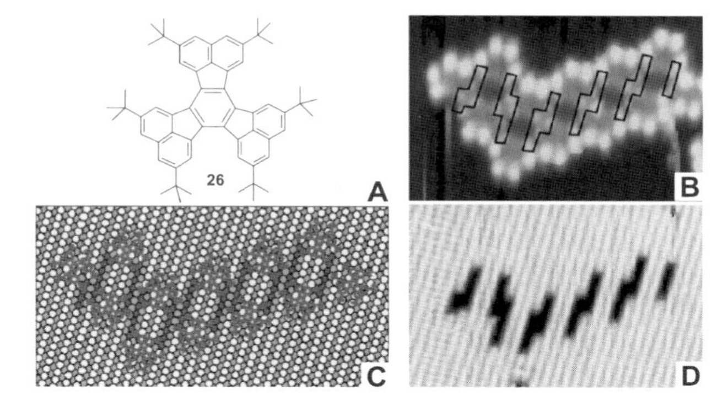


**Fig. 13** **a** Molecular structure of **26**. **b**, **d** Constant-current STM images ($10.5 \times 6.9\ nm^2$) on Cu(110) under UHV conditions. **b** **26** double-row structure. The trenches in the underlying surface are sketched. **c** Ball model of the double-row structure. The substrate atoms are shaded darker the deeper the layers lie. The molecules are shown in *gray*. **d** The trenches in the surface layers are disclosed after manipulating the molecules aside. (Reproduced with permission from Ref. [70])

teractions. In the case of *tert*-butyl-free decacyclene molecules, no ordered structures are observed at low coverage and no restructuring of the Cu(110) surface is induced; this is attributed to the strong interaction of the aromatic $\pi$ system with the substrate [72].

## 2.6 Lander-Type Molecules

A surface reconstruction was also observed for another type of molecule, the so-called Lander molecules: these are large aromatic molecules containing a large $\pi$ system substituted with di-*tert*-butylphenyl groups (Fig. 14, **27**) [73, 74]. These molecules act as templates accommodating metal atoms at the step edges of the copper substrate, forming nanostructures (two Cu atoms wide and eight Cu atoms long) that are adapted to the dimensions of a single molecule. The process of step restructuring is thermally activated and does not occur at low temperatures (150 K) [75]. **27** can be "forced" to align in 1D rows at low surface coverage by the formation of a striped periodic supergrating created by the controlled oxidation of a Cu(110) surface. The molecules adsorb only on the bare Cu regions in between the Cu – O areas, as the attractive van der Waals interaction between the central $\pi$ board and the surface is decreased on the oxidized areas. The conformations of the molecules are identical to those found on untreated Cu(110): as a result, the long axis of the molecules is perpendicular to the direction of the molecular chain (Fig. 14). In contrast, a longer analogue **28**, which does not fit in between the nanotem-

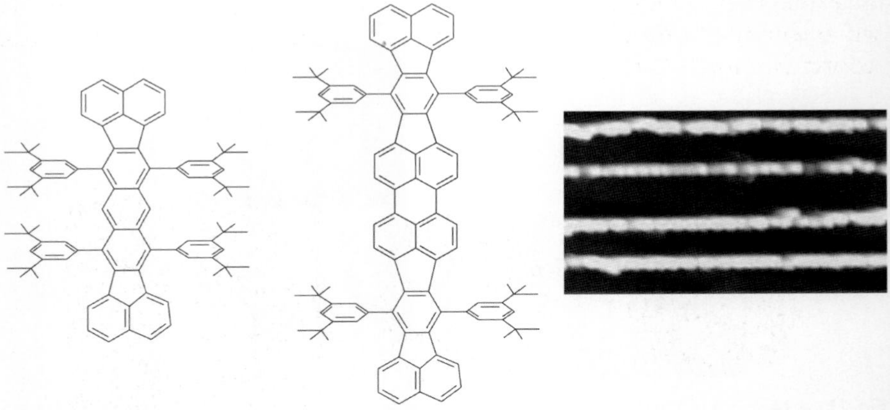

27

28

**Fig. 14** Lander molecules. STM image recorded under UHV conditions. Molecules are trapped in rows on the nanostructured Cu – O surface of Cu(110). The molecules are stacked in rows with their long axis perpendicular to the growth direction of the rows. (Reproduced with permission from Ref. [76])

plate stripes, is preferentially aligned with its long axis along the stripes. The supergrating not only provides specific adsorption sites for the molecules but it also steers the orientation of the adsorbed molecules [76].

## 2.7 Fullerenes

Fullerene overlayers at solid surfaces have attracted much interest in recent years because of their possible future applications. Detailed STM images of $C_{60}$ (Fig. 15, **29**) overlayers have been obtained on several metal surfaces under UHV conditions [77, 78]. On threefold symmetry surfaces of noble metals and aluminum, the $C_{60}$ packing is found to present a hexagonal or a quasihexagonal lattice. The molecule packing is often governed by the $C_{60} - C_{60}$ van der Waals interaction though strong $C_{60}$–substrate interactions can lead to the formation of stressed $C_{60}$ monolayers or even lack of monolayers [79, 80]. At the liquid–solid interface, hexagonal monolayer formation is observed on Au(111), though the choice of solvent has an effect on the self-assembly. At saturated conditions, the poorer solvent tetradecane leads to a rough overlayer [81], while in 1,2,4-trichlorobenzene, which is a good solvent for $C_{60}$, crystalline monolayers were formed [82]. The latter is due to a more balanced competition between physisorption and dissolution. $C_{70}$ (**30**), which presents a reduced symmetry and an ellipsoidal shape, forms a close-packed hexagonal and quasihexagonal lattice on the threefold symmetry surface of noble metals, both under UHV conditions [83] and at the liquid–solid interface [84]. The coexistence of two types of domains is related to the molecular shape: the molecules lie on the surface either with their long axes or with their short axes parallel to the substrate. Recently, the self-assembly of exotic $C_{60} - C_{60}$ and $C_{60} - C_{70}$ dimers and $C_{60}$ trimers was reported on Au(111) [85, 86].

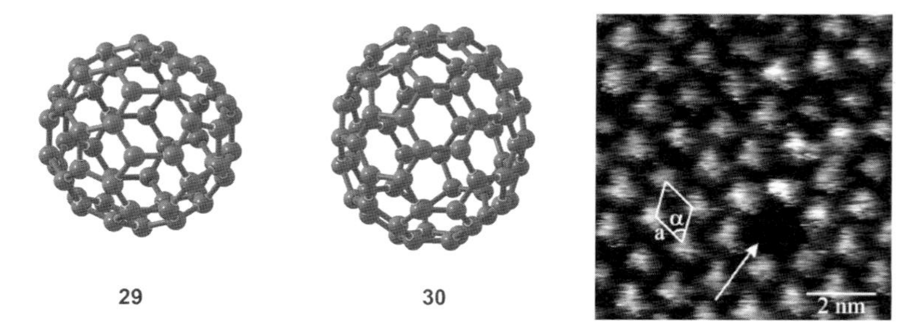


**Fig. 15** Models of $C_{60}$ (**29**) and $C_{70}$ (**30**). STM height image of hexagonally packed $C_{60}$ molecules at the 1,2,4-trichlorobenzene–Au(111) interface. The *arrow* indicates a missing molecule within the 2D crystal. (Reproduced with permission from Ref. [82])

## 2.8
## Squaraine Dyes

Symmetric and asymmetric squaraine dyes (Fig. 16) which easily undergo intermolecular charge transfer (**31**) form 2D monolayers when adsorbed from 1-phenyloctane or liquid-crystalline solvents on HOPG [87, 88]. The central $C_4O_2$ unit acts as the electron acceptor, while the aniline moieties act as electron donors, giving rise to squaraine molecules with a donor–acceptor–donor internal structure. This specific structure leads to strong attractive intermolecular interactions. Squaraine molecules appear as elongated bright structures. For short alkyl chains, various polytypes of herringbone packings were observed arising from the strong donor–acceptor interactions and Coulomb forces between oppositely polarized areas of adjacent squaraine molecules. The longer the alkyl chains, the more important the effect of the van der Waals interactions between the alkyl chains in the overall stabilization of the monolayer. In addition to the formation of polymorphous structures multilayers were often formed and imaged. The donor–acceptor interactions between adjacent molecules within the same layer and between adjacent molecules in adjacent molecular layers provide the electronic coupling needed for facile charge transport.

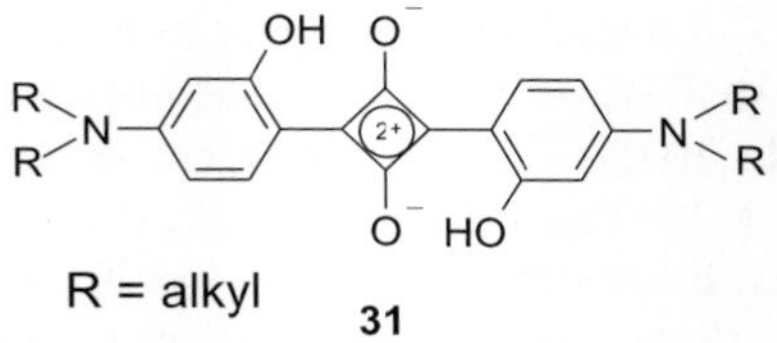

**Fig. 16** Squaraine dye

## 2.9
## Other Dyes

Those dyes mentioned in the previous sections are only a selection of the dye systems which have been investigated by means of STM. Many more systems have been successfully investigated, including alkylated quinacridone [89], coumarine [90] and tetrathiofulvalene [91–93] derivatives on HOPG, rhodamine B on I-Au(111) [94], eosin molecules [95] and xanthene dyes [96] on Au(111), tetrachlorothioindigo and thioindigo on HOPG and $MoS_2$ [97], and carbocyanine dyes on silver bromide surfaces [98] to name a few. The latter example relates to the use of dye molecules as spectral sensitizers in the photographic industry.

# 3
# Conjugated Oligomers and Polymers

Organic semiconducting materials possess unique electronic properties which open up a wide range of applications both in optoelectronics and photonics. The spatial arrangement of the molecules determines the properties of the devices; therefore, it is an important goal to drive the ordering of conjugated oligomers and polymers on flat solid surfaces to highly ordered and thermodynamically stable supramolecular structures.

## 3.1
## Conjugated Oligomers

In view of their processability, oligomers are often studied as model systems for conjugated polymers.

### 3.1.1
### Oligo-*p*-phenyleneethynelene

End-functionalized phenyleneethynylene oligomer (Fig. 17, **32**) served as a model for $\alpha,\omega$-thiol functionalized poly(*p*-phenyleneethynylene)s [99]. A regular arrangement was observed, with trimers lying with their conjugated backbone flat on the HOPG plane. The spacing between the parallel backbones was smaller than that modeled for completely extended hexyl groups, suggesting that they are disordered. In addition, a shape-persistent macrocycle (Fig. 18, **33**) based on the phenyl–ethynyl backbone containing various extraannular alkyl side chains has been adsorbed from a 1,2,4-trichlorobenzene solution at the surface of highly oriented pyrolitic graphite,

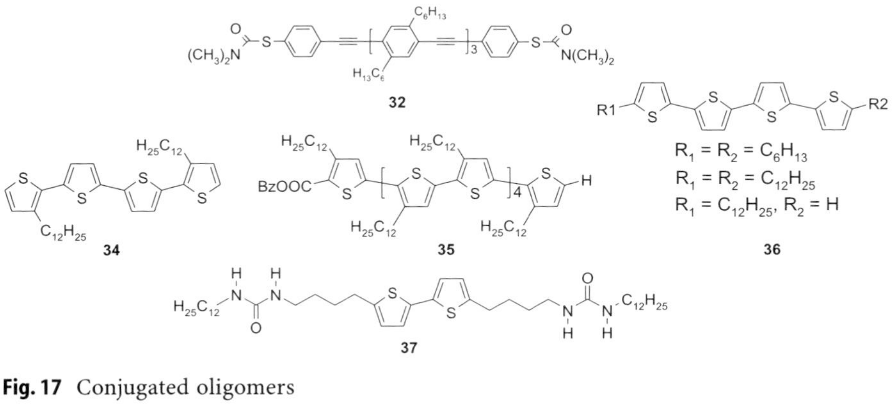

**Fig. 17** Conjugated oligomers

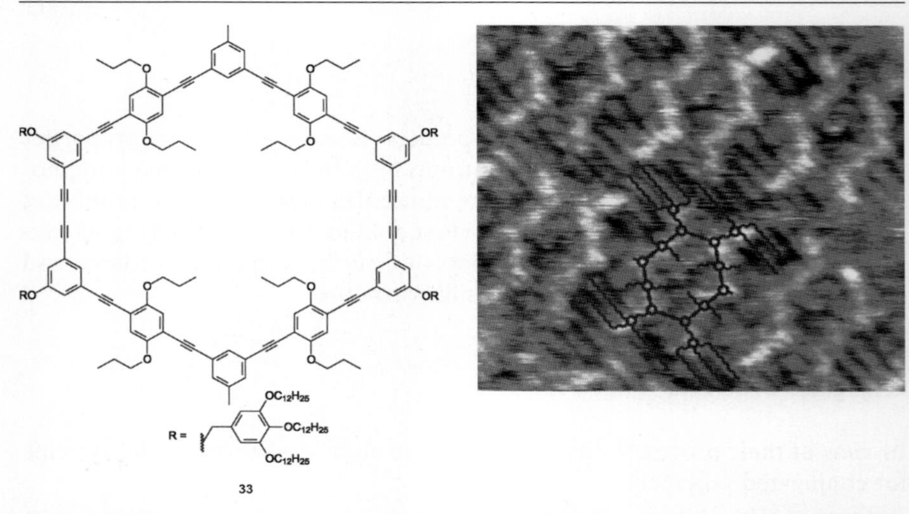


**Fig. 18** Chemical structure of macrocycle **33** and STM image at the 1,2,4-trichlorobenzene–graphite interface. (Reproduced with permission from Ref. [100])

leading to an ordered 2D pattern. Upon adsorption, the interaction between the substrate and the adsorbates caused a flattening of the shape-persistent macrocycle. These macrocycles could be used for templating purposes and the formation of 3D supramolecular structures (Fig. 18) [100].

### 3.1.2 $\alpha$-Oligothiophenes

In particular, oligothiophenes have gained considerable interest for their electronic properties and several studies have appeared on the ordering of oligomers, studied by STM. A study of the self-assembly at the liquid–solid interface of a homologue series of $\beta$-dodecylated oligothiophenes containing 4, 8, 12 or 16 thiophenes, the shortest one of which (**34**) is shown in Fig. 17, revealed that the 2D organization depends on the length of the oligothiophene backbone where the molecules assemble in a lamella-type arrangement [101]. The length of overlap (side-by-side) between the aromatic backbones of adjacent molecules decreases linearly with the oligothiophene length (Fig. 19).

A change of the alkyl chain length of a $\beta$-alkylated quaterthiophene (dodecyl, hexyl, propyl) also affects the ordering, the shortest one showing a "herringbone-like" structure (**34**). In all three cases, the 2D arrangement of the adsorbates coincides well with the molecular order in one layer of the 3D crystal [102]. Overall only limited interaction between adjacent $\pi$ systems is possible. In the case of $\beta$-alkylated oligothiophenes where each thiophene group carries a dodecyl group (Fig. 17, **35**), the oligothiophene molecules are separated from each other by the interdigitated alkyl chains [103]. The influ-

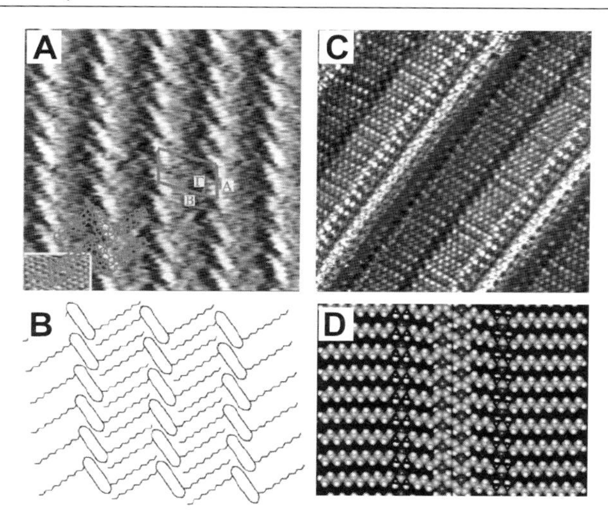


**Fig. 19** STM images and models of **a**, **b** **34** ($10 \times 10\,nm^2$) at the 1,2,4-trichlorobenzene–graphite interface and **c**, **d** **37** ($10.4 \times 10.4\,nm^2$) at the 1-octanol–graphite interface. Only in the case of **37** is there significant $\pi$–$\pi$ stacking. (Reproduced with permission from Ref. [102])

ence of the increasing polarity of the substituents on the monolayer structure of oligothiophenes physisorbed on graphite has been investigated [104]. Other studies on physisorbed oligothiophenes and the influence of length and substituents on the 2D organization have also been reported [105–107].

$\alpha,\alpha'$-Dialkylated quaterthiophenes Fig. 17, **36**) form 2D crystals on $MoS_2$ in which one of the alkyl side chains of a molecule is sandwiched between two thiophene units within a row, so hindering $\pi$–$\pi$ stacking [106]. Lack of $\pi$–$\pi$ interaction was also observed for other $\alpha$-functionalized oligothiophenes [104, 105] and steroid-bridged thiophenes [107]. STM experiments of monolayers of endcapped quaterthiophene and quinquethiophene on Ag(111) show the formation of a highly ordered 2D crystalline monolayer with flat-lying molecules with limited $\pi$–$\pi$ interactions [108, 109]. Highly efficient $\pi$–$\pi$ stacking was achieved by substituting the oligothiophenes (mono, bi, ter) in the $\alpha$-position with long alkyl chains carrying at both sides a urea group (Fig. 17, **37**). The ordering of the molecules at the liquid–HOPG interface is determined by the hydrogen bonding between the urea groups along the stacking axis [110]. As a result, the distance between two oligothiophene units is 0.46 nm. The oligothiophene rings do not lie flat on the surface but are tilted, leading to $\pi$–$\pi$ interactions (Fig. 19) as revealed by STS [111].

In addition to linear oligomers, cyclic $\alpha$-conjugated macrocyclic oligothiophenes (Fig. 20, **38**) have also been investigated. Spontaneous ordering was observed. The energy minimum does not correspond to a fully planar molecule but rather to a spiderlike conformation in which the alkyl side

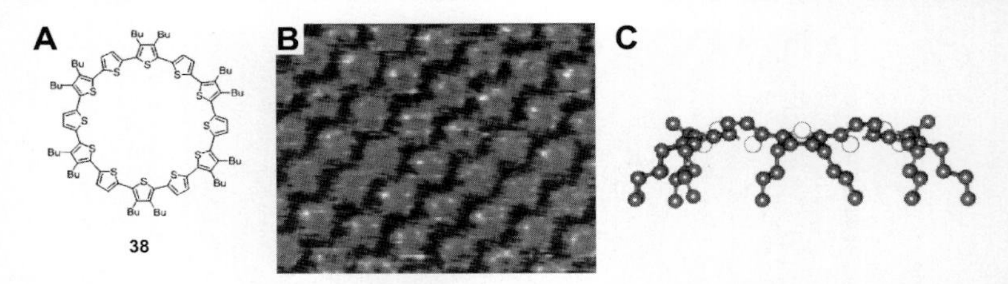


**Fig. 20** **a** Structure of a cyclic $\alpha$-conjugated oligothiophene macrocycle **38** at the liquid–HOPG interface. **b** STM image ($15 \times 12\,nm^2$) of the ordering and 2D crystal of **38** adsorbed on graphite. **c** Calculated minimum-energy conformation of an individual macrocycle. (Reproduced with permission from Ref. [112])

chains are bent downward as a consequence of the uniform distortion of each second thiophene ring [112, 113].

### 3.1.3
### Oligo-*p*-phenylenevinylenes

*p*-Phenylenevinylenes form another important type of conjugated oligomers. Three main classes of chiral *p*-phenylenevinylenes have been investigated [114–116]. They all carry (*S*)-2-methyl butoxy groups along the backbone. In the first class, oligo-*p*-phenylenevinylenes which are functionalized with three dodecyl chains at both termini (Fig. 21, **39**) can be organized in highly organized 2D crystals on graphite by spontaneous self-assembly (Fig. 22). A profound effect of the length of the $\pi$-conjugated backbone on the adlayer structure was uncovered. While for the longer oligomer systems the $\pi$ system has a strong impact on the organization of the molecules, shortening the $\pi$-conjugated moiety leads to monolayer structures controlled by the alkyl chains. In the second and the third classes, the molecules are functionalized by hydrogen-bonding groups such as ureido-*s*-triazine (class 2) (Fig. 21, **40**) or 2,5-diaminotriazine groups (class 3) (Fig. 21, **41**). The ureido-*s*-triazine derivatized oligo-*p*-phenylenevinylenes show dimerization via self-complementary hydrogen bonding (Fig. 22). The 2,5-diaminotriazine derivatized oligo-*p*-phenylenevinylenes show cyclic hexamer formation induced by the hydrogen bonding between the 2,5-diaminotriazine units (Fig. 22). By functionalizing the oligomers the ordering can be controlled, which might affect their properties. In none of the systems is intermolecular $\pi$–$\pi$ stacking achieved. Moreover, as these compounds are chiral and enantiopure, their molecular chirality is also expressed at the liquid–solid interface. Typically, those molecules stacked in a row are in every domain rotated in the same direction with respect to the stack normal (not the mirror image) or the cyclic hexamers formed by the diaminotriazine derivatives always show the same virtual rotation direction.

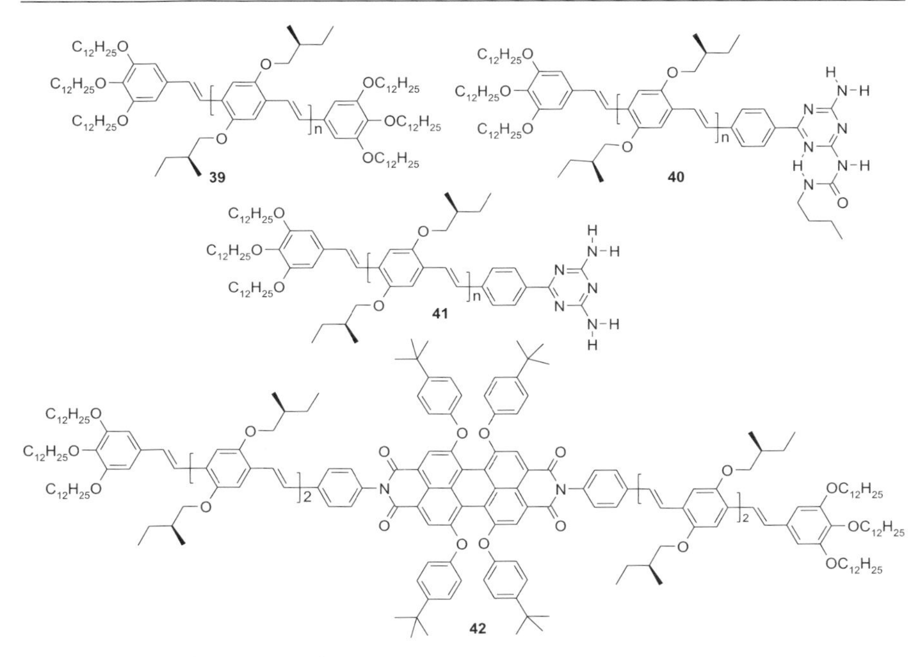


**Fig. 21** Oligo-*p*-phenylenevinylenes

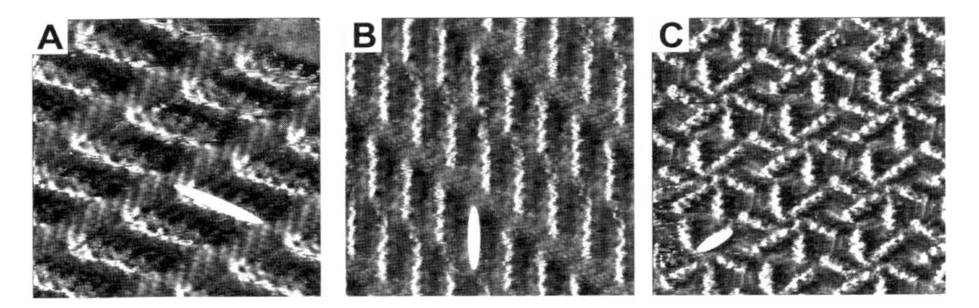


**Fig. 22** STM images of **a** **39** ($n = 4$) ($10.7 \times 10.7\ nm^2$) (1,2,4-trichlorobenzene–graphite interface), **b** **40** ($n = 2$) ($12.1 \times 12.1\ nm^2$) (1,2,4-trichlorobenzene–graphite interface) and **c** **41** ($n = 2$) ($18.4 \times 18.4\ nm^2$) at the 1-phenyloctane–graphite interface. Hydrogen bonding has a strong effect on the supramolecular architecture. (Reproduced with permission from Ref. [114])

### 3.1.4
### Composite Conjugated Oligomers

Of interest are also compounds which combine different properties. **42** (Fig. 21) contains two oligo-*p*-phenylenevinylene units symmetrically linked to a central perylenediimide core [117]. The oligo-*p*-phenylenevinylene units and perylenediimides are typical electron donor and electron acceptor moieties, respectively. These molecules are stacked in rows as observed for **39**. Noticeably, though both units are aromatic and should be characterized by a higher tunneling current than the alkyl chains, by bias-dependent imaging, one can differentiate between the *p*-phenylenevinylene units and the perylenediimide core (see Sect. 8).

## 3.2
## Conjugated Polymers

In addition to conjugated oligomers, conjugated polymers can also be studied under ambient conditions or at the liquid–solid interface. The self-assembly of $\pi$-conjugated macromolecules on insulating solid substrates offers a strategy for the construction of well-defined and stable nanometer-sized structures with chemical functionalities and physical properties that are of potential use as active components in electronic devices [118, 119].

### 3.2.1
### Poly-*p*-phenyleneethynylene

Exposure of a HOPG substrate to an almost saturated solution of poly(*p*-phenyleneethynylene) (Fig. 23, **43**) results in the spontaneous self-assembly of the conjugated polymer in a nematic-like molecularly ordered monolayer. The conjugated skeletons are aligned along preferred directions, according to the threefold symmetry of the HOPG lattice. The achievement of molecular imaging was attributed to the stiffness and low polydispersity of the molecules. The reduced spacing between neighboring parallel backbones indicates a disordered arrangement of the hexyl chains. The difference in the 2D structure of the polymer and the oligomer [99] is attributed to polydispersity which prevents the assembly of the macromolecules in perfect 2D crystals [120].

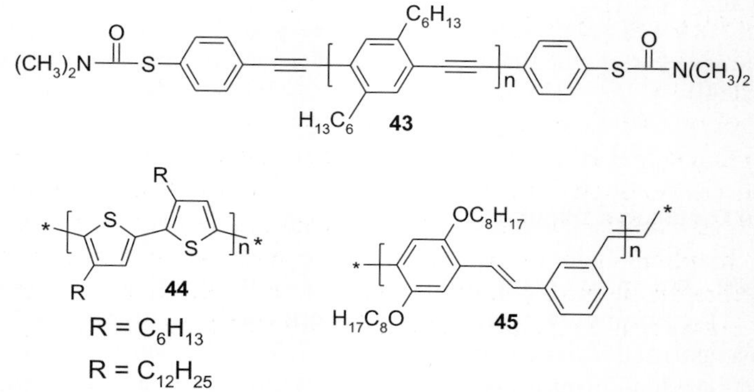

**Fig. 23** Conjugated polymers

### 3.2.2
### Polythiophenes

The self-assembly of head-to-tail coupled poly(3-alkylthiophene)s (Fig. 24, **44**) has also been visualized [121–123]. The polymer strands are oriented according to the three crystallographic axes of the HOPG substrate. The linear arrangement of the conjugated chains over large distances invokes an all-*trans* conformation of the thiophene repeating units. In addition, chain folding was observed in which *cis* conformations of the thiophene units are prerequisite for the fold (Fig. 24). Owing to the epitaxial effect of the graphite surface, the alkyl chains are fully interdigitated, which is in contrast to the case of polycrystalline bulk material. The observation of a second layer of adsorbed macromolecules shows that conductivity increases and more so for conjugated molecules stacked parallel, indicating a better electronic coupling than for the crossed ones [121]. Also casting from chloroform solutions on HOPG leads to the formation of polymer polycrystals and gives insight into the mesoscopic ordering. The average size of the 2D polycrystals is 20 nm in both parallel (along the $\pi$-conjugated backbone) and transverse directions [122, 123].

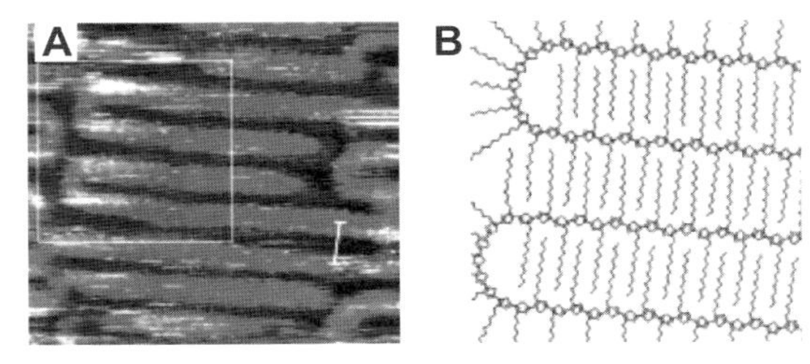


**Fig. 24** **a** STM image of **44** at the 1,2,4-trichlorobenzene–HOPG interface ($20 \times 20\ \mathrm{nm}^2$). Folding can clearly be observed. **b** Model. (Reproduced with permission from Ref. [121])

### 3.2.3
### *p*-Phenylenevinylene Copolymers

Monolayer films of poly[(*m*-phenylenevinylene)-*co*-(2,5-dioctoxy-*p*-phenylenevinylene) (Fig. 23, **45**), a regioregular polymer whose structure is related to that of the commoner poly(*p*-phenylenevinylene), have been investigated on HOPG upon deposition from a toluene solution. Over a month, it was observed that the degree of order increases. High-resolution images reveal that the phenyl groups are planar to the substrate and that the polymer strands adopt an all-*trans* conformation. Diverse molecular foldings all result from the presence of *cis* conformations in different sequences [124].

### 3.2.4 Polydiacetylenes

Other types of conjugated polymers, the polydiacetylenes, have also been extensively investigated. In contrast to the polymers discussed before, the polydiacetylenes are synthesized after treatment of the preordered diacetylene monomers which will be discussed in Sect. 7.3.

## 4 STM Imaging of Mixed Monolayers

One of the goals of investigating molecules at surfaces is to gain insight into and to control 2D monolayer formation. Applying mixtures of molecules on surfaces can lead to unique patterns, with potential novel properties. In addition, it can give insight into the balance between molecule–substrate and intermolecular interactions. Of particular interest is the formation of regular 2D co-crystals.

Host–guest interactions can also be used to stimulate and control the adsorption of large molecules such as phthalocyanines [125], coronene [125] and fullerenes [126–128] under UHV [126, 127] and ambient conditions [125] and under potential control [128].

### 4.1 Mixtures Including PTCDA and Phthalocyanine

Phase separation of the dyes in mixtures is a common behavior, which is not that surprising given that the formation of alternating periodic structures often requires directional and strong heteromolecular interactions. In the early 1990s, STM measurements were conducted on the dyes PTCDA and **5** in the presence of liquid crystals on HOPG in order to immobilize the dye molecules under ambient conditions [129]. The outcome was phase separation: only one of the compounds could be observed or a domain of a given compound was surrounded by the other compound. PTCDA formed the typical herringbone structure, while **5** exhibits a hexagonal symmetry.

### 4.2 Mixture Including PTCDI, Melamine and $C_{60}$

Hydrogen bonding under UHV conditions has been exploited to design open honeycomb networks formed when PTCDI is coadsorbed with melamine (1,3,5-triazine-2,4,6-triamine) on a silver-terminated silicon surface (Fig. 25) [126]. Melamine, which has threefold symmetry, forms the vertices of the network, while the straight edges correspond to PTCDI. The

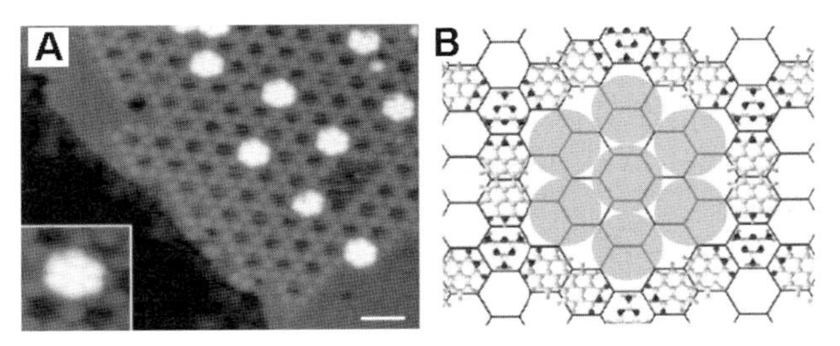


**Fig. 25** **a** Image of $C_{60}$ heptamers trapped in "nanovessels" formed by the **2**-melamine network. The *scale bar* is 5 nm. **b** Model. (Reproduced with permission from Ref. [126])

choice of these two compounds was motivated by the fact that in this case heteromolecular hydrogen bonding is stronger than homomolecular hydrogen bonding: three hydrogen bonds per PTCDI–melamine pair versus two for a PTCDI or a melamine pair. The substrate plays an important role as it allows molecules to diffuse freely on the surface. Stepwise introduction of the two components is vital for the formation of the network. Only if PTCDI is adsorbed first on the substrate, forming close-packed islands and short chains, followed by deposition of melamine and annealing, is an open network formed. The pores of the network are large enough to host guest molecules such as $C_{60}$: heptameric $C_{60}$ clusters with a compact hexagonal arrangement of the individual molecules form within the pores. Increase of the $C_{60}$ coverage leads to the formation of a second-layer $C_{60}$, on top of the PTCDI–melamine network: a new fullerene surface phase is formed that is controlled and templated by the underlying hydrogen-bonded network.

## 4.3 Mixtures Including a PTCDI or a Merocyanine Derivative

Perylene bisimides (**46**) and merocyanines (**47**) with hydrogen-bonding receptor sites are little soluble pigment dyestuffs which are the subject of intensive research owing to their interesting optoelectronic and supramolecular organization properties. Well-defined heterocomplex formation based upon multitopic hydrogen bonding between their imide functions and the 1,3-diaminotriazine moiety of an alkylated compound (**48**) at the liquid–solid interface has been realized and correlated to the composition and structure of the observed heterocomplexes and the extent and quality of their long-range ordering with the orientation of the hydrogen-bonding sites in the molecules [130]. The formation of termolecular complexes by mixtures of **46** and **48**, and **47** and **48**, respectively, and their composition and geometry which is in line with predictions are clear indications that hydrogen bonding is indeed involved. Though in both cases the hydrogen-bonding interactions are similar (identical hydrogen-bonding sets are involved) and termolecular complexes are formed, there are important differences at the monolayer level

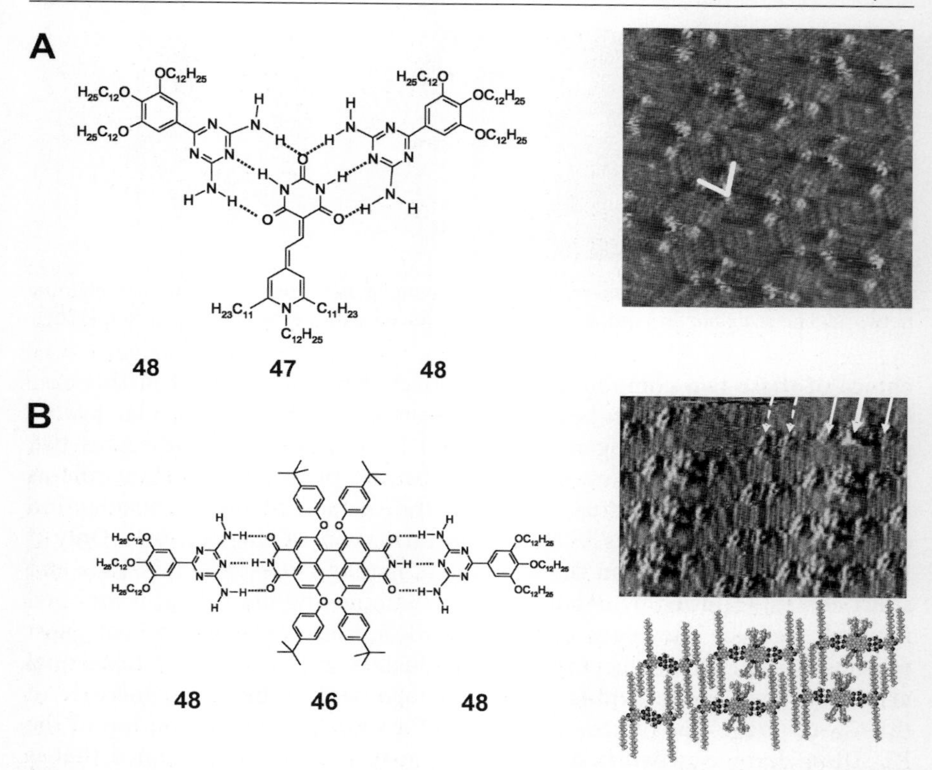


**Fig. 26** Chemical structures of the hydrogen-bonded trimers and the respective STM images at the 1-phenyloctane–graphite interface. **a 48–47–48** ($11.5 \times 11.5\,\mathrm{nm}^2$), **b 48–46–48** ($12.3 \times 12.3\,\mathrm{nm}^2$). (Reprinted with permission from Ref. [130])

(Fig. 26). A **48–47** mixture leads to the formation of true 2D crystals: heterocomplexes are exclusively formed and they organize in large domains. In contrast, **48–46–48** complexes are formed only locally, they assemble in rows, they do not cover complete domains, and they coexist with noncomplexed dimers of **48** within the same domain. These differences can at least in part be attributed to the symmetry of the hydrogen-bonding interactions involved.

## 4.4 Mixtures of Phthalocyanines and Porphyrins

In addition to hydrogen bonding, other types of noncovalent interactions, such as perfluorophenyl–phenyl interactions, are very well suited for self-assembly purposes. The perfluorinated cobalt phthalocyanine (F16Co5) molecules do not form an ordered structure when deposited on Au(111) under UHV [131, 132]. This disorder and the lack of submolecular resolution are in contrast to the behavior of the protonated complex, which is

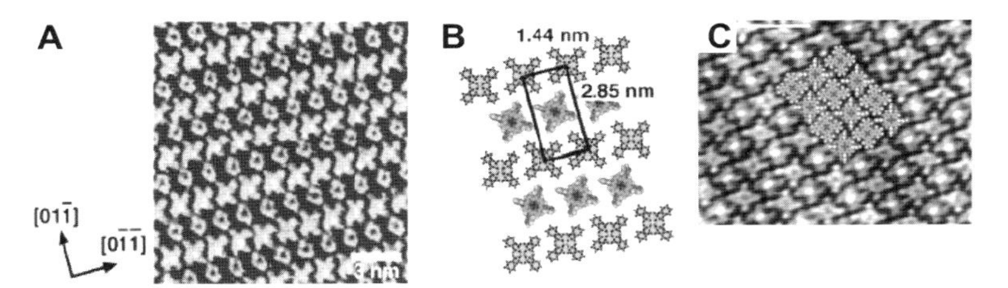


**Fig. 27** STM image **a** of Co**5** and Cu**6** mixed adlayer on a Au(100)–(hex) surface in 0.1 M $HClO_4$ ($15 \times 15\ nm^2$). **b** Molecular model. **c** STM image of a 1 : 1 mixed monolayer of Ni**6** and F16Co**5** under UHV conditions on Au(111). The F16Co**5** molecules can easily be recognized by the bright Co atoms. The *scale bar* is 5 nm. (Adapted with permission from Refs. [131, 133])

attributed to differences in van der Waals attraction and electron affinities. Ni tetraphenyl porphyrin (Ni**6**) forms a simple 2D crystal structure when adsorbed on gold. A 2:1 mixture of Ni**6** and F16Co**5** forms in addition to disordered regions and ordered regions of Ni**6**, well-ordered regions of an entirely new structure with 1 : 1 ratio (Fig. 27). The phthalocyanine molecules can easily be identified by the Co ions, which appear bright.

Using the EC-STM approach, a mixture of Co**5** and Cu**6** has been investigated on Au(100) and Au(111) in 0.1 M $HClO_4$ [133]. In contrast to the case for Au(111), where a disordered phase and a highly ordered phase of Cu**6** is observed, on Au(100) stripes composed of alternate bright and dark lines are observed, revealing highly ordered molecular rows (Fig. 27). The binary molecular arrays are well-organized 2D. The Co**5** molecules have a propeller-shaped image with a central brightest spot and four additional spots at the corners. The Cu**6** molecules appear as a ring, the center of which is a dark spot. The difference in brightness at the centers of Cu**6** and Co**5** molecules is explained by the difference in the mode of occupation of *d* orbitals. The central cobalt ion in each Co**5** molecule appears as a bright spot, whereas the copper ion in each Cu**6** molecule appears as a dark spot.

## 4.5 Mixtures Including Subphthalocyanine and $C_{60}$ or Calixarenes and $C_{60}$

Depending on the relative surface coverage of subphthalocyanine **7** and $C_{60}$ on Ag(111), periodic intermixed monolayers consisting of 1D chains of 1-nm width or 2D hexagonal patterns were formed [127]. Also under potential control well-defined host–guest complexes are formed. Empty calixarenes and $C_{60}$–calixarene complexes can be visualized at the electrolyte Au(111) interface. In this case, the $C_{60}$–calixarene complex is preformed before formation of the monolayer [128].

## 4.6
## Entrapment of Large Molecules

A 2D network with tetragonal cavities with inner widths of 2.3 × 1.3 nm using 1,3,5-tris(10-carboxydecyloxy)benzene [125] has been used as the template for the inclusion of CuPc (**Cu5**) and coronene (**18**) on the surface of HOPG. A cavity is formed by the dimerization of the carboxylic acid functional groups of 1,3,5-tris(10-carboxydecyloxy)benzene. When **Cu5** molecules are entrapped in the cavities, highly uniform arrays are formed. One **Cu5** fits in a cavity. In contrast, inclusion of **18** in the cavities leads to domains with one or two molecules adsorbed per cavity (Fig. 28). The expected size of dimers of **18** fits perfectly within the cavity size. Each **18** corresponds to a bright ringlike structure. The distance between the coronenes in the dimer is in agreement with the lattice constant of a closely packed coronene monolayer adsorbed on Ag(111) and HOPG.

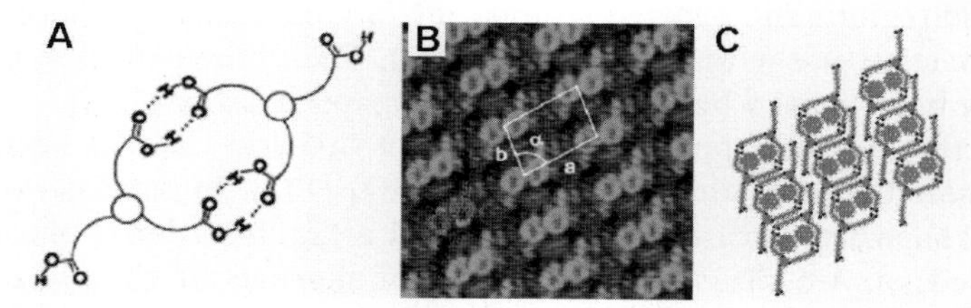


**Fig. 28** **a** Dimerization of 1,3,5-tris(10-carboxydecyloxy)benzene forming a cavity. **b** STM image of coronenes trapped in the hydrogen-bonded network (11.8 × 11.8 nm$^2$) at the air–HOPG interface. The *bright "doughnuts"* correspond to the coronenes. **c** Model. (Reproduced with permission from Ref. [125])

## 4.7
## Complexes of Phthalocyanines and Porphyrins With (Functionalized) Alkanes

Other types of complexes have also been reported under ambient conditions. Alkane lamellae ($C_{34}H_{70}$) have been used as a molecular buffer layer to immobilize **Cu5** and alkylated **Cu5** [134]. **Cu5** forms nanometer-sized domains and adsorbs on top of the alkane layer, revealing well-resolved submolecular features at different tunneling conditions. Nonalkylated **Cu5** molecules could not be imaged on HOPG in the absence of the alkanes. **Cu5** could be removed from the alkane layer by tip manipulation, indicating that the **Cu5** adsorbed on top of the monolayer. The ordering of **Cu5** is independent of that of the alkane buffer layer. Despite the fact that the interaction between **Cu5** and graphite is much stronger than between **Cu5** and the alkane layer, the lateral corrugation barrier of **Cu5** on the alkane layer is much larger than on the graphite substrate, which decreases the motion of **Cu5** by thermal diffusion or tip manipulation [135]. Also functionalized alkanes can act as a buffer layer such as stearic acid, 1-octadecanol and 1-iodooctadecane (C18I) on HOPG. The alkane derivatives form templates which provide different adsorption sites for Cu**5**. Cu**5** was found to exclusively adsorb as a monomer or as a dimer on the hydrocarbon-chain parts of the alkane derivatives [136]. Tridodecylamine has a strong effect upon the adsorption, diffusion and assembly of Cu**5**. Isolated Cu**5** and clusters could be stabilized at the alkane part of the lamellae (Fig. 29). The lateral diffusion of single Cu**5** as well as clusters of adsorbed Cu**5** molecules is exclusively along the direction of the tridodecylamine lamellae. With higher Cu**5** coverage, well-ordered bimolecular Cu**5** bands were observed as a result of molecular templating of the tridodecylamine lamellae [137]. Adsorption of the molecules on top of the alkyl derivatives is not the only adsorption mode. When mixtures of **5** or Cu**5** with *n*-octadecyl mercaptan (C18SH) and mixtures of **5** with C18I, 1-bromooctadecane (C18Br), 1-chlorooctadecane (C18Cl) and octadecyl cyanate (C18CN) were investigated, under optimized experimental conditions, alternating rows of **5** or Cu**5** and C18*X* could be observed [138]. Increasing the ratio of C18*X* leads to the formation of various phases, including the formation of large areas of uniform single molecular arrays on the submicrometer scale, where the ordering of **5** is consistent with its adsorption structure reported on HOPG [139]. The (Co)**5** units appear as the bright structures, while the alkyl chains appear darker. The groove between C18*X* lamellae appears to be a favorable site for adsorption of **5**. As this kind of assembly was not observed with stearic acid and octadecanol, the interaction between **5** and the functional groups of the alkyl derivatives appears to play an important role in the assembling process. Also a phthalocyanine with short alkyl chains (eight butyl groups) (Cu**5**–Bu8) self-assembles in the presence of C18I forming a regular pattern of alternating rows of Cu**5**–Bu8 and C18I (Fig. 29) [140]. The intermolecular distance of Cu**5**–Bu8 is larger than for Cu**5**, most likely owing to the butyl groups. On the other hand, a similar analogue with octyl chains **12** was found not to assemble in this alternating fashion. Phase separation took place and this could be related to unbalanced intermolecular interactions.

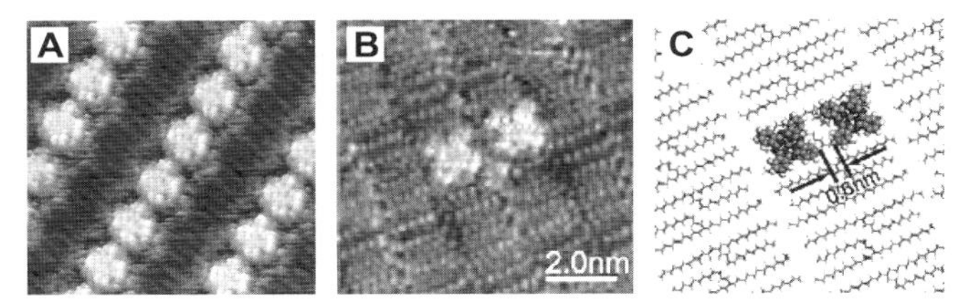


**Fig. 29** **a** STM image of mixed monolayer of Cu5–Bu8 and 1-iodooctadecane ($12 \times 12\ \text{nm}^2$) at the air–HOPG interface. **b**, **c** STM image and model of Cu**5** dimer on top of a tridodecylamine monolayer at the air–HOPG interface. (Reproduced with permission from Refs. [137, 140])

## 4.8
## Controlling Phase Behavior

In order to reveal the important factors in the phase behavior of structurally related compounds where the main mode of intermolecular interaction is hydrogen bonding, **37** was mixed with analogous compounds (Fig. 30) [141]. The bithiophene unit was replaced by a linear alkyl group. The distance between the two urea groups is identical if the alkyl spacer contains 14 carbon atoms. Study of the assembly of the pure compounds before mixing reveals that they indeed form rows where the main intermolecular interaction is hydrogen bonding via the urea groups. Mixing of **37** with compounds having dodecyl terminal groups but an alkyl spacer of different length (6, 9, 12, 14, 15 or 16 carbon atoms) shows that optimal mixing is obtained with C12-u-C14-u-C12 (C*X* refers to length alkyl chain, u is urea), while complete phase separation is obtained for C12-u-C6-u-C12. More important than the difference in molecular size is the fact that for those compounds with an alkyl spacer length different from 14 methylene units the urea groups cannot form ideal hydrogen bonds. This was confirmed by mixing **37** with a molecule having a tetradecyl spacer though with terminal hexyl chains instead of dodecyl

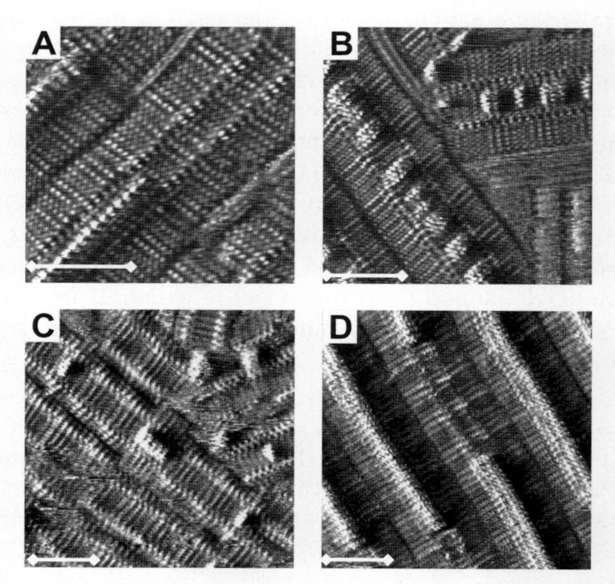


**Fig. 30** STM images of mixtures of **a** **37** and C12-u-C9-u-C12 (complete phase separation), **b** **37** and C12-u-C14-u-C12 (optimal mixing), **c** **37** and C6-u-C14-u-C6 (in addition to clusters, isolated **37** molecules are observed) and **d** **37** and C15-u-C9-u-C15 (complete phase separation). **37** can be recognized by the as *bright bithiophene units* in the center of a row. The *scale bar* indicates 4 nm. Experiments were carried out at the 1-octanol–HOPG interface. (Reproduced with permission from Ref. [141])

chains (C6-u-C14-u-C6). Though this molecule is much shorter than C12-u-C6-u-C12, both molecules mix very well as the identical spacer length still allows hydrogen bonding along a stack. The importance of hydrogen bonding as the most important interaction in this system is finally confirmed by mixing **37** with bisurea derivatives which are identical in length to **37** but which have the urea groups located at different positions along the molecular axis, such as C15-u-C9-u-C15. Despite their identical dimensions the molecules completely phase separate.

# 5
# Dyes and Chirality

Since the late 1990s, the expression of molecular chirality at surfaces has received increasing attention. STM provides a convenient approach to investigate (sub)monolayers, and the chirality aspects thereof, with submolecular resolution. It is one of the very few techniques that allow identification of the (sub)monolayer chirality and even absolute conformation. It is not the purpose of this review to give an extensive overview of all the studies that have been reported on chirality at surfaces but this section is intended to give a flavor of the relation between self-assembly directed noncovalent interactions and chirality at surfaces for dye molecules [142]. Chiral discrimination between mirror-image stereoisomers can lead to the spontaneous spatial separation of a racemic mixture into the enantiomerically pure phases. In 3D systems the formation of these so-called conglomerates is rather the exception than the rule: most racemic mixtures crystallize as racemates with the unit cell composed of an equal number of molecules with opposite handedness or as random solid solutions. Owing to the confinement of the molecules in a plane and the interaction with the substrate, conglomerate formation becomes more likely.

Mirror-image molecules form always mirror-image supramolecular structures on a surface. That is indeed what is found for enantiopure heptahelicene (Fig. 31, **49**). When enantiopure heptahelicene molecules are forced into a close-packed monomolecular layer, molecular chirality is transferred to monolayer chirality [143]. When the molecules are squeezed together repulsive forces dominate the lateral interaction. The self-assembly is obviously not only governed by the interaction between the molecules but also by adsorbate–substrate interactions, which determine the mobility of the molecules on the surface. On Ni(111) and Ni(100) the low mobility of helicene did not allow the observation of chiral effects. On Cu(111) the molecules were observed to diffuse readily at coverages below 95%. At 95% coverage, a long-range ordered structure is observed, apparently built up from clusters containing six molecules and clusters containing three molecules. At 100% surface coverage, the unit cell of the adsorbate lattice contains a group of three

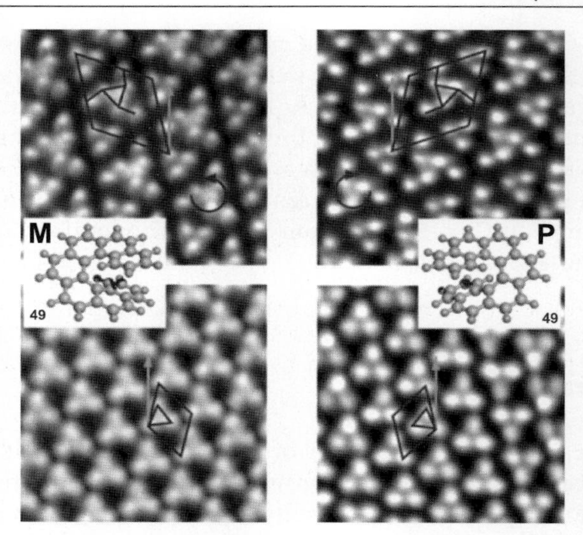


**Fig. 31** STM images of (*M*)-heptahelicene (*left*) and (*P*)-heptahelicene (*right*) ($10 \times 10\,nm^2$) on Cu(111) recorded under UHV conditions. Monolayer coverage is 0.95 (*top*) and 1.00 (*bottom*). Both enantiomers clearly form mirror-image structures. (Reproduced with permission from Ref. [143])

molecules. The observed adsorbate lattice structures show enantiomorphism: adsorption of the (*P*)-enantiomer of the helicene leads to structures which are mirror images of those observed for the (*M*)-enantiomer. In addition to the chiral shape of the unit cell, the arrangement of the molecules within the unit cell is also chiral. The hexamer (95% coverage) has a pinwheel-like shape and the pinwheel's wings point either anticlockwise as found for (*M*)-**49** or clockwise as found for (*P*)-**49**. In the case of the trimers (100% coverage) the mirror symmetry is expressed by tilts of the three-molecule cloverleaf units into opposite directions with respect to the adsorbate lattice vectors. The adsorption of racemic helicene led to the formation of enantiomorphous mirror domains in which the enantiomers are partially separated [144].

The adsorption of enantiopure molecules gives rise to the formation of chiral surfaces and both enantiomers form each other's mirror image. However, most achiral molecules become 2D chiral upon adsorption on a substrate. Consider, for instance, STM images of naphtho[2,3-*a*]pyrene (Fig. 32, **50**), which is a simple aromatic system. STM images reveal equal numbers of the 2D enantiomers on the surface [145]. The ordered molecular domains contained only one 2D enantiomer: segregation occurred of homochiral molecules into domains with chiral unit cells. The chiral nature of the domains is revealed by the fact that in a given domain all molecules can be superimposed by rotation of the molecules within the plane of the Au(111) surface. Again, both 2D enantiomers form mirror-image domains.

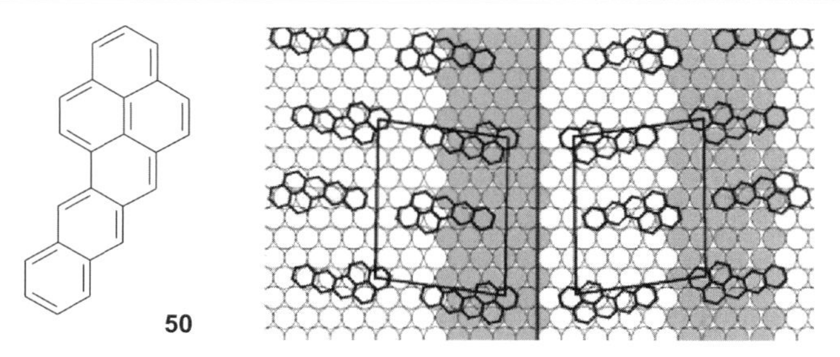


**Fig. 32** Chemical structure of **50** and model of the reconstructed Au(111) surface (*dark substrate atoms*) and the intersection of two chiral domains of molecules of **50**. Experiments were performed under UHV conditions. (Reproduced with permission from Ref. [145])

All achiral systems presented which become 2D chiral upon adsorption on a surface show conglomerate formation: a domain is composed of molecules with the same side facing the substrate [146]. Adsorption of a prochiral molecule to an achiral surface creates a racemic adlayer if the adsorption geometry establishes different interactions between the two (formerly) enantiotopic groups on the surface. Chiral domains with chiral unit cells form when interactions between identical 2D enantiomorphs are more favorable than between opposite 2D enantiomorphs. Mixed domains with unit cells containing both enantiomorphs form when interactions between opposite isomers are more favorable. The stereochemical morphology of monolayers formed from prochiral 1,5-bis(3′-thiaalkyl) anthracenes (Fig. 33, **51**) on HOPG switches from a 2D racemate to a 2D conglomerate by the addition of a single methylene unit to each side chain.

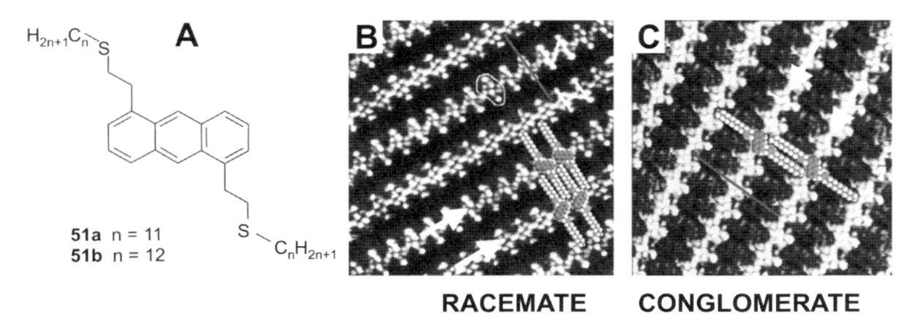


**Fig. 33** **a** Structures of **51a** and **51b**. **b** STM image ($12 \times 12\,nm^2$) of **51a** adsorbed onto HOPG. From row to row, the anthracene units (indicated by *arrows*) are twisted. A 2D racemate is formed. **c** STM image ($11 \times 11\,nm^2$) of **51b** adsorbed on HOPG. From row to row, the anthracene units are parallel. A 2D conglomerate is formed. The location of the anthracene units is indicated by *arrows*. Experiments were performed at the 1-phenyloctane–graphite interface. (Reproduced with permission from Ref. [146])

# 6 Dynamics

## 6.1 Translation

Diffusion is an important process on surfaces. There are only a few reports which deal with the diffusion of large molecules on surfaces [147–149] and this has been recently reviewed [14]. According to a very simple model, surface diffusion can be viewed as a random 2D walk of an adsorbate, hopping from adsorption site to adsorption site on the substrate. The diffusion process can be described in terms of three parameters: (1) the *activation energy for diffusion* is the barrier an adsorbate has to overcome on the potential energy surface to reach at least the neighboring adsorption site; (2) the *prefactor* indicates the attempt frequency to escape from its adsorption site; (3) the *root-mean-square jump length* contains information on the average distance of a jump. With high-speed STM imaging, quantitative data can be obtained and modeled.

For decacyclene (DC) and hexa-*tert*-butyl decacyclene (**26**) adsorbed on an atomically clean Cu(110) surface, it was shown that molecular diffusion is strongly 1D, occurring parallel to the close-packed ⟨110⟩ direction of the Cu(110) substrate (Fig. 34) [147]. Arrhenius plots of the hopping rates and tracer diffusion constants show a lower activation energy for diffusion of **26** (0.60 eV) compared with DC (0.72 eV), which is attributed to the six *tert*-butyl spacer groups of **26**. The spacer groups increase the distance between the aromatic core and the substrate, which as a result leads to a decrease in adsorbate–substrate interactions. As a result, the diffusion constant is 4 times higher than for DC. Jumps are not restricted to nearest-neighbor sites on the Cu substrate. Surprisingly, long jumps play a dominant role in the diffusion

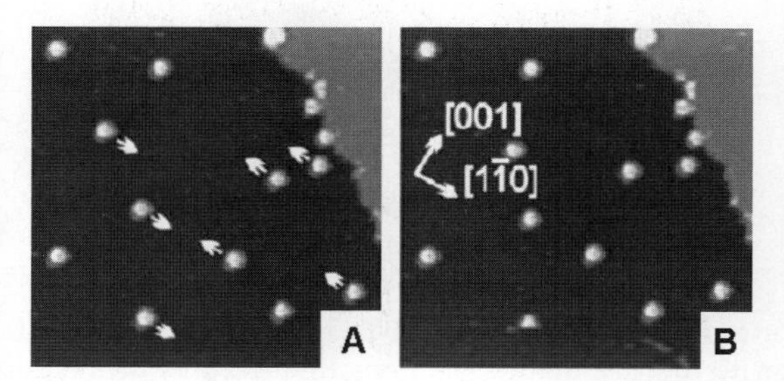


**Fig. 34** Images form a constant-current movie of **26**, imaged as bright spots, on Cu(110) at 194 K (UHV). Molecular displacements can clearly be discerned; *arrows* indicate the direction in which molecules will have moved in the successive image. (Reproduced with permission from Ref. [147])

of these molecules. The root-mean-square jump lengths are determined to be as large as approximately four and approximately seven Cu nearest-neighbor distances for DC and **26**, respectively. Large molecules occupy a significantly larger area on the substrate and they might find it more difficult to find a low-energy site. The longer jumps for **26** might be related to the decreased interaction with the surface. For $C_{60}$ on Pd(110) an activation energy of 1.4 eV was determined along the close-packed direction ⟨110⟩. This high value indicated strong interactions and directional bonding between $C_{60}$ carbon ring systems and Pd surface atoms [148].

Diffusion is also observed for higher monolayer coverages. Subphthalocyanine **7** self-assembles into a 2D crystalline overlayer on Ag(111) with a honeycomb pattern characterized by a low packing density (2D-condensed phase) [39]. Next to a condensed island, noisy streaks appear in the scan direction, indicating that they exhibit mobility on the time scale of one scan line (2D-gas phase) (Fig. 35). These domains evolve in time as shown by time-lapse imaging sequences. The borders of the 2D-condensed phase move as a function of time, indicating that the condensed phase is in dynamic coexistence with the gas phase. These two phases coexist in a 2D thermodynamic equilibrium at room temperature. The molecules in the gas phase are stably adsorbed for a certain time but tend to hop to nearby adsorption sites. A simple model can account for the observations. If the thermal energy of the adsorbed molecule is small compared with the corrugation of the surface potential, one can consider the adsorbed molecule as localized in a minimum of the surface potential. Each molecule has to overcome the energy barrier

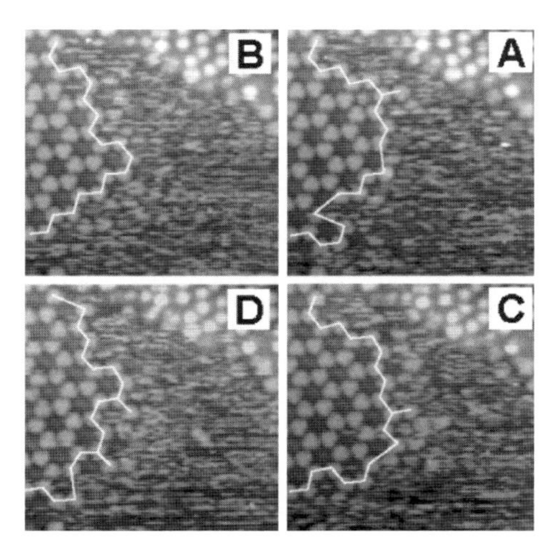


**Fig. 35** Sequence of STM images showing the time evolution of a condensed molecular island of 7. [Ag(111): UHV conditions]. One *bright spot* is a molecule. The interval time between two images is 3 min 26 s. The scan range is 25.8 × 25.8 nm. The *white line* connects the outermost molecules. (Reproduced with permission from Ref. [39])

imposed by the corrugation of the surface potential in order to hop to an adjacent minimum and thus participate in the diffusion process.

## 6.2 Rotation

Translation is not the only motion at surfaces. At monolayer coverage, **26** is immobilized and forms a 2D van der Waals crystal on atomically clean Cu(100) surfaces under UHV conditions [150]. The internal structure of each molecule consists of six lobes arranged in a hexagonal lattice with alternating distances of 0.6 and 0.8 nm between the lobes. Each of the lobes can be assigned to a *tert*-butyl appendage. At submonolayer coverages, these molecules diffuse around owing to the weak adsorption of the H atoms of the *tert*-butyl groups with the Cu surface and the molecules cannot be imaged. At close-to-complete substrate coverage, the images reflect the structures obtained at full monolayer coverage except for the presence of voids, which appear in a random way. Surprisingly, in those voids and only in those voids, molecules often appear with a torus-type shape and these were always out of registry with the surrounding supramolecular lattice (Fig. 36). Molecules in registry with the supramolecular lattice always revealed the typical six lobes of the *tert*-butyl groups. The torus shape is a result of molecular rotation with speeds higher than the scan rate for imaging. The experimental results were supported by calculations of the rotational activation energy.

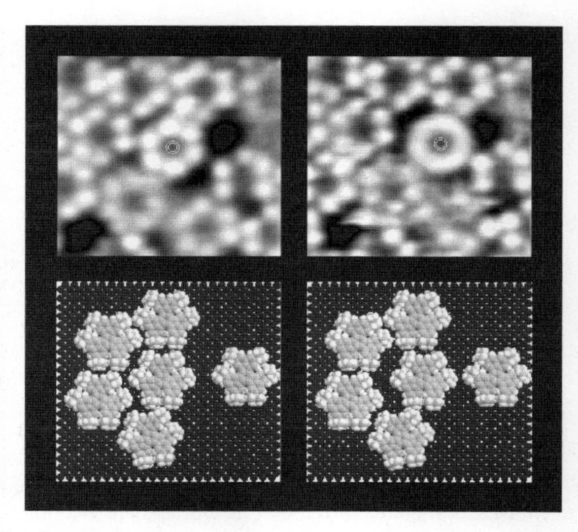

**Fig. 36** Immobilized **26** (*left*). Rotating **26** in the void (*right*). [Cu(100): UHV conditions]. (Reproduced with permission from Ref. [150])

## 6.3 Ostwald Ripening

Dynamic phenomena are of course not restricted to UHV conditions. Ostwald ripening describes the growth of larger domains at the expense of smaller domains. The thermodynamic driving force is the reduction of the circumference-to-area ratio and thereby the lowering of the interfacial or line energy. For example, application of unsaturated solutions of 2-hexadecyl-9,10-anthraquinone in phenyloctane on HOPG leads to the formation of domains [151]. On a time scale of seconds a small domain shrinks until it disappears, while the larger domains grow at its expense. After several minutes to a few hours the 2D polycrystals become single crystals on a micrometer scale. It was shown for this system that Ostwald ripening is determined by a reorientation of molecules at the domain boundaries of the 2D polycrystals. At a domain boundary, molecules are often not ideally close packed, and the free volume within the 2D polycrystal is significantly increased. Therefore, individual molecules or lamella fragments at the domain boundary can change their orientation with respect to the underlying substrate without diffusing within the single crystals and without a transition to the supernatant solution [152].

At the liquid–solid interface, the adlayer is in equilibrium with the supernatant solution and several examples show that there is exchange of molecules. These exchange processes are important for the formation of supramolecular structures: the exchange of molecules in two and in three dimensions (monolayer–solution) provides a "healing" mechanism to repair defects.

# 7 Tip-Induced Manipulation

In addition to imaging, the scanning tunneling microscope tip can also be used to manipulate molecules, to move them around and even to induce reactions.

## 7.1 Tip-Induced Translational and Conformational Changes

Lateral manipulation at constant current allows molecular adsorbates to be moved on metal surfaces by STM under UHV conditions. The scanning tunneling microscope tip is kept close to the surface by increasing the tunneling current (or decreasing the tunneling voltage, or both) and is then moved laterally over the target molecule. This allowed at room temperature the displacement of single molecules of **14** on different Cu substrates [48]. At low temperatures, a stronger repulsive force must be exerted to be effective and using a modified approach, manipulation of single molecules of **14** was also proven to be successful [153]. It even turned out to be possible to change the orientation of one leg of **14** in a controlled manner [154].

Not only single molecules but also assemblies of molecules could be manipulated, which was illustrated for crown-ether-functionalized phthalocyanines which form a gel on HOPG. Tip-induced manipulation affected the orientation of the molecules on the surface (edge-on or face-on) [155].

## 7.2 Tip-Induced Orientational Switching

In addition to scanning tunneling microscope tip induced conformational and translational manipulation, the scanning tunneling microscope tip can also be used to induce a reversible orientational switching, as demonstrated for subphthalocyanine molecules **7** [41]. The asymmetric, polar molecules adsorbed in an epitaxial array on a Cu(100) surface, initially with the axial chlorine atom either upward or downward. After scanning at a negative bias changes in image contrast were attributed to the upside down turning of the upward molecules, while all molecules switched to the upward orientation at a positive bias.

## 7.3 Tip-induced Reactivity

Recently, the formation was reported of conjugated polymer 1D structures (polydiacetylenes) by initiating the chain polymerization of diacetylene containing carboxylic acid derivatives using a scanning tunneling microscope tip with a spatial precision on the order of 1 nm at the solid (HOPG)–air interface (Fig. 37) [156, 158]. The molecules are aligned to form straight chains and the chains are arranged in a manner such that the COOH end groups of a chain are opposite those of a neighboring chain. The controlled fabrication of nanowires is of importance in the field of nanotechnology. In analogy with light excitation, the scanning tunneling microscope tip initiates the reaction by applying a pulsed sample bias. A very bright line appeared, starting from the point of stimulation, which indicates that the reaction initiated by the scanning tunneling microscope tip has propagated and a polydiacetylene chain is formed. The length of the polydiacetylene chain is determined by the size of the domains of diacetylene monomer molecules having the same orientation. The initiation by the scanning tunneling microscope tip leads to polymerization in both directions. Interestingly, these authors were able to control the maximum length of the polydiacetylene chains within a domain. In order to achieve that, they created an artificial defect in the form of a 6-nm-wide hole at a predetermined position within the monolayer using a scanning

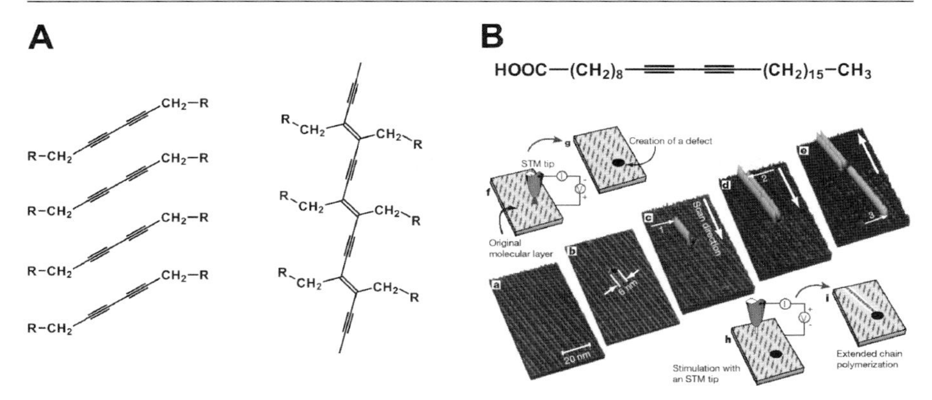


**Fig. 37** **a** Model of stacked diacetylene monomers (*left*); Model of a polydiacetylene chain (*right*). **b** STM images and diagrams showing the process of controlling the initiation and termination of linear chain polymerization with a scanning tunneling microscope tip. STM images were obtained at the solid (HOPG)–air interface. *a* Original monolayer of 10,12-nonacosadiynoic acid. *b* Creation of an artificial defect. *c* First chain polymerization, initiated at the *arrow*. *d* Second chain polymerization, initiated at the *arrow*. *e* Third chain polymerization, initiated at the *arrow*. (Reproduced with permission from Ref. [156])

tunneling microscope tip. Such a defect terminates the polymerization reaction. Molecular modeling shows that the polymer backbone is raised from the graphite substrate and suggests that organic molecules with various novel conformations could be created on 2D solid surfaces [157]. STS experiments on isolated polydiacetylene chains clearly reveal the $\pi$ band expected in the 1D conjugated polymer system [158].

## 8 Dyes and Spectroscopy

Substantial research has been performed on bias-dependent imaging and STS of dye-related systems. Here, only a few selected examples will be highlighted. In bias-dependent imaging, the images are recorded at different bias voltages and polarity. In STS, the voltage is ramped while the feedback loop is turned off and an $I$–$V$ graph is obtained at a certain position on the substrate.

This research is often motivated by the miniaturization of electronics and the potential molecules may have in devices, even at the level of a single molecule. The electron tunneling through redox single porphyrin molecules in aqueous solution has been investigated. In this system, the substrate and tip Fermi levels can be flexibly adjusted relative to the molecular orbitals by controlling the substrate potential versus a reference electrode in the solution, at a constant sample-tip bias. By tuning the substrate Fermi level to an un-

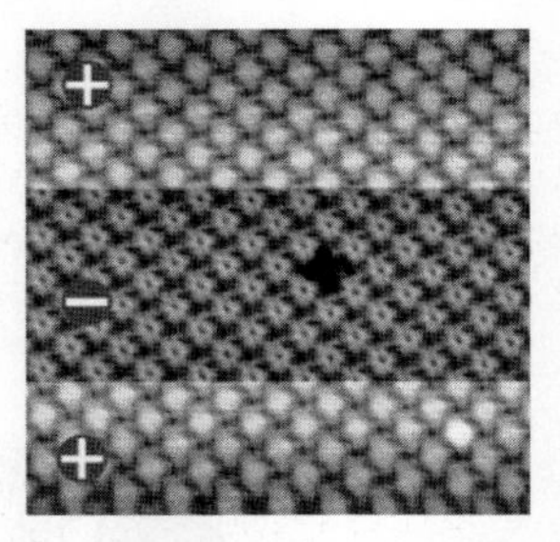


**Fig. 38** STM constant-current topograph of a monolayer of **8** on HOPG ($25 \times 25\,nm^2$) obtained at 50 K (UHV). The polarity of the tunneling voltage was switched twice as indicated. (Reproduced with permission from Ref. [44])

occupied molecular orbital, a nearly tenfold increase in the tunneling current was observed owing to resonant enhancement [8].

The bias-dependent STM imaging and spectroscopy of a naphthalocyanine dye (Fig. 38, **8**) has been investigated. Compared with phthalocyanine, the more extended, conjugated electron system of **8** influences the energies of the molecular orbitals and is expected to lead to a smaller HOMO–lowest unoccupied molecular orbital (LUMO) gap. For negative sample bias, i.e., tunneling out of occupied sample states to the tip, the center of **8** appears dark, while at positive sample bias, i.e., tunneling from the tip into unoccupied sample states, all molecules appear as bright humps, without any indication of a decrease in the apparent height at the center. In the case of weak interactions between the adsorbate and the substrate (physisorption), the observed effect can be explained in terms of molecular orbitals. The HOMO shows no electron density in the middle of the molecule, in contrast to the LUMO. So the HOMO is considered to contribute to the image contrast at negative sample bias while the LUMO (and/or LUMO + 1) affects the tunneling at positive bias voltages. Further investigation of the electronic structure has been performed by STS. The STS data indicate semiconducting behavior: no current is measured in the gap region from – 0.9 to + 0.9 V [44].

A covalently bound donor–acceptor–donor derivative **42** showed a bias-dependent contrast (Fig. 39) [118]. The donor molecule is an oligo-*p*-phenyl-

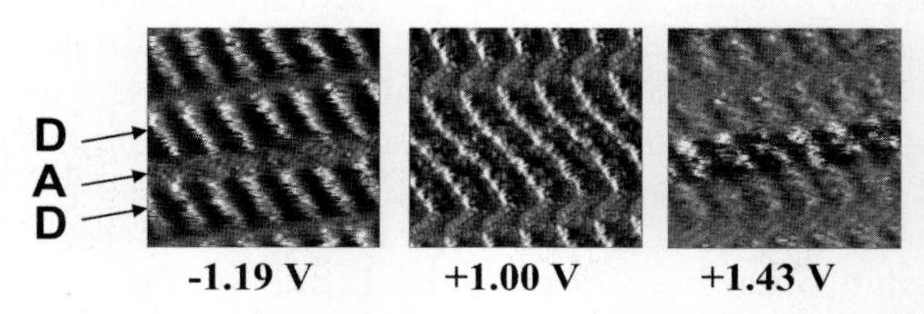


**Fig. 39** Bias-dependent STM constant-height images of **42** ($15.5 \times 15.5\,nm^2$) at the 1-phenyloctane–graphite interface. The sample bias is indicated. *D* refers to the donor and *A* to the acceptor moiety, respectively. (Reproduced with permission from Ref. [117])

enevinylene derivative and the acceptor unit is a perylenediimide derivative. At high negative bias voltages, the tunneling current through the donor moieties was much higher than through the acceptor moiety, while the opposite was observed at high positive bias voltages. The results were interpreted in terms of a resonant tunneling model which will be discussed in the following example.

The tunneling microscopy and spectroscopy of an electron donor (hexaperi-hexabenzocoronene) covalently linked to six acceptors (anthraquinone) in self-assembled layers at the graphite–liquid interface **23** has been investigated [159]. The tunneling current through the HBC cores is larger than through the anthraquinones. This difference is bias-dependent and is much smaller at positive sample bias. In the $I$–$V$ experiments, the current is larger for negative sample bias, while over anthraquinone regions larger currents are observed at sufficiently large positive sample bias. Thus, HBC regions exhibit higher tunneling probability at negative than at positive sample bias, whilst anthraquinone regions exhibit the opposite behavior. The break in symmetry of the $I$–$V$ curves can be due to different distances between molecule and tip or substrate and/or to different energetic gaps between the molecule's HOMO and LUMO and the Fermi level of HOPG. By assuming resonantly enhanced tunneling through molecular states as the dominant mechanism for contrast formation in STM/STS experiments, we can explain the aforementioned observations by the simple model of resonant tunneling (Fig. 40) [160]. The probability for electrons to tunnel from occupied states in the substrate into empty states of the tip (or vice versa) strongly increases if there are molecular states of the adsorbate energetically close to the states involved in the tunneling process. The different asymmetries of the $I$–$V$ curves of HBC and anthraquinone, respectively, can be explained by resonant tunneling through the HOMO of HBC and the LUMO of anthraquinone, respectively.

Stimulated by these results, a prototypical single-molecule field-effect transistor with nanometer-sized gates was realized. Adsorption of dimethoxyanthracene (an electron donor) on anthraquinone (an electron acceptor), which is covalently linked to HBC **23**, induces an interface dipole which shifts the substrate work function by approximately 120 meV, as revealed by STS experiments (Fig. 41) [161].

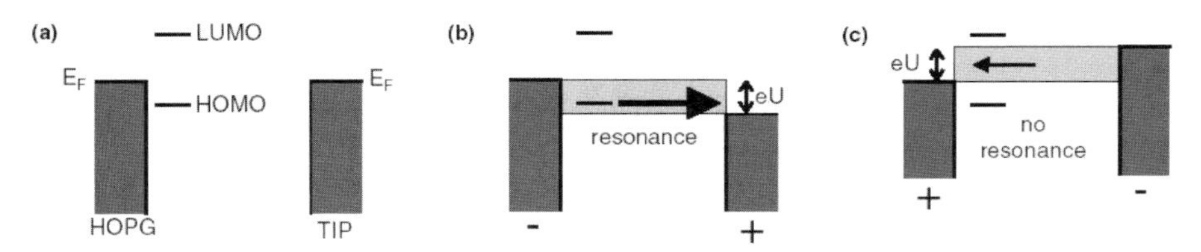


**Fig. 40** Schematics illustrating the origin of asymmetry in the $I$–$V$ curves and energy levels. (Reproduced with permission from Ref [159])

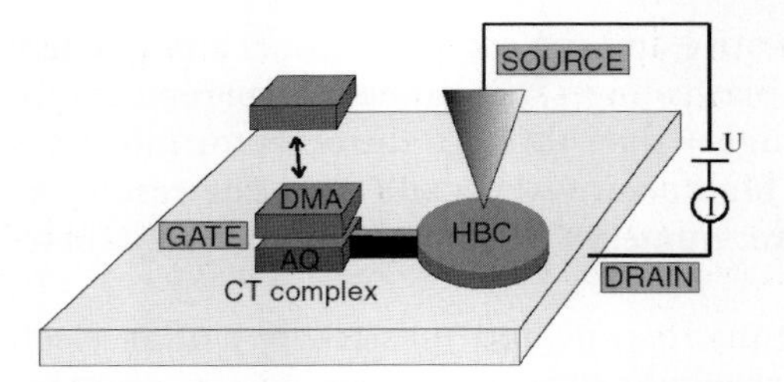


**Fig. 41** Schematic drawing of a prototypical single-molecule chemical-field-effect transistor. (Reproduced with permission from Ref. [161])

The scanning tunneling microscope tip can also be used to induce light emission. Light is emitted out of the junction of a scanning tunneling microscope when an energetic tunneling electron excites a radiative process. This phenomenon provides a light source of nanometer-scale dimensions and enables an optical study of nanoscopic excitations. Spectral analysis of the STM-induced luminescence in the presence of molecules adsorbed on the surface can add chemical specificity. A few examples include the STM-induced luminescence from a $C_{60}$ monolayer on Au(110) [162] and molecular spectroscopy of dye aggregates [163]. Intrinsic molecular fluorescence from porphyrin molecules on Au(100) has been realized by using a nanoscale multimonolayer decoupling approach with nanoprobe excitation in the tunneling regime [164]. The molecular origin of luminescence is established by the observed well-defined vibrationally resolved fluorescence spectra. The molecules fluoresce at low turn-on voltages for both bias polarities, suggesting an excitation mechanism via hot electron injection from either tip or substrate. The excited molecules decay radiatively through Franck–Condon $\pi$–$\pi^*$ transitions. Crucial is the decoupling of the emitter from the substrate—molecular fluorescence is quenched close to the metal substrate—by using the molecules both as a spacer (formation of multilayers) and an emitter. Increase of the layer thickness leads in addition to a plasmon-mediated emission band to STM-induced emission of molecular origin.

## 9
## Conclusions

STM is an established technique to unravel the secrets of the ordering of molecules, in general, and dyes, in particular, on surfaces. Much information has been gathered on the balance between molecule–molecule and molecule–adsorbate interactions. The strength of STM is that it goes beyond mere imaging. It gives information on dynamic processes, it can be used as a manipulation tool and, importantly, it is also a spectroscopic tool. This aspect will become increasingly important in view of the popularity of "molecular electronics", eventually on the single-molecule scale.

**Acknowledgements** The authors thank the Federal Science Policy through IUAP-V-03, the Institute for the promotion of innovation by Science and Technology in Flanders (IWT), and the Fund for Scientific Research-Flanders (FWO). S.D.F. is a postdoctoral fellow of FWO.

## References

1. (a) Binnig G, Rohrer H, Gerber C, Weibel E (1982) Phys Rev Lett 49:57; (b) Binnig G, Rohrer H (1982) Helv Phys Acta 55:726
2. Ohtani H, Wilson RJ, Chiang S, Mate CM (1988) Phys Rev Lett 60:2398
3. Forrest SR (1997) Chem Rev 97:1793
4. Foster JS, Frommer JE (1988) Nature 333:542
5. (a) Spong JK, Mizes HA, LaComb LJ Jr, Dovek MM, Frommer JE, Foster JS (1989) Nature 338:137; (b) Smith DPE, Horber H, Gerber C, Binning G (1989) Science 245:43; (c) Smith DPE, Horber JKH, Binnig G, Nejoh H (1990) Nature 344:641
6. McGonigal GC, Bernhardt RH, Thomson DJ (1990) Appl Phys Lett 57:28
7. Rabe J, Buchholz S (1991) Science 253:424
8. Tao NJ (1996) Phys Rev Lett 76:4066
9. Itaya K (1998) Prog Surf Sci 97:1129
10. Kunitake M, Akiba U, Batina N, Itaya K (1997) Langmuir 13:1607
11. He Y, Ye T, Borguet E (2002) J Am Chem Soc 124:11964
12. Chiang S (1997) Chem Rev 97:1086
13. Barlow SM, Raval R (2003) Surf Sci Rep 50:210
14. Rosei F, Schunack M, Naitoh Y, Jiang P, Gourdon A, Laegsgaard E, Stensgaard I, Joachim C, Besenbacher F (2003) Prog Surf Sci 71:95
15. Möbius M, Karl N (1992) J Cryst Growth 116:492
16. Hoshino A, Isoda S, Kurata H, Kobayashi T (1994) J Appl Phys 76:4113
17. Ludwig C, Gompf B, Glatz W, Petersen J, Eisenmenger W, Möbius M, Zimmermann U, Karl N (1992) Z Phys B 86:397
18. Ludwig C, Gompf B, Petersen J, Strohmaier R, Eisenmenger W (1994) Z Phys B 93:365
19. Seidel C, Awater C, Liu XD, Ellerbrake R, Fuchs H (1996) Surf Sci 397:123
20. Schmitz-Hübsch T, Fritz T, Staub R, Back A, Armstrong NR, Leo K (1999) Surf Sci 437:163
21. Gloeckel K, Seidel C, Soukopp A, Sokolowski M, Umbach E, Boehringer M, Berndt R, Schneider WD (1998) Surf Sci 405:1
22. Staub R, Toerker M, Fritz T, Schmitz-Hübsch, Sellam F, Leo K (2000) Surf Sci 445:368
23. Eremtchenko M, Schaefer JA, Tautz FS (2003) Nature 425:602
24. Uder B, Ludwig C, Petersen J, Gompf B, Eisenmenger WZ (1995) Phys B 97:389
25. Keeling DL, Oxtoby NS, Wilson C, Humphry MJ, Champness NR, Beton PH (2003) Nano Lett 3:9
26. Kaneda Y, Stawasz ME, Sampson DL, Parkinson BA (2001) Langmuir 17:6185
27. (a) Gimzewski JK, Stoll E, Schlittler RR (1987) Surf Sci 181:267; (b) Lippel PH, Wilson RJ, Miller MD, Wöll C, Chiang S (1989) Phys Rev Lett 62:171
28. Lu X, Hipps KW, Wang XD, Mazur U (1996) J Am Chem Soc 118:7197

29. Hipps KW, Lu X, Wang XD, Mazur U (1996) J Phys Chem 100:11207
30. Lu X, Hipps KW (1997) J Phys Chem B 101:11207
31. Strohmaier R, Ludwig C, Petersen J, Gompf B, Eisenmenger W (1996) J Vac Sci Technol B 14:1079
32. Lackinger M, Hietschold M (2002) Surf Sci 520:619
33. Scudeiro L, Barlow DE, Hipps KW (2000) J Phys Chem B 104:11899
34. Scudeiro L, Balrow DE, Mazur U, Hipps KW (2001) J Am Chem Soc 123:4073
35. Yoshimoto S, Tada A, Suto K, Narita R, Itaya K (2003) Langmuir 19:672
36. Yoshimoto S, Tada A, Suto K, Itaya K (2003) J Phys Chem B 107:5836
37. Yoshimoto S, Tada A, Suto K, Yau SL, Itaya K (2004) Langmuir 20:3159
38. Yanagi H, Schlettwein D, Nakayama H, Nishino T (2000) Phys Rev B 61:1959
39. Berner S, Brunner M, Ramoino L, Suzuki H, Güntherodt HJ, Jung TA (2001) Chem Phys Lett 348:175
40. Suzuki H, Berner S, Brunner M, Yanagi H, Schlettwein D, Jung TA, Guentherodt HJ (2001) Thin Solid Films 393:325
41. Yanagi H, Ikuta K, Mukai H, Shibutani T (2002) Nano Lett 2:951
42. Suzuki H, Miki H, Yokoyama S, Mashiko S (2003) J Phys Chem 107:3659
43. Berner S, de Wild M, Ramoino L, Ivan S, Baratoff A, Güntherodt HJ, Suzuki H, Schlettwein D, Jung TA (2003) Phys Rev B 68:115410
44. Lackinger M, Müller T, Gopakumar TG, Müller F, Hietschold M, Flynn GW (2004) J Phys Chem B 108:2279
45. Yokoyama T, Yokoyama S, Okuno Y, Mashiko S (2001) Nature 413:619
46. Lei SB, Wang C, Yin SX, Wang HN, Wu F, Liu HW, Xu B, Wan LJ, Bai CL (2001) J Phys Chem B 105:10838
47. Qiu X, Wang C, Zeng Q, Xu B, Yin S, Wang H, Xu S, Bai C (2000) J Am Chem Soc 122:5550
48. Jung TA, Schlittler RR, Gimzewski JK, Tang H, Joachim C (1996) Science 271:181
49. Jung TA, Schlittler RR, Gimzewski JK (1997) Nature 386:696
50. Moresco F, Meyer G, Rieder KH, Ping J, Tang H, Joachim C (2002) Surf Sci 499:94
51. Yokoyama T, Yokoyama S, Kamikado T, Mashiko S (2001) J Chem Phys 115:3814
52. Batina N, Yamada T, Itaya K (1995) Langmuir 11:4568
53. Askadskaya L, Boeffel C, Rabe JP (1993) Ber Bunsen-Ges Phys Chem 97:517
54. Charra F, Cousty J (1998) Phys Rev Lett 80:1682
55. Wu P, Zeng Q, Xu S, Wang C, Yin S, Bai CL (2001) Chem Phys Chem 12:750
56. Katsonis N, Marchenko A, Fichou D (2003) J Am Chem Soc 125:13682
57. Walzer K, Sternberg M, Hietschold M (1998) Surf Sci 415:376
58. Lackinger M, Griessl S, Heckl WM, Hietschold M (2002) J Phys Chem B 106:4482
59. Yoshimoto S, Narita R, Itaya K (2002) Chem Lett 356
60. Yoshimoto S, Narita R, Wakisaka M, Itaya K (2002) J Electroanal Chem 532:331
61. Stabel A, Herwig P, Müllen K, Rabe J (1995) Angew Chem Int Ed Engl 34:1609
62. Ito S, Wehmeier M, Brand JD, Küble C, Epsch R, Rabe JP, Müllen K (2000) Chem Eur J 6:4327
63. Samori P, Fechtenkötter A, Jäckel F, Böhme T, Müllen K, Rabe JP (2001) J Am Chem Soc 123:11462
64. Iyer VS, Yoshimura K, Enkelmann V, Epsch R, Rabe JP, Müllen K (1998) Angew Chem Int Ed Engl 37:2696
65. Ito S, Herwig PT, Böhme T, Rabe JP, Rettig W, Müllen K (2000) J Am Chem Soc 122:7698
66. Samori P, Severin N, Simpson CD, Müllen K, Rabe JP (2002) J Am Chem Soc 124:9454

67. Samori P, Yin X, Tchebotareva N, Wang Z, Pakula T, Jäckel F, Watson MD, Venturini A, Müllen K, Rabe JP (2004) J Am Chem Soc 126:3567
68. Tchebotareva N, Yin X, Watson MD, Samori P, Rabe JP, Müllen K (2003) J Am Chem Soc 125:9734
69. Watson MD, Jäckel F, Severin N, Rabe JP, Müllen K (2004) J Am Chem Soc 126:1402
70. Schunack M, Petersen L, Kühnle A, Lægsgaard E, Stensgaard I, Johannsen I, Besenbacher F (2001) Phys Rev Lett 88:456
71. Schunack M, Lægsgaard E, Stensgaard I, Johannsen I, Besenbacher F (2001) Angew Chem Int Ed Engl 40:2623
72. Schunack M, Lægsgaard E, Stensgaard I, Besenbacher F (2002) J Chem Phys 117:8493
73. Langlais VJ, Schlittler RR, Tang H, Gourdon A, Joachim C, Gimzewski JK (1999) Phys Rev Lett 83:2809
74. Schunack M, Rosei F, Naitoh Y, Jiang P, Gourdon A, Joachim C, Besenbacher F (2002) J Chem Phys 117:6259
75. Rosei F, Schunack M, Jiang P, Gourdon A, Lægsgaard E, Stensgaard I, Joachim C, Besenbacher F (2002) Science 296:328
76. Otero R, Naitoh Y, Rosei F, Jiang P, Thostrup P, Gourdon A, Lægsgaard E, Stensgaard I, Joachim C, Besenbacher F (2004) Angew Chem Int Ed Engl 116:2144
77. Wilson JR, Meijer G, Bethune DS, Johnson RD, Chambliss DD, de Vries MS, Hunziker HE, Wendt HR (1990) Nature 348:621
78. Sakurai T, Wang XD, Xue QK, Hasegawa Y, Hashizume T, Shinohara H (1996) Prog Surf Sci 51:263
79. Kuk Y, Kim DK, Suh YD, Park KH, Oh HP, Kim SK (1993) Phys Rev Lett 70:1948
80. Jehoulet C, Obeng YS, Kim YT, Zhou F, Bard AJ (1992) J Am Chem Soc 114:4237
81. Marchenko A, Cousty J (2002) Surf Sci 513:233
82. Uemura S, Samori P, Kunitake M, Hirayama C, Rabe JP (2002) J Mater Chem 12:3366
83. Chen T, Howells S, Gallagher M, Darid D, Lamb LD, Huffman D, Workman RK (1992) Phys Rev B 45:14441
84. Katsonis N, Marchenko A, Fichou D (2004) Adv Mater 16:309
85. Matsumoto M, Inukai J, Tsutsumi E, Yoshimoto S, Itaya K, Ito O, Fujiwara K, Murata M, Murata Y, Komatsu K (2004) Langmuir 20:1245
86. Kunitake M, Uemura S, Ito O, Fujiwara K, Murata Y, Komatsu K (2002) Angew Chem Int Ed Engl 41:969
87. Stawasz ME, Sampson DL, Parkinson BA (2000) Langmuir 16:2326
88. Takeda N, Stawasz ME, Parkinson BA (2001) J Electroanal Chem 498:19
89. De Feyter S, Gesquière A, De Schryver FC, Keller U, Müllen K (2002) Chem Mater 14:989
90. Pan LL, Zeng QD, Lu J, Wu DX, Xu SD, Tan ZY, Wan LJ, Wang C, Bai CL (2004) Surf Sci 559:70
91. Abdel-Mottaleb MMS, Gomar-Nadal E, De Feyter S, Zdanowska M, Veciana J, Rovira C, Amabilino DB, De Schryver FC (2003) Nano Lett 3:1375
92. Gomar-Nadal E, Abdel-Mottaleb MMS, De Feyter S, Veciana J, Rovira C, Amabilino DB, De Schryver FC (2003) Chem Commun 906
93. Lu J, Zeng QD, Wang C, Wang LJ, Bai CL (2003) Chem Lett 32:856
94. Wang D, Wan LJ, Wang C, Bai CL (2002) J Phys Chem B 106:4223
95. Su GJ, Yin SX, Wan LJ, Zhao JC, Bai CL (2003) Chem Phys Lett 370:268
96. Su GJ, Yin SX, Wan LJ, Zhao JC, Bai CL (2004) Surf Sci 551:204
97. Petersen J, Strohmaier R, Gompf B, Eisenmenger W (1997) Surf Sci 389:329
98. Janssens G, Touhari F, Gerritsen JW, van Kempen H, Callant P, Deroover G, Vanenbroucke D (2001) Chem Phys Lett 344:1

99. Samori P, Francke V, Enkelmann V, Müllen K, Rabe JP (2003) Chem Mater 15:1032
100. Höger S, Bonrad K, Mourran A, Beginn U, Möller M (2001) J Am Chem Soc 123:5651
101. Bäuerle P, Fischer T, Bidlingmaier B, Stabel A, Rabe JP (1995) Angew Chem Int Ed Engl 34:303
102. Azumi R, Götz G, Debaerdemaeker T, Bäuerle P (2000) Chem Eur J 6:735
103. Kirschbaum T, Azumi R, Mena-Osteritz E, Bäuerle P (1999) New J Chem 241
104. Stecher R, Gompf B, Muenter JRS, Effenberger F (1999) Adv Mater 11:927
105. Stecher R, Drewnick F, Gompf B (1999) Langmuir 15:6490
106. Azumi R, Götz G, Bäuerle P (1999) Synt Met 101:569
107. Vollmer MS, Effenberger F, Stecher R, Gompf B, Eisenmenger W (1999) Chem Eur J 5:96
108. Soukopp A, Glöckler K, Bäuerle P, Sokolowski M, Umbach E (1996) Adv Mater 8:902
109. Seidel C, Soukopp A, Li R, Bäuerle P, Umbach E (1997) Surf Sci 374:17
110. Gesquière A, Abdel-Mottaleb MMS, De Feyter S, De Schryver FC, Schoonbeek F, van Esch J, Kellogg RM, Feringa BL, Calderone A, Lazzaroni R, Brédas JL (2000) Langmuir 16:10385
111. Gesquière A, De Feyter S, De Schryver FC, Schoonbeek F, van Esch J, Kellogg RM, Feringa BL (2001) Nano Lett 1:201
112. Krömer J, Rios-Carreras Idoia, Fuhrmann G, Musch C, Wunderlin M, Debaerdemaeker T, Mena-Osteritz E, Bäuerle P (2000) Angew Chem Int Ed Engl 39:3481
113. Mena-Osteritz E, Bäuerle P (2001) Adv Mater 13:243
114. Gesquière A, Jonkheijm P, Schenning APHJ, Mena-Osteritz E, Bäuerle P, De Feyter S, De Schryver FC, Meijer EW (2003) J Mater Chem 13:2164
115. Gesquière A, Jonkheijm P, Hoeben FJM, Schenning APHJ, De Feyter S, De Schryver FC, Meijer EW (2004) Nano Lett 4:1175
116. Jonkheijm P, Miura A, Zdanowska M, Hoeben FJM, De Feyter S, Schenning APHJ, De Schryver FC, Meijer EW (2004) Angew Chem Int Ed Engl 43:74
117. Miura A, Chen Z, De Feyter S, Zdanowska M, Uji-i H, Jonkheijm P, Schenning APHJ, Meijer EW, Würthner F, De Schryver FC (2003) J Am Chem Soc 125:14968
118. Lehn JM (1993) Science 260:1762
119. Bäuerle P (1993) Adv Mater 5:879
120. Samori P, Francke V, Müllen K, Rabe JP (1999) Chem Eur J 5:2312
121. Mena-Osteritz E, Meyer A, Langeveld-Voss BMW, Janssen RAJ, Meijer EW, Bäuerle P (2000) Angew Chem Int Ed Engl 39:2680
122. Grévin B, Rannou P, Payerne R, Pron A, Travers JP (2003) J Chem Phys 118:7097
123. Grévin B, Rannou P, Payerne R, Pron A, Travers JP (2003) Adv Mater 15:881
124. Lei SB, Wan LJ, Wang C, Bai CL (2004) Adv Mater 16:828
125. Lu J, Lei SB, Zeng QD, Kang SZ, Wang C, Wan LJ, Bai CL (2004) J Phys Chem 108:5161
126. Theobald J, Oxtoby NS, Phillips MA, Champness NR, Beton PH (2003) Nature 424:1029
127. de Wild M, Berner S, Suzuki H, Yanagi H, Schlettwein D, Ivan S, Baratoff A, Guentherodt HJ, Jung TA (2002) Chem Phys Chem 10:881
128. Pan GB, Lu JM, Zhang HM, Wan LJ, Zheng QY, Bai CL (2003) Angew Chem Int Ed Engl 42:2747
129. Freund J, Probst O, Grafström S, Dey S, Kowalski J, Neumann R, Wörtge M, zu Pulitz G (1994) J Vac Sci Technol B 12:1914
130. De Feyter S, Miura A, Yao S, Chen Z, Würthner F, Jonkheijm P, Schenning APHJ, Meijer EW, De Schryver FC (2005) Nano Lett 5:77
131. Hipps KW, Scudiero L, Barlow DE, Cooke MP Jr (2002) J Am Chem Soc 124:2126

132. Scudiero L, Hipps KW, Barlow DE (2003) J Phys Chem B 107:2903
133. Suta K, Yoshimoto S, Itaya K (2003) J Am Chem Soc 125:14976
134. Xu B, Yin S, Wang C, Qiu X, Zeng Q, Bai CL (2000) J Phys Chem B 104:10502
135. Yin S, Wang C, Xu B, Bai CL (2002) J Phys Chem B 106:9044
136. Lei SB, Yin SX, Wang C, Wan LJ, Bai CL (2004) J Phys Chem B 108:224
137. Lei SB, Wang C, Wan LJ, Bai CL (2004) J Phys Chem B 108:1173
138. Lei SB, Wang C, Yin SX, Bai CL (2001) J Phys Chem B 105:12272
139. Ludwig C, Strohmaier R, Peterson J, Gompf B, Eisenmenger W (1994) J Vac Sci Technol B 12:1963
140. Lei SB, Yin SX, Wang C, Wan LJ, Bai CL (2002) Chem Mater 14:2837
141. De Feyter S, Larsson M, Schuurmans N, Verkuijl B, Zoriniants G, Gesquière A, Abdel-Mottaleb MM, van Esch J, Feringa BL, van Stam J, De Schryver F (2003) Chem Eur J 9:1198
142. Pérez-Garcia L, Amabilino DB (2002) Chem Soc Rev 31:342
143. Fasel R, Parschau M, Ernst KH (2003) Angew Chem Int Ed Engl 42:5178
144. Ernst KH, Kuster Y, Fasel R, Müller M, Ellerbeck U (2001) Chirality 13:675
145. Brian France C, Parkinson BA (2003) J Am Chem Soc 125:12712
146. Wei Y, Kannappan K, Flynn GW, Zimmt MB (2004) J Am Chem Soc 126:5318
147. Schunack M, Linderoth TR, Rosei F, Laegsgaard E, Stensgaard I, Besenbacher F (2002) Phys Rev Lett 88:156102
148. Weckesser J, Barth JV, Kern K (2001) Phys Rev B 64:161403
149. Weckesser J, Barth JV, Kern K (1999) J Chem Phys 110:5351
150. Gimzewski JK, Joachim C, Schlitter RR, Langlais V, Tang H, Johannsen I (1998) Science 281:531
151. Stabel A, Heinz R, De Schryver FC, Rabe JP (1995) J Phys Chem 99:505
152. Rabe JP, Buchholz S (1991) Phys Rev Lett 66:2096
153. Moresco F, Meyer G, Rieder KH, Tang H, Gourdon A, Joachim C (2001) Appl Phys Lett 78:306
154. Moresco F, Meyer G, Rieder KH, Tang H, Gourdon A, Joachim C (2001) Phys Rev Lett 86:672
155. Samori P, Engelkamp H, de Witte P, Rowan AE, Nolte RJM, Rabe JP (2001) Angew Chem Int Ed Engl 40:2348
156. Okawa Y, Aono M (2001) Nature 409:683
157. Miura A, De Feyter S, Abdel-Mottaleb MMS, Gesquière A, Grim PCM, Moessner G, Sieffert M, Klapper M, Müllen K, De Schryver FC (2003) Langmuir 19:6474
158. Akai-Kasaya M, Shimizu K, Watanbe Y, Saito A, Aono M, Kuwahara Y (2003) Phys Rev Lett 91:255501
159. Jäckel F, Wang Z, Watson MD, Müllen K, Rabe JP (2004) Chem Phys Lett 387:372
160. Mizutani W, Shigeno M, Kajimura K, Ono M (1992) Ultramicroscopy 42-44:236
161. Jäckel F, Watson MD, Müllen K, Rabe JP (2004) Phys Rev Lett 92:188303
162. Berndt R, Gaisch R, Gimzewski JK, Reihl B, Schlitter RR, Schneider WD, Tschudy M (1993) Science 262:1425
163. Touhari F, Stoffels EJAJ, Gerritsen JW, van Kempen H (2001) Appl Phys Lett 79:527
164. Dong ZC, Guo XL, Trifonov AS, Dorozhkin PS, Miki K, Kimura K, Yokoyama S, Mashiko S (2004) Phys Rev Lett 92:86801

Top Curr Chem (2005) 258: 257–313
DOI 10.1007/b135682

Published online: 8 August 2005

# Self-Assembled Monolayers of Chromophores on Gold Surfaces

Volker Kriegisch · Christoph Lambert (✉)

Universität Würzburg, Institut für Organische Chemie, Am Hubland, 97074 Würzburg, Germany
*lambert@chemie.uni-wuerzburg.de*

**Abstract** In the last two decades the field of self-assembled monolayers (SAMs) has received much attention both in research and for applications because of the stability and the simple handling of SAMs. In this account we present a short overview about the morphology and preparation of the different gold surfaces used for SAMs and about self-assembled alkanethiol monolayers on Au(111). The synthetic strategies and the aggregation of self-assembled chromophores on gold is described in more detail. The most important chromophores are introduced along with their spectroscopic UV/vis, fluorescence, electron and energy transfer characteristics. Applications of self-assembled dyes on gold as sensors, photoswitchable materials, catalysts and for photocurrent generation are outlined.

**Keywords** Chromophore · Self-assembled monolayer (SAM) · Gold · Electron transfer

**Abbreviations**

| | |
|---|---|
| AFM | Atomic force microscopy |
| ARXPS | Angle resolved X-ray photoelectron spectroscopy |
| COx | Cytochrome oxidase |
| EN | Energy transfer |
| ET | Electron transfer |
| FET | Field-effect transistors |
| FT-IRRAS | Fourier transform infrared reflection absorption spectroscopy |
| GOD | Glucose oxidase |
| IRRAS | Infrared reflection absorption spectroscopy |
| ITO | Indium tin oxide |
| LB | Langmuir–Blodgett |
| $MRH^+$-GOD | Protonated merocyanine-glucose oxidase |
| NEXAFS | Near-edge X-ray absorption fine structure spectroscopy |
| NLO | Non-linear optical |
| OLED | Organic light-emitting diode |
| OPV | Oligophenylenevinylene |
| RAIR | Reflection absorption infrared spectroscopy |
| SAM | Self-assembled monolayer |
| SP-GOD | Nitrospiropyran-glucose oxidase |
| STM | Scanning tunnelling microscopy |
| TPP | Tetraphenylporphyrin |
| XPS | X-ray photoelectron spectroscopy |

# 1 Introduction

The research topic of self-assembled monolayers (SAMs) has grown tremendously in synthetic sophistication and depth of characterisation over the past 20 years [1]. Functionalised self-assembled monolayers [2, 3] have attracted the attention of many scientists because of their potential applications in chemical sensors [4–6], biosensors [7, 8], NLO-systems [9–13], molecular switches [8, 14–19], molecular electronics [20–25], photovoltaic devices [26–28], active surfaces for patterning [29–33] and chemical architecture [34, 35]. Although chromophore SAMs attached to gold nanoparticles [36–55] and to other surfaces, such as $SiO_2$ [1, 3, 56–59] or ITO (indium tin oxide) [17, 60, 61], play an increasingly important role and even might surpass bulk gold in many technological aspects we restrict ourselves here to chromophore SAMs on bulk gold surfaces.

It is well known that organosulfur compounds react with gold surfaces. In 1982 Taniguchi et al. observed the spontaneous formation of pyridine disulfide SAMs on a gold surface [62]. In 1983, Nuzzo and Allara showed that SAMs of alkanethiolate on gold could be prepared by adsorption of *n*-alkyl disulfides from dilute solutions [63]. This report initiated a rapid growth in SAM research.

The reasons for the large interest in SAMs are their great advantages over Langmuir–Blodgett (LB) films [64, 65]. Although there have been numerous attempts to organise molecules in LB films, the instability of the physisorbed layers and the defects caused by previously adsorbed material are inherent limitations [66]. On the other hand SAMs are ordered molecular assemblies formed by the adsorption of an active surfactant on a solid surface, which includes the formation of a covalent bond of thiol, e.g. to gold or silver. This simple process makes SAMs inherently manufacturable and thus technologically attractive. Chemisorption involves relatively large heats of bond formation (40–160 kJ/mol) and has two advantages: firstly, the chemical reaction displaces any previously formed physically attached adsorbates or impurities from the surface. Secondly, the adsorbed species, once bonded, are difficult to remove from the surface. However, there are three disadvantages of chemisorption: the uncertain degree of coverage, the possibility of further chemical reactions (e.g. thiolates on gold slowly oxidise to sulfoxides in high-humidity environment) and the formation of surface dipoles [66].

While the majority of publications deal with thiols on gold, silanes on hydroxylated surfaces like silicon, glass, mica or ITO are important systems for many potential applications [3, 55, 67, 68], although high quality SAMs of alkyltrichlorosilane derivatives are difficult to produce. SAMs of fatty acids are an important link between the LB and the self-assembly techniques [3].

While the early days of SAM research were almost exclusively devoted to the investigation of the formation, structure and physical properties of alkanethiols on gold, in the past decade the function of such SAMs has moved into the foreground. In many cases the functionality of SAMs is associated with chromophores or $\pi$-systems in the broadest sense. Although one can hardly speak of dyes in the common sense because the colour of these SAM chromophores is invisible to the human eye owing to the low absorptivity of the monolayer, changes of absorptivity, orientation, redox state, electron transfer behaviour etc. can readily be measured by a variety of physical methods and can be used to build up functional units on a monolayer scale.

Therefore, if we speak of "chromophores" in this context we simply mean more or less extended functional $\pi$-systems.

The purpose of this review is to give an overview of the different chromophore-carrying thiols used for self-assembly on gold surfaces, their spectroscopic behaviour and their possible applications. Because of their stability and easy preparation, chromophores on gold may become important systems for many technological applications. Efforts continue to achieve better reproducibility in monolayer preparation. We attempt to provide a general picture about the structure and preparation of gold surfaces for SAMs and to give the basics of alkanethiol monolayers on gold. Structural factors depending on the chromophores bound to gold surfaces and the different preparation pathways are also discussed.

## 2 Morphology of Gold Surfaces for SAMs

Considering crystalline gold for SAM preparation, one has to distinguish two different morphologies, i.e. polycrystalline and single-crystal substrates. If the substrate is a single crystal, a particular face must be selected. Although many studies have been reported on bulk single gold crystals, most "single-crystal" work has been done on evaporated or sputtered gold films on cleaved mica, polished single-crystal silicon or glass. The gold films exhibit a strong Au(111) character when deposited on glass or silicon, and they are essentially 100% Au(111) when deposited on freshly cleaved mica (Fig. 1) [69–72].

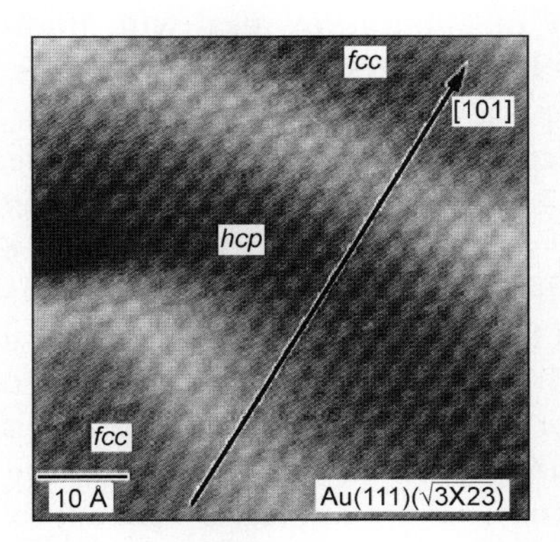


**Fig. 1** Constant-current STM topograph of Au(111) surface showing quasihexagonal arrangement of Au atoms. Reprinted with permission from [106]. Copyright (1997) American Chemical Society

Creager et al. showed that SAMs on polycrystalline gold electrodes are superior in terms of defects of the alkanethiol monolayers to SAMs on evaporated gold films on smooth substrates (silicon wafer or cleaved mica) if the polycrystalline gold is annealed at 1000 °C, polished, and then etched with dilute aqua regia [73].

Electrodes using sputtered and annealed gold films on glass support SAMs that are apparently free of pinholes [74, 75] even though the grain structure is much smaller on sputtered films than on evaporated films [69]. However, an atomically smooth plane of gold atoms is still anticipated to yield the highest ordered and most regularly packed monolayer [2].

One effective characterisation tool for the investigation of SAMs, apart from the most widely used characterisation methods (AFM, STM and X-ray diffraction [76]) is UV/vis spectroscopy. The thickness of the gold films on glass substrates used for UV/vis transmittance measurements varies between 5 and 20 nm. A major drawback of this technique is the intense absorption of the gold substrate [77]. To minimise the self-absorption of the gold substrate, ultrathin gold films of $\leq 10$ nm thickness can be used [78]. It is well known that the substrate may influence the electronic and, hence, the optical properties of the adsorbate and vice versa. Such effects are mostly studied in colloid systems [79–81], although it is quite clear that the interaction may also be noticeable in ultrathin gold films, where the "island" structure resembles that of a colloid (Fig. 2).

Rubinstein et al. showed that these discontinuous and inhomogeneous films, composed of nanometer-sized gold clusters (islands), exhibit absorption in the visible or near-infrared region, which is attributed to the surface plasmon resonance of gold (Fig. 3) [78] comparable to other ultrathin metallic films [82–84].

The sizes and optical properties of these Au(111) textured islands can be controlled by the evaporation conditions and subsequent annealing (see Fig. 2). The influence of the evaporation parameters on the gold layer texture were extensively studied [85–88].

A very simple method to get a single-crystal electrode is to melt a gold wire in a hydrogen flame; Au(111) crystal planes appear in small steps on the surface of the molten ball [89]. "Annealing" a gold ball prepared according to this procedure by cycling the electrode potential using dilute perchloric acid and traces of chloride removes atomic-scale defects from single crystal planes [90].

Another very important carrier for SAMs are gold nanoparticles [37, 54] because of their stability and their fascinating aspects associated with individual particles, size-related electronic, magnetic and optical properties (quantum size effects), and their applications in sensors [91] and for biomolecular labelling or as immunoprobes [53, 92, 93].

Owing to the roughness of the surface the actual surface area of gold is always larger than the geometrical surface area. The actual surface area ($A$) of

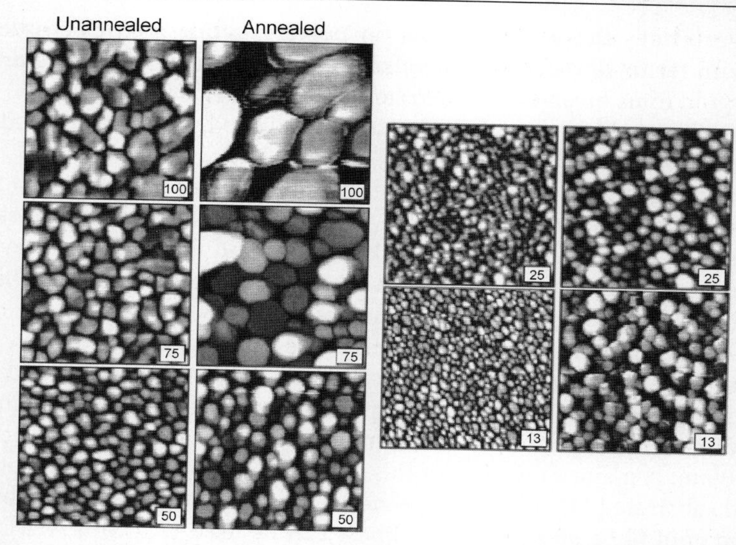


**Fig. 2** Tapping-mode STM topographic images of ultrathin gold films (300 × 300 nm) of 100, 75, 50, 25 and 13 Å thickness on mica. Compared to the unannealed films annealing leads to larger separation between individual islands, increase of the island diameter, and flattening of the upper surface, exposing the Au(111) crystallographic face. Reprinted with permission from [78]. Copyright (2000) American Chemical Society

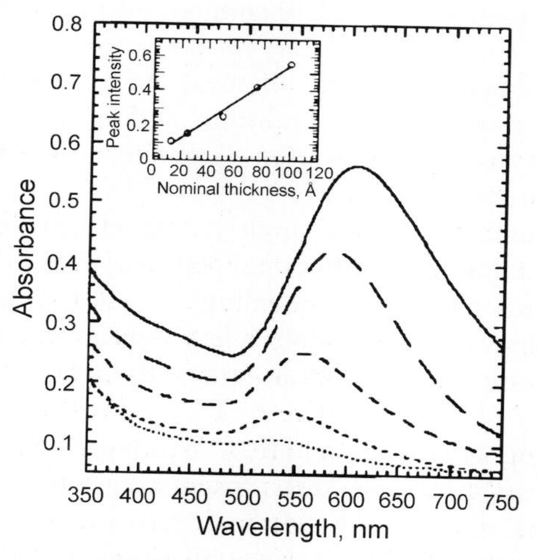


**Fig. 3** UV/vis spectra of ultrathin gold films on mica, showing the gold surface plasmon band for annealed films. Thicknesses: 13, 25, 50, 75 and 100 Å (from *bottom* to *top*). *Inset*: Dependence of the band intensity on the gold nominal thickness. Reprinted with permission from [78]. Copyright (2000) American Chemical Society

gold electrodes is important to know for the calculation of the surface coverage of gold electrodes with thiols. Soriaga et al. introduced a reliable method for determination of $A$ using chemisorbed iodine [94].

# 3 Basic Aspects of SAMs

Owing to the easy preparation, to the well-defined order and also to the relative inertness of the substrate, thiols on gold have become a model system for SAMs. The most studied and probably best understood SAMs are those of alkanethiolates on Au(111) surfaces [3, 95–103]. The following description gives a brief overview of the basics of alkanethiol-SAMs; more information can be obtained from the reviews mentioned above and related literature.

## 3.1 Structure of Alkanethiol-SAMs

A gold substrate can be coated with an organised monolayer simply by immersing the substrate into a dilute solution of thiol or disulfide. The formation of the SAMs may be considered formally as an oxidative addition of the S – H bond to the gold surface, followed by a reductive elimination of hydrogen. Extensive X-ray photoelectron spectroscopic (XPS) experiments suggest that chemisorption of alkanethiols on gold(0) surfaces yields the gold(I) thiolate (R– S⁻) species [104]. The presumed adsorption chemistry is:

$$\mathrm{R-SH + Au_n{}^0 \rightarrow R-S^- - Au^+ + \frac{1}{2}H_2 + Au_{n-1}{}^0}$$

The bonding of the thiolate group to the gold surface is very strong (the homolytic bond strength is approximately 160 kJ/mol) [105].

Even though the mechanism of SAM formation is still under debate the structure of the monolayer is well understood. In general the arrangement of sulfur atoms is hexagonal and commensurate with the underlying gold lattice (Fig. 4a). This imposes a S – S distance of 4.97 Å [3]. Depending on the length of the alkyl chains and their relative orientation superlattices may form as exemplified by the c(4 × 2) superlattice of a $(\sqrt{3} \times \sqrt{3})R30°$ overlay structure of octanethiol on Au(111) (Fig. 4b) [106]. While it is generally believed that the stability of the alkanethiol SAMs increases with increasing chain length owing to increasing van der Waals interactions, Ulman et al. suggested two chemisorption modes with 180° surface-S – C angle ($sp$-hybridisation of S) and 104° surface-S – C angle ($sp^3$-hybridisation of S), the latter being slightly more stable [107]. Indeed, Zharinkov et al. demonstrated that hybridisation and thus spatial orientation of the bonding orbitals of sulfur is the deter-

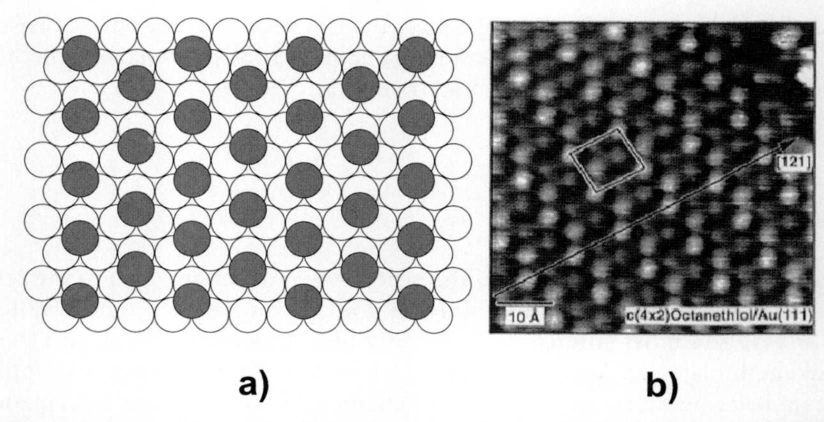


**Fig. 4** (**a**) Hexagonal coverage scheme for alkanethiolates on Au(111). The *open circles* are gold atoms and the *grey circles* are sulfur atoms [3]; (**b**) Constant-current STM topograph of an octanethiol monolayer on Au(111) which shows a $c(4 \times 2)$ superlattice of a $(\sqrt{3} \times \sqrt{3})R30°$ overlay structure. Reprinted with permission from [106]. Copyright (1997) American Chemical Society

mining factor for the orientation and density of the monolayers rather than intermolecular interactions [108].

Because in SAMs the sulfur–sulfur interactions are believed to be minimal and the sulfur–sulfur distance is greater than the minimal distance of the alkyl chains, the latter are tilted in order to maximise the interchain alkyl–alkyl van der Waals interactions. This tilt angle is found to be $\sim 30°$ with respect to the surface normal towards the nearest neighbour direction (Fig. 5) [95].

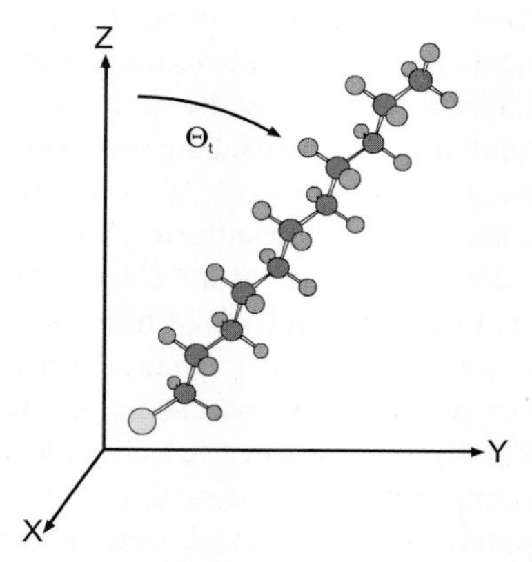


**Fig. 5** The tilt angle $\Theta_t$ of the alkanethiol chain relative to the surface normal [95]

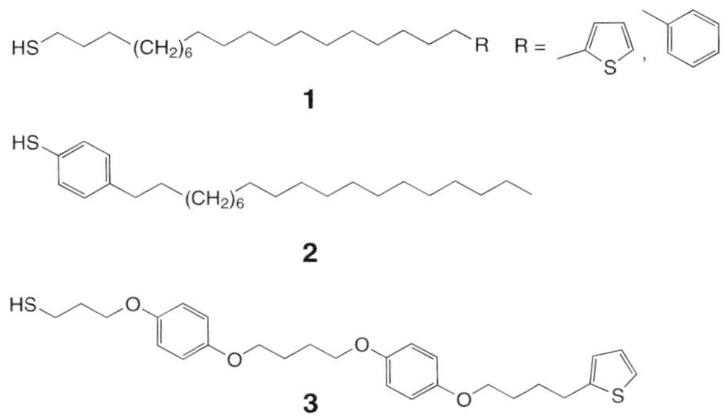

**Fig. 6** Structure of alkanethiols **1–3** [109]

The tilt angles of alkanethiols towards the surface normal can be varied by incorporation of aromatic groups into the alkyl chains (Fig. 6). Larger tilt angles (37–40°) can be observed by introducing aliphatic and aromatic moieties, which are directly connected (Fig. 6, compounds **1** and **2**). By connecting the aromatic and the aliphatic groups via an oxygen bridge smaller tilt angles, with a nearly perpendicular orientation of the chromophores to the goldsurface, are found (Fig. 6, compound **3**). As derived from RAIR-spectra, the tilt angle of SAMs on gold can be influenced in a desired manner by appropriate modification of the alkyl chains [109].

## 3.2 Adsorption Kinetics of Alkanethiol-SAMs

Kinetic studies of alkanethiol adsorption from ethanol solutions onto Au(111) surfaces have shown that at relatively low concentrations ($10^{-3}$ M) two distinct adsorption kinetic steps can be observed: a very fast step, which takes a few minutes and leads to about 80–90% of maximal coverage; and a slow step, which lasts several hours and at the end of which the thickness and contact angles reach their final values [110]. The initial step, described by the diffusion-controlled Langmuir adsorption, was found to be strongly dependent on the thiol concentration. At 1 mM solution the first step was over after ~ 1 min, while it required over 100 min at 1 μM concentration. The second step can be described as a surface crystallisation process where the alkyl chains move from disordered states into unit cells, thus forming a two-dimensional crystal [110]. A more detailed study of Buck et al. involved different types of adsorption sites and showed that the rate-determining step strongly depends on the type of solvent [111–113].

Concerning the chain length of alkanethiol monolayers, Bain et al. noted that the monolayer properties of longer-chain *n*-alkanethiols ($n > 8$) were consistent but that shorter-chain thiol monolayers were qualitatively different, displaying greater disorder [110]. Comparing alkanethiols with different chain length Peterlinz and Georgiadis found that for the initial fast adsorption step (25 min for a 1 mM solution) the rate decreased with increasing chain length whereas the opposite trend was observed in the second step, which lasted many hours [112]. Monolayers composed of $C_8$- and $C_{12}$-chain molecules reached > 80% of the final thickness in the first step while the longer-chain films ($C_{16}$ and $C_{18}$) only reached 40–50%. Not withholding that the literature is full of dramatically conflicting reports regarding the effects of chain length on thiol SAM growth kinetics [99, 110–116], we sum up by saying that in general alkanethiols with short chains (< $C_6$) form disordered monolayers, while in longer alkanethiol chains (> $C_7$) the van der Waals interactions are strong enough to build well-ordered SAMs.

# 4 Structural Aspects of Chromophore SAMs

## 4.1 Synthesis of SAMs

Generally, two different strategies are employed for the formation of SAMs of functional components on gold surfaces. One method is the direct adsorption of a functional component that carries a thiol or disulfide group. This sulfur group is attached to the functional component by a synthetic procedure prior to the SAM formation step (Fig. 7) [27, 117–119]. The sulfur group itself may either be a thiol, a disulfide, an acetyl protected thiol or [1, 2]-dithiolane.

The alternative method is covalent attachment of the functional components onto a preformed SAM of $\omega$-functionalised (e.g. – COOH, $NH_2$) alkanethiols (Fig. 8) [120–122].

For an easy variation of the spacer length in mixed SAMs, Schuhmann and Gorton developed a three-step modification procedure (Fig. 9) [121].

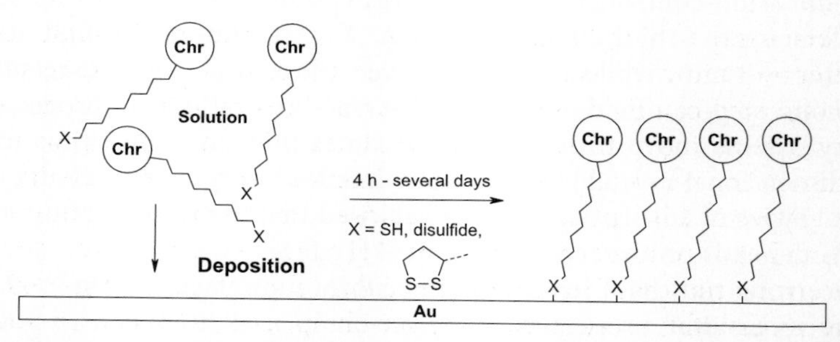


**Fig. 7** Direct adsorption method of thiolated compounds onto gold (Chr = chromophore)

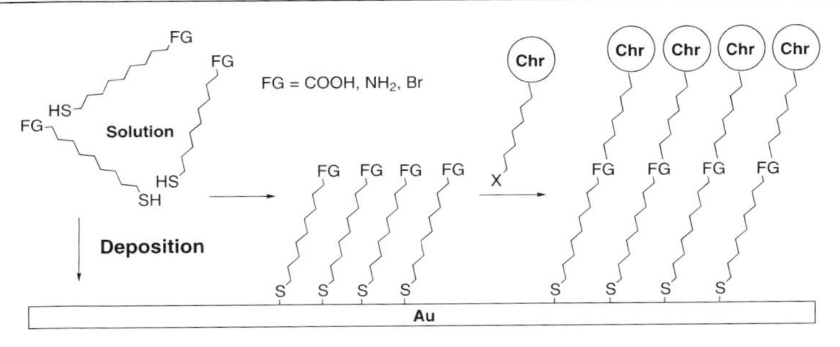


**Fig. 8** Covalent attachment of the functional components onto preformed SAM of $\omega$-functionalised alkanethiols (Chr = chromophore)

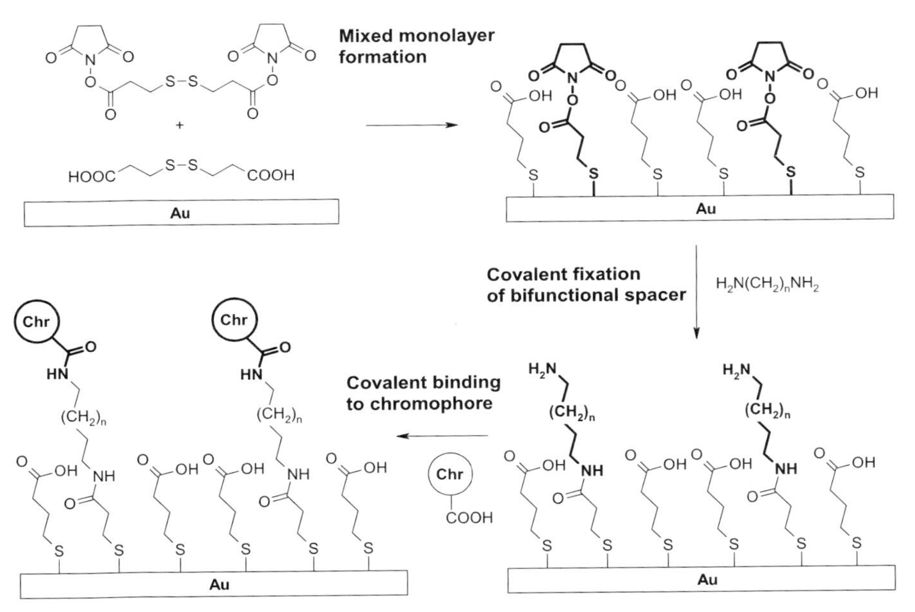


**Fig. 9** Schematic representation for the immobilisation of a chromophore on a dilute spacer [121]

In the first step, a monolayer consisting of an activated disulfide (3,3′-dithiopropionic acid di-(*N*-succinimidyl ester); Lomant's reagent) mixed with 3-carboxypropyl-disulfide (1 : 99) is self-assembled on the electrode surface. Addition of high excess of 1, $\omega$-diaminoalkanes to the monolayer, which contains the activated head groups, yields a monolayer terminated with amino groups. Activation of the carboxylic residues of the chromophore leads to amide-bonded chromophore monolayers. This approach has the advantage that the concentration of functionalised molecules can be determined by the initial ester/disulfide ratio. In this way, chromophore-SAMs can be studied in which the chromophore units are well separated from each other.

## 4.2 Structures of Basic $\pi$-Systems in SAMs

No general statement concerning the structure of self-assembled chromophores on gold surfaces can be made. Two reasons account for this problem: the first is that the chromophore moieties are usually much larger than alkyl chains, which inhibits a well-ordered arrangement of chromophore-appended alkanethiol-SAMs. The second problem arises because of specific interactions between the chromophore units, such as $\pi$–$\pi$ stacking etc. We will briefly describe the structures of some simple $\pi$-systems attached to gold. Although these systems can hardly be viewed as chromophores they may serve to illustrate some features of $\pi$-system-SAMs, which also play a role in more complex chromophore-SAM assemblies.

Thioaromatic compounds were found to form highly oriented and densely packed SAMs on gold. The molecular orientation and orientational order of the adsorbed thioaromatic molecules depends on the number of aromatic rings and the rigidity of the aromatic systems. Comparison of the chemical homologue TP (thiophenol), BPT (*p*-biphenylthiol) and TPT (*p*-terphenylthiol) monolayers shows that the molecular orientation becomes less tilted with increasing length of the aromatic chain (Fig. 10A) [123]. This tendency can be explained by the assumption that the intermolecular forces responsible for the self-assembling become stronger with an increasing number of aromatic rings. Investigations of AnT (anthracene-2-thiol), which has more rigidity than TPT, show that the size of the conjugated system affects the final molecular orientation within the layer to a larger extent than the rigidity of the molecule. In contrast to TP with only one phenyl ring, AnT forms a highly oriented SAM, which is presumably a consequence of the stronger intermolecular interaction between the conjugated anthracene backbones of the adjacent AnT systems [123]. In contrast to the above mentioned BPT, biphenyldithiol (BPDT) and biphenyldimethylthiol (BPDMT) do not form well-oriented monolayers because of oligomerisation reactions in which two to four units are connected by disulfide bridges due to their air sensitivity (Fig. 10D). However, the increase of the aromatic backbone interaction in terphenyldimethylthiol (TPDMT) leads to the formation of SAMs with a high degree of molecular orientation (Fig. 10C) [124]. Depending on the $\omega$-functionalisation, carboxy terphenylmethanethiols (CTPMT) form highly oriented bilayers where the second layer of COOH-terminated CTPMT is bound to the first layer by hydrogen bonds in a head-to-head fashion (Fig. 10B) [125].

Independent of their position, the incorporation of aryl groups in alkyl chains influences the chain molecular orientation to the surface, but does not prevent dense packing. Depending on the bonding between the aryl groups and alkyl chains (alkyl-Ph versus alkyl-O-Ph) the molecules possess smaller or larger tilt angles than normal alkanethiols (see Sect. 3.1, Fig. 6) [109].

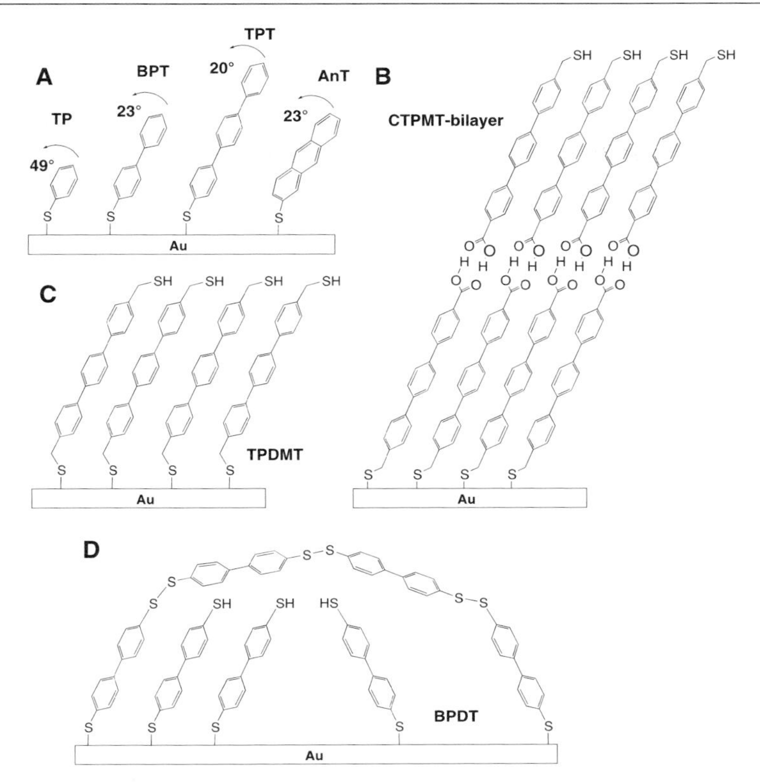


**Fig. 10** Thioaromatic compound used for structural investigations [123–125]

Comparison of the benzyl mercaptans **4** and 4-biphenylmethanethiols **5** with the thiophenols **6** and the biphenylthiols **7** (Fig. 11) shows that the chromophores **4** and **5**, with a methylene unit between the thiol and the aromatic group, form closely packed and well-ordered monolayers independent of the length of the *p*-alkoxy substituents. For **6** and **7**, ordered monolayers can only be achieved by increasing the length of the alkoxy chains, indicating that the bond angle between the thiolate head group and the $\pi$-system influences the packing of the monolayers [126, 127].

Zharnikov et al. investigated the even–odd effects of biphenyl- (**8**) and terphenyl-substituted (**9**) alkanethiols with XPS, FT-IRRAS and NEXAFS in terms of the spacer length between the aromatic backbone and the thiol function and the packing density on different substrates. While on Au(111) the chromphores with even numbers of methylene units are loosely packed and the aromatic unit is tilted towards the gold surface, the tilt angle behaviour changes in the opposite way by using Ag(111) surfaces (Fig. 12) [128–130].

Oligo(phenylene-ethynylene)benzenethiols **11** and **12** also form the more highly ordered SAMs the longer the conjugated backbone is [131]. While STM

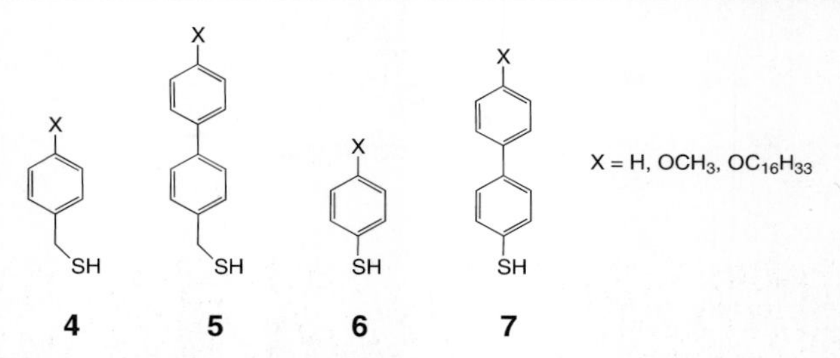


**Fig. 11** Molecules **4**–**7** for structural investigations in SAMs [126]

images indicate that **10** forms islands or pits on the gold surface, **11** exhibits a certain degree of order and **12** consists of highly ordered domains, which have a row structure consisting of dimeric subunits (Fig. 13). Compound **12** and the nitro-substituted compound **13** posses tilt angles on Au(111) surfaces between 33–39° with a $(\sqrt{3} \times \sqrt{3})R30°$ superlattice structure [132].

Oligothiophenes are also attractive conjugated molecules for SAM formation [60]. In the case of oligothiophene-derivatised thiols, both the thiophene rings and the thiol group can directly adsorb onto the Au surface: in thienylalkanethiol-SAMs with short alkyl chains between the oligothiophene and the thiol, the thiophenes directly adsorb onto the gold surface because of the small van der Waals interactions of the alkyl chains. In contrast, increasing the chain length of the alkyl bridges of the oligothiophenes leads to well-ordered monolayers.

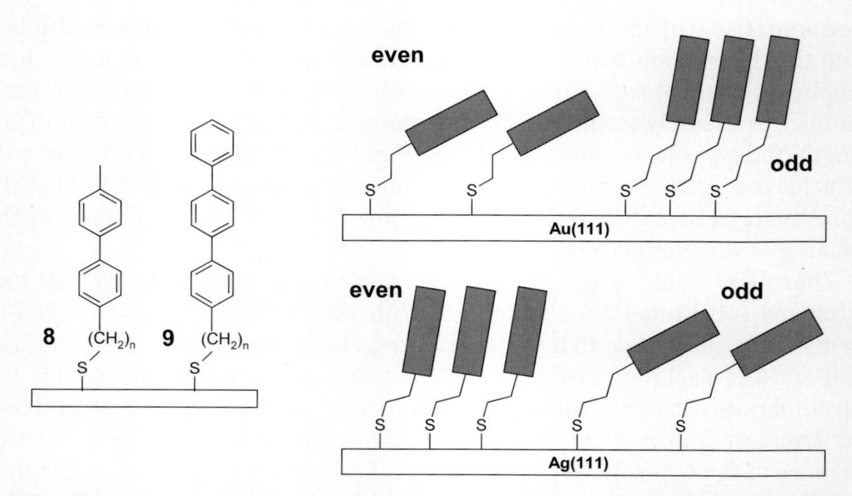


**Fig. 12** Schematic drawing of the orientation and packing of the biphenyl- (**8**) and terphenyl-substituted (**9**) alkanethiols on Au(111) and Ag(111) [128, 130]

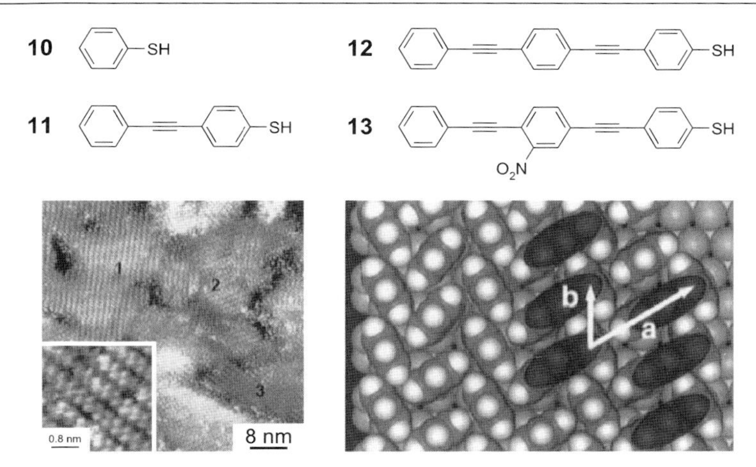


**Fig. 13** *Top*: Structures of oligo(phenylene-ethynylene)benzenethiols **10–13** [131, 132]. *Bottom left*: STM image of **12** on Au(111)/mica. *Bottom right*: top-down view of a $2(\sqrt{3} \times \sqrt{3})R30^\circ$ model for SAMs of **12** on Au(111). Ovals indicate a possible dimerisation scheme. Reprinted with permission from [131]. Copyright (1996) American Chemical Society

Oligothiophenes carrying an alkanethiol group on one of the internal $\beta$-positions adsorbed on gold are not very well organised. Although the SAMs are quite dense they contain many defects due to the T-like geometry of the single molecules (Fig. 14, **14**, **15**) [133]. FTIR-spectroscopy indicates loosely interacting alkyl chains in a densely packed structure in which the thiophene-oligomere is slightly tilted up towards the surface normal.

In contrast, ter- and quarterthiophene-alkanethiols bearing alkanethiol chains at the $\alpha$-position are densely packed and highly ordered (Fig. 14,

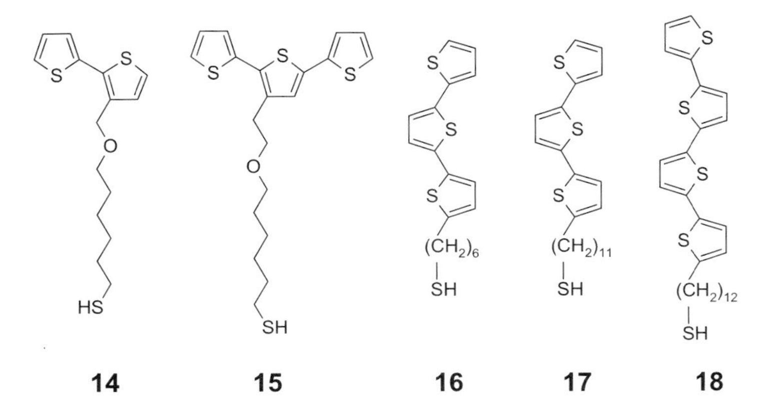


**Fig. 14** Different oligothiophenes for SAMs [133–136]

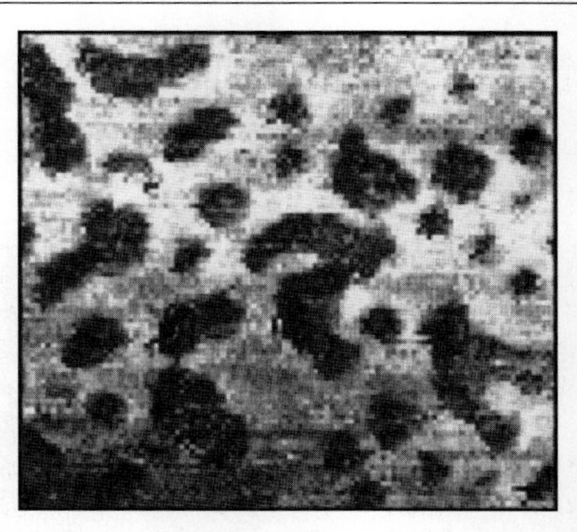

**Fig. 15** STM image showing a $44 \times 41\ nm^2$ area of a SAM of a mixture of 25% $CH_3(CH_2)_{15}SH$ and 75% $CH_3O_2C(CH_2)_{15}SH$. Brighter regions are $CH_3O_2C(CH_2)_{15}SH$ domains and darker regions are $CH_3(CH_2)_{15}SH$ domains. Reprinted with permission from [141]. Copyright (1994) American Chemical Society

**16, 18**). IRRAS spectroscopy indicates that the quarterthiophene is almost parallel to the surface normal [134, 135] while the terthiophene-alkanethiol **17** is tilted about 14° away from the surface normal [136].

SAMs consisting of a mixture of non-functionalised alkanethiols and $\omega$-functionalised alkanethiols provide the chance to study "isolated" functionalised sites in SAMs. In these mixed SAMs the non-functionalised alkanethiols serve as "dummy" molecules to dilute the functionalised compounds and to isolate the active sites in order to minimise their interactions. These mixed SAMs play an important role, e.g. in the development of biorecognition surfaces [137–139]. One obstacle in producing mixed monolayers is the different rate of adsorption of the non-functionalised and the end-functionalised thiols. Consequently, the final composition of the mixed SAMs not only depends on the initial ratio of the two thiols but also on their individual as well as cooperative adsorption and desorption kinetics [140]. Another problem is the tendency for similar molecules to aggregate into islands on the surface (Fig. 15) [141, 142]. Therefore, the final SAM composition cannot be predicted easily.

## 4.3 Selected Important Chromophores in SAMs

### 4.3.1 Porphyrin and Metalloporphyrin SAMs

In recent years organised assemblies of porphyrins on gold surfaces have attracted considerable attention, mainly directed towards the development of photovoltaic devices, catalysis and photonic sensors [27, 143].

The structure of porphyrins in SAMs has been widely studied because of the strong and characteristic UV/vis absorption features of porphyrin chromophores (see Sect. 5.1) [78, 142, 144]. Porphyrins may form two different types of aggregation: a face-to-face porphyrin $\pi$-aggregation (sandwich-type H-aggregate) and a side-by-side porphyrin $\pi$-aggregation (J-aggregate) (Fig. 16). The J-type aggregation is often found with cyanine dyes [145, 146].

Systematic studies of the photoelectrochemical properties of SAMs of porphyrin disulfide dimers clarified the effects of the photocurrent on the spacer length (Fig. 17) [27, 142, 147–150]. The photocurrents decrease dramatically with a decrease in the spacer length, indicating that there are two competitive deactivation pathways for the excited porphyrin, i.e. quenching by the electrode and by the electron carrier. The porphyrin monolayers show a more highly ordered structure the longer the methylene spacer is. The adjacent porphyrin rings take on a J-aggregate-like partially stacked structure in the monolayer [142]. The porphyrin ring planes in the monolayer with an even number ($n = 2, 4, 6, 10$) of methylene groups are tilted significantly to the gold surface, while porphyrins with an odd number ($n = 1, 3, 5, 7, 11$) of methylene spacers take on nearly perpendicular orientation to the gold surface [142].

Metalloporphyrin SAMs also play an important role in catalysis. Several thiol-derivatised metalloporphyrins have been prepared that contain zinc [151, 152], manganese [153] or cobalt [152, 154–156]. Such metalloporphyrin systems bound to gold electrodes have been used to investigate the reduction of oxygen to hydrogen peroxide [154–156]. The electrocatalytic activity of this process depends on the interfacial architecture of the metalloporphyrin SAMs [155]. Metalloporphyrins in which the porphyrin rings are coplanar to the Au-surface show a higher electrocatalytic activity than

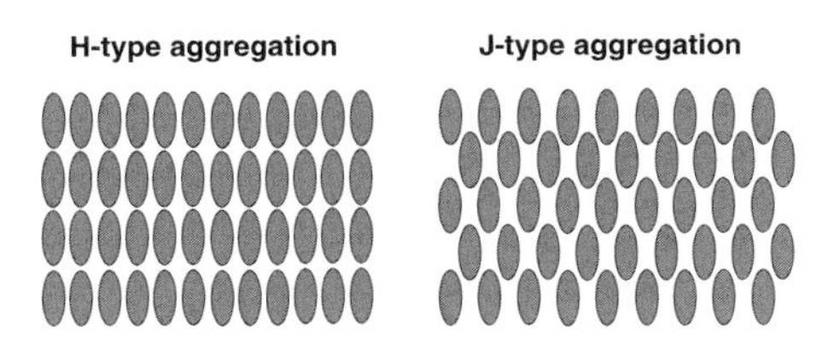


**Fig. 16** Schematic structures of H- and J-type aggregate of porphyrins

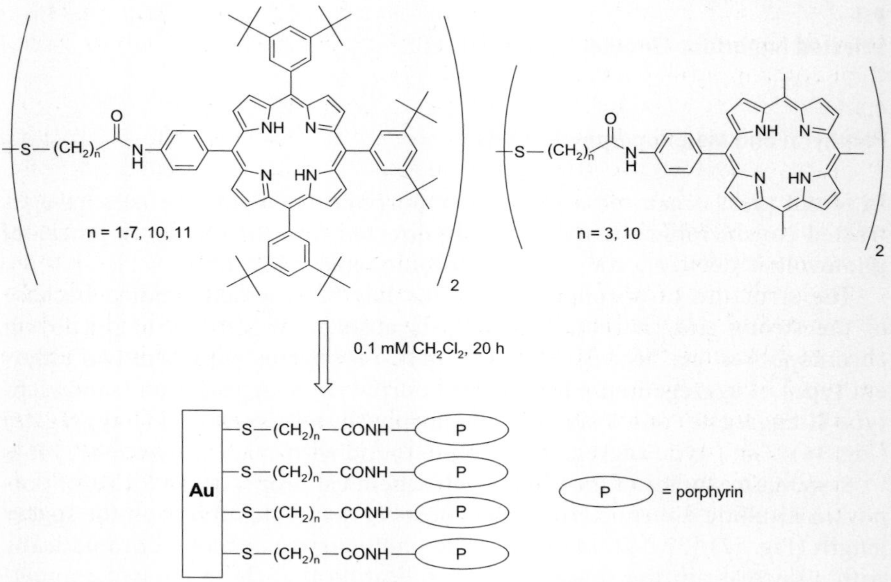

**Fig. 17** Gold electrodes modified with self-assembled monolayers of porphyrin disulfide dimers with different chain length of spacers [142, 241–243]

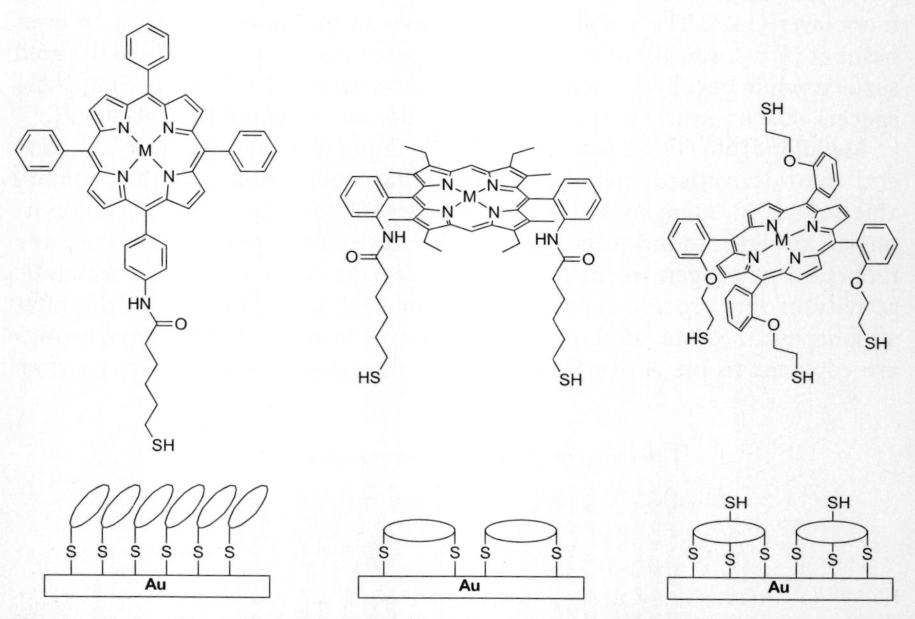


**Fig. 18** SAMs of metalloporphyrins with different number of thiol-anchors and their bonding onto a gold surface (M = $H_2$, Co) [154–156]

porphyrins with a nearly perpendicular orientation. The coplanar orientation towards the gold electrode could be achieved by increasing the number of thiol-containing anchors (Fig. 18) [154, 155].

Ratner et al. introduced a new type of tetraphenylporphyrin (TPP) that forms SAMs on gold. These porphyrins are novel in that the sulfur-containing anchors are attached to the porphyrin via a sixfold-coordinated phosphorus centre (Fig. 19a), which leads to a parallel orientation of porphyrin and gold surface [144, 157]. A similar parallel orientation can also be achieved by chemisorption via axial ligand substitution of the metal centre (ruthenium, osmium) of metalloporphyrins in mixed SAMs containing imidazole-terminated adsorbates (Fig. 19b) [158]. The same strategy was used by Rubinstein and Shanzer to build up SAMs by connecting cobalt and iron TPPs to various N-donor ligand monolayers [78, 159].

For the investigation of pH-dependent photoinduced electron transfer Uosaki et al. used a tetraphenylporphyrin coupled to a mercaptoquinone (Fig. 20) [160, 161]. In electrolyte solutions containing methylviologen as an electron acceptor and EDTA as an electron donor, anodic (cathodic) pho-

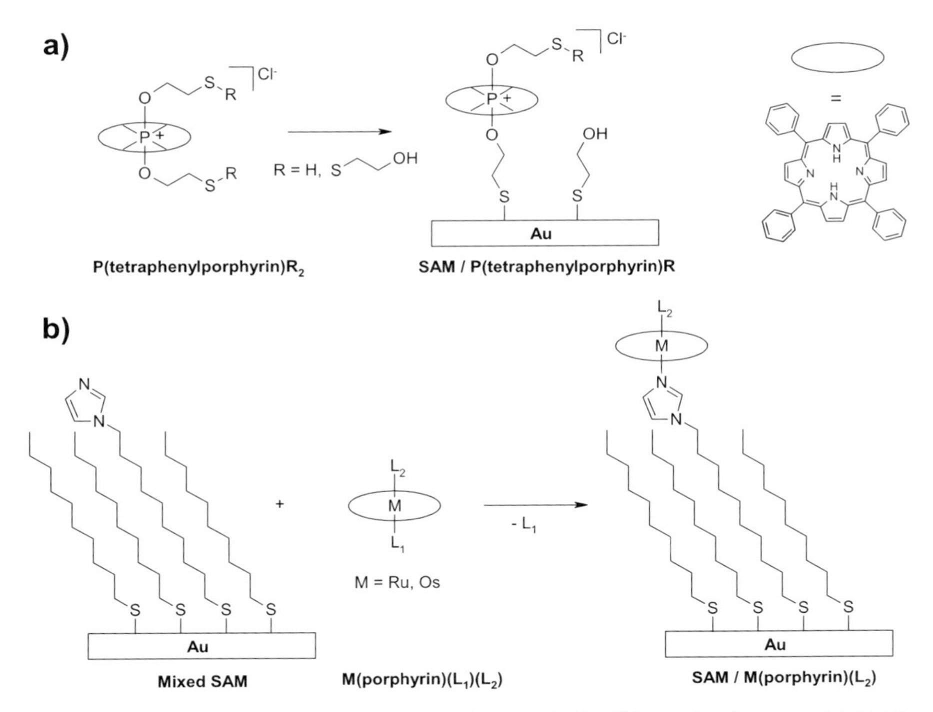


**Fig. 19** (**a**) Attachment of P(TPP)-thiol and P(TPP)-disulfide molecules to gold [157]; (**b**) Deposition of M(porphyrin)($L_2$) to a mixed SAM of $C_{10}$ – SH and imidazol-$C_{10}$ – SH on a planar gold substrate [158]

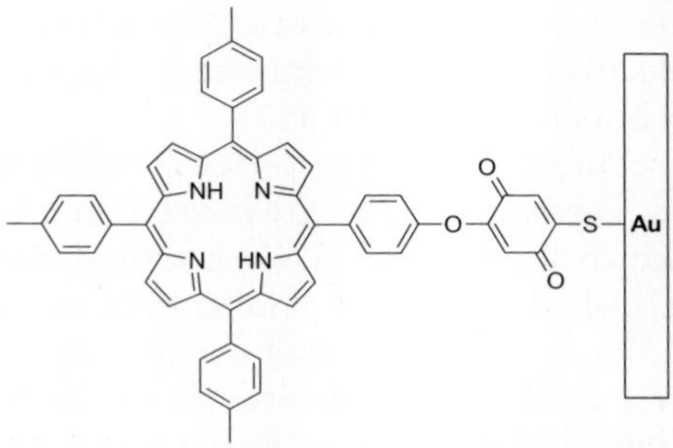

Fig. 20 Porphyrin-mercaptoquinone SAMs used for investigation of pH-dependent photoinduced ET [160, 161]

tocurrents were observed at potentials more positive (more negative) than the redox potential of the quinone moiety. By increasing the pH of the solution from pH 3.5 to 5.5 the redox potential of the quinone moiety decreased. The shift of the potential dependence of the photocurrents to the negative direction also indicates that the photoinduced electron transfer direction can be controlled by the pH of the electrolyte solution.

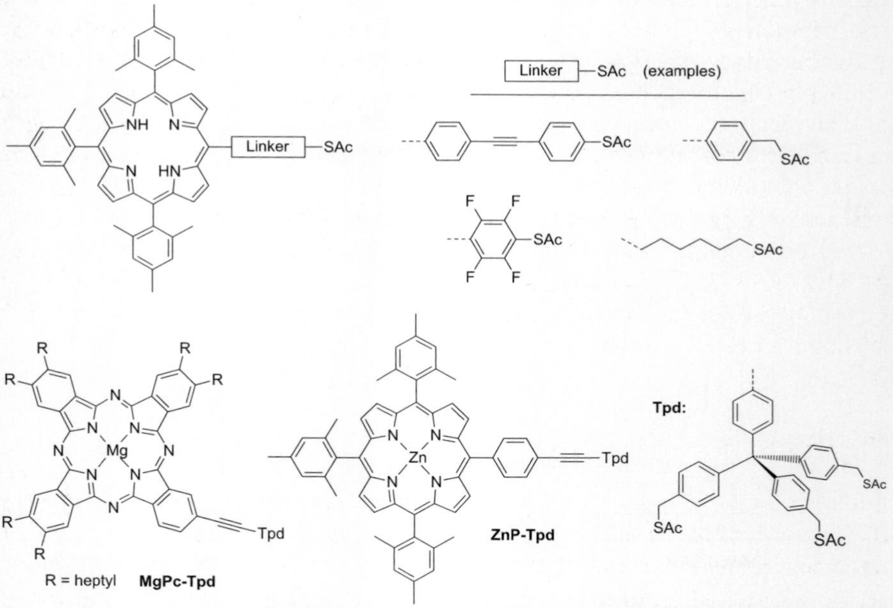

Fig. 21 Different redox-active porphyrins and phthalocyanines for studies of heterogeneous ET and of molecular information storage [162, 163, 165, 166]

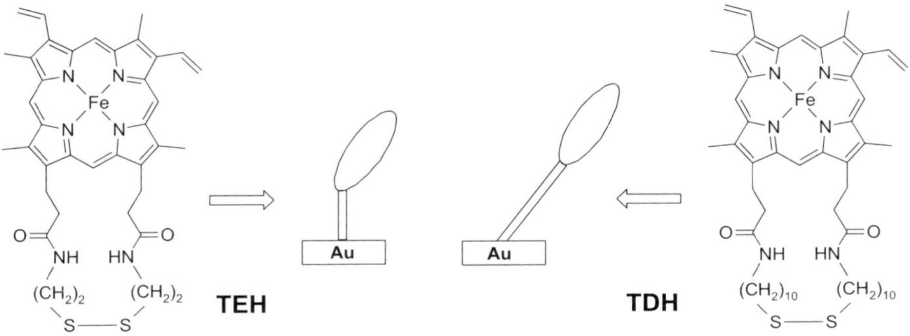

**Fig. 22** Heme derivatives and structural models of SAMs (TEH = thioethylated heme derivative; TDH = thiodecylated heme derivative) [119]

Lindsey et al. developed new routes for the preparation of thiol-derivatised porphyrin monomers and porphyrin building blocks that require no handling of free thiols for measuring the rate of heterogeneous electron transfer in SAMs with regard to molecular information storage [162–166]. Acetyl-protected thiols were used because they are relatively stable and can be used directly for SAM formation [167]. Electrochemical studies of SAMs indicate that a tripodal tether provides a more robust anchor to the Au surface than does a tether with a single site thiol attachment. The ET and charge-dissipation characteristics of the different tethers are generally similar, but the tripodal anchor offers superior stability characteristics (Fig. 21) [166]. A further advantage of the tripodal anchor is that it ensures a 90° orientation of porphyrin chromophores towards the Au surface.

The reconstitution reaction of heme derivatives with apo-protein on gold surfaces is a convenient method for fabricating well-ordered protein monolayers on solid substrates as a first step for the structural analysis of proteins and also for device applications. Kobayashi and coworkers synthesised thiolated heme compounds to investigate the self-assembly behaviour on gold (Fig. 22) [119, 168]. UV- and IRRAS-spectroscopy indicate that the tilt angle of the adsorbed molecules depends on the alkyl chain length between the surface and the chromophore.

### 4.3.2 Phthalocyanine-SAMs

Phthalocyanines have a high potential as the chemically active component in both conductometric and optical sensors, in photovoltaic cells and in optical data storage devices [169–171].

Russell et al. synthesised a series of phthalocyanine alkanethiols and disulfides with different length of alkyl spacers (Fig. 23). IRRAS spectroscopy indicates that the orientation of the phthalocyanine macrocycles towards the gold

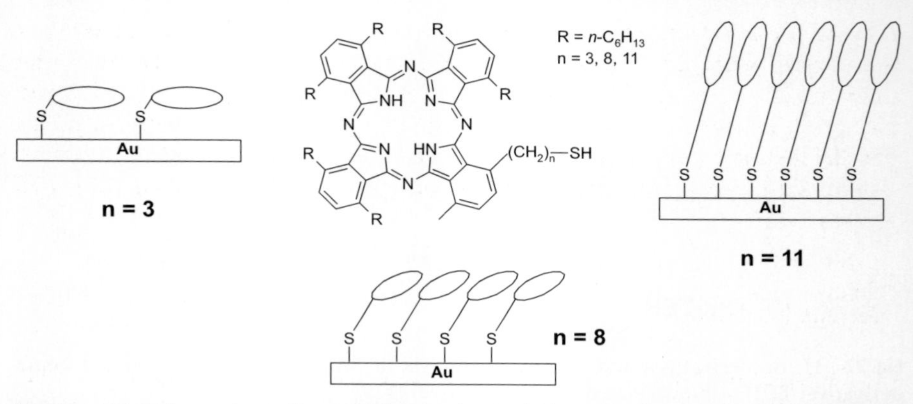


**Fig. 23** Phthalocyanine alkanethiols synthesised by Russell et al. and schematic representation of the possible orientation of the phthalocyanine SAMs on the gold substrate [172–174]

surface depends upon the length of the alkanethiol anchor chain. When a $C_{11}$ hydrocarbon chain was used the phthalocyanine macrocycles form a densely packed and highly ordered SAM with the phthalocyanine rings arranged in a nearly perpendicular configuration with respect to the gold surface. In contrast, a $C_3$ alkyl anchor chain results in a less closely packed monolayer in which the phthalocyanine macrocycles are arranged parallel to the gold surface [172–174]. This emphasises the structure-forming influence of alkyl chain interactions with increasing chain length.

To investigate the heterogeneous ET rates in phthalocyanine SAMs a series of monomers, dimers, trimers and oligomers of triple-decker eu-

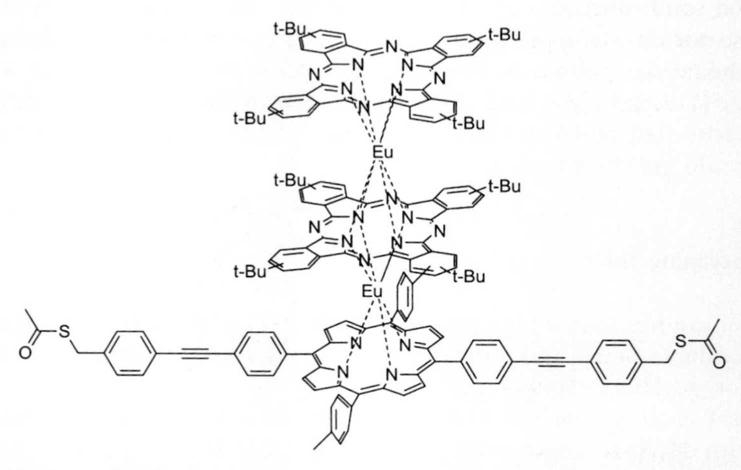


**Fig. 24** Bis(*S*-acetylthiol)-derivatised europium triple-decker monomer [175]

ropium phthalocyanine complexes bearing *S*-acetylthiol groups at the termini were prepared (Fig. 24). In SAMs, the complexes are oriented with their linkers/macrocycle planes parallel to the surface. This contrasts with monothio-derivatised analogues which prefer perpendicular geometry. The parallel geometry of the dithiol-derivatised triple-deckers is qualitatively consistent with a covalent attachment to the gold surface via both thiol end groups [175].

### 4.3.3
### Fullerene ($C_{60}$)-SAMs

$C_{60}$ is an important SAM building block because of its remarkable electronic, spectroscopic and structural properties [26–28, 176]. To explore the relationship between the structure and the photochemical properties of $C_{60}$, Imahori et al. prepared different kinds of $C_{60}$ alkanethiols by systematically changing the linking positions of the pyrrolidine ring fused to the $C_{60}$ moiety at the phenyl group from *ortho*, *meta* to *para* (Fig. 25) [26, 27, 177, 178]. Electrochemical measurements showed that well-ordered SAM structures are formed on gold electrodes. A stable anodic photocurrent was observed in the presence of an electron acceptor when the electrode was illuminated with monochromatic light. Depending on the linking position, the intensity of the photocurrent increases gradually in the order *o*-, *m*- to the *p*-isomer.

Stabilisation of the $C_{60}$ monolayers and alteration of the electronic properties can be achieved by introducing oligothiophenes between the gold surface and $C_{60}$ instead of the alkyl spacer [28, 179]. Surface coverage studies show that the packing of the monolayer depends on the chain length of the oligothiophenes, i.e. for $n = 1$ the quarterthiophene chain is oriented coplanar to the gold surface while for $n = 2$ the octithiophene is tilted. One disadvantage of these oligothiophene-$C_{60}$ SAMs for photocurrent generation is that the excited states are quenched by close oligothiophene–oligothiophene interactions and by the gold surface [180]. The deactivation pathways are suppressed if tripodal anchors are used. This ensure a perpendicular alignment of the

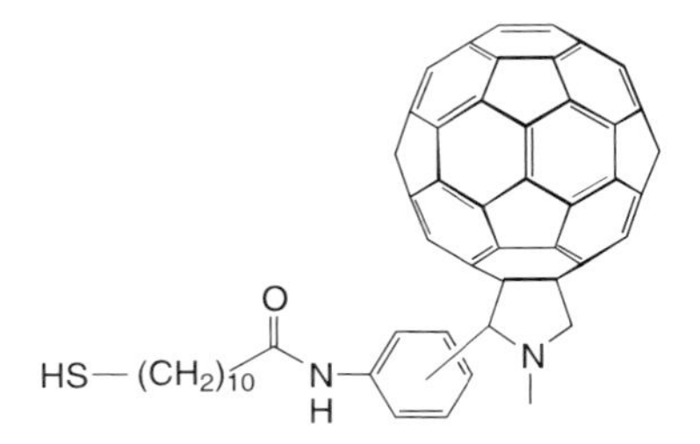


**Fig. 25** $C_{60}$ alkanethiols with different pyrrolidine linking position (*o*, *m*, *p*-isomer) [26, 27, 177, 178]

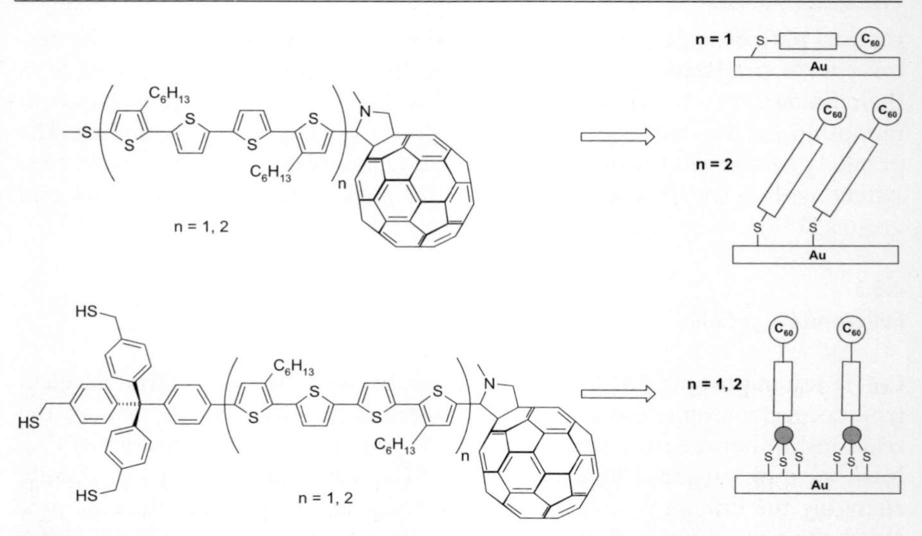


**Fig. 26** Oligothiophene-fullerene systems containing disulfide or tripodal anchors [28]

oligothiophene chain to the gold surface, which ensures non-aggregated chromophores (Fig. 26) [179].

### 4.3.4 SAMs of Aromatic Azo Compounds

Azobenzenes are frequently used as molecular switches because of their reversible photoinduced *trans* to *cis* isomerisation [181, 182].

Fujishima et al. found that SAMs prepared of azobenzene-terminated long chain alkanethiols on gold surfaces show a herring-bone structure in which the long axes of the azobenzene moieties are parallel to each other and the short axes of neighbouring molecules are perpendicular to each other (Fig. 27) [183, 184]. In contrast, phenylazonaphthol alkanethiol monolayers are densely packed as a J-aggregate and well oriented with a tilt angle of 28° [185].

The light-driven interconversion of *trans* to *cis* isomers of azobenzene was used to control the heterogeneous ET of a SAM consisting of a ferrocene-alkanethiol that was substituted by azobenzene. Electrochemical investigation of a mixed SAM containing 20% of the above mentioned *cis* azobenzene-ferrocene derivative and 80% of the *trans* isomer can be converted to pure *trans* isomer SAM by electroreduction and converted back to the 20/80 mixture by UV irradiation [18].

Cook et al. studied the photoinduced *trans* to *cis* isomerisation of a (pyridylazo)phenol alkanethiol SAM and its interaction with transition metal ions by spectroscopic and electrochemical methods [19]. The SAM in the *trans*

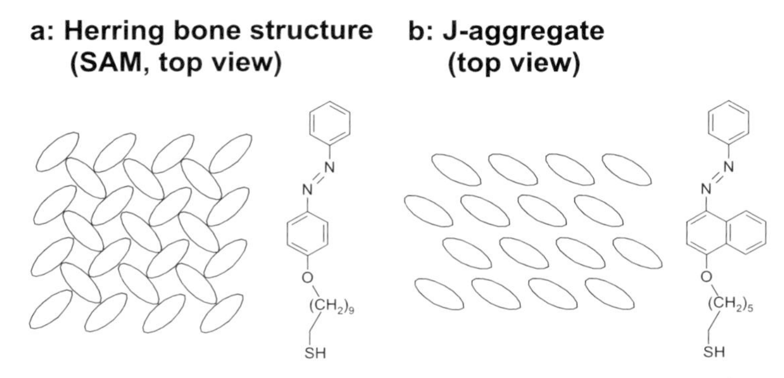


**Fig. 27** Molecular packing modes in SAM films of azobenzene alkanethiols (**a**) and of phenylazonaphthol alkanethiols (**b**) [183, 185]

isomeric form provides a bidendate ligand that can chelate metal ions such as $Ni^{2+}$ and $Co^{2+}$ (Fig. 28). Following irradiation with UV light ($\lambda = 365$ nm) the SAM photoisomerises to the *cis* form, which is unable to chelate metals. This photoisomerisation can be reversed by irradiating the *cis* monolayer with UV light at 439 nm.

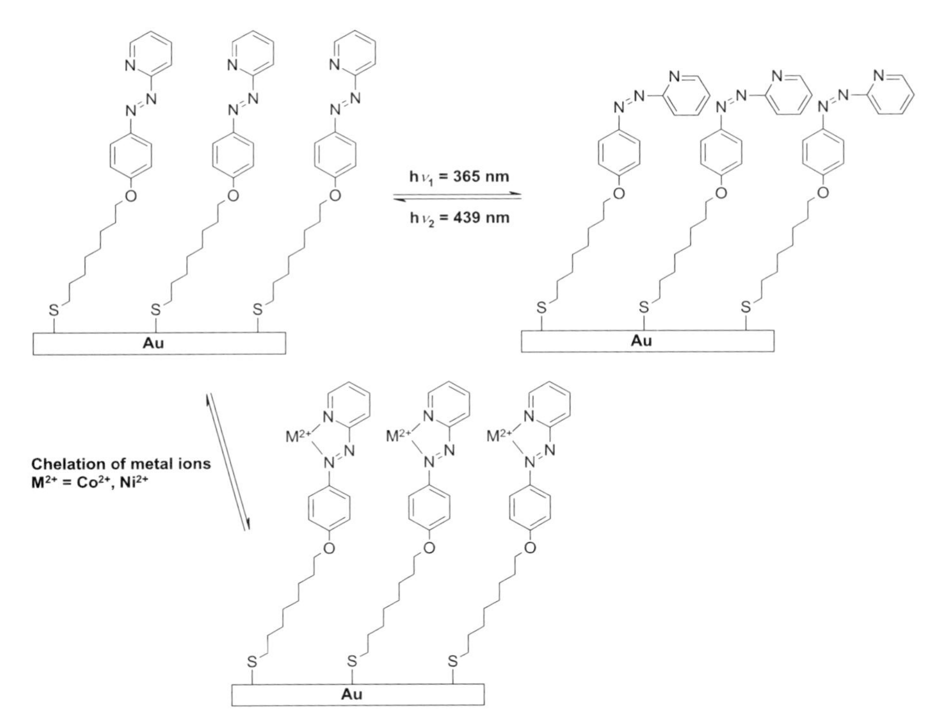


**Fig. 28** Photoswitching and metal-ion chelation by 8-[4-(2-pyridylazo)phenoxy]octanethiolate-SAM [19]

## 4.3.5 SAMs of Pyrene Chromophores

Pyrene is one of the most prominent standard chromophores for probing fluorescence in relation to excimer formation [186]. Fox and Whitesell et al. used a tripodal anchor group to attach a pyrene chromophore in a stable and well-ordered monolayer (Fig. 29, **19**) [122, 187]. Although attached to a conducting surface, these monolayers display significant fluorescence from a pyrene excited state. The SAMs obtained are so stable that extensive washing did not result in any loss of fluorescence intensity.

The assembly of molecular components for light harvesting and charge separation in artificial photosynthetic systems is of current interest. To mimic the multistep electron transfer in natural systems, Imahori et al. prepared a mixed SAM that combines an artificial antenna system (pyrene) with an artificial reaction centre (porphyrin), to examine the possibility of photoinduced energy transfer in two-dimensional assemblies (Fig. 29, **20**) [27, 147]. The ratio of porphyrin : pyrene in the mixed SAM estimated from the absorption spectra on the gold surface is significantly lower than that of the solution from which the SAM was prepared. The strong $\pi$–$\pi$ interaction of

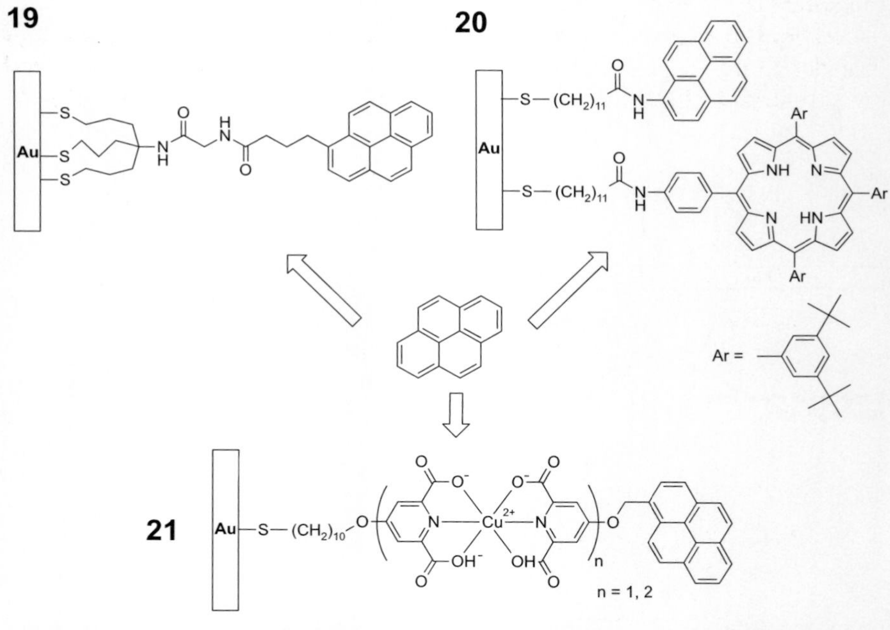

**Fig. 29** SAMs containing pyrene chromophores: **19**: pyrene bound to the gold surface via a tripodal anchor group [122, 187]; **20**: mixed SAM of pyrene and porphyrin [27, 147]; **21**: multilayer system consisting of pyrene and Cu(II)-complexes [188]

the pyrene moieties as compared to the relatively weak interaction between the porphyrin moieties due to the bulky *tert*-butyl groups may be responsible for the preference of the adsorption of the pyrene over the porphyrin onto the gold surface.

Recently, MacDonald and coworkers followed a strategy for the assembly of supramolecular photocurrent-generating systems in which the light-absorbing group (pyrene) is non-covalently coupled to a gold surface via metal–ligand complexation (Fig. 29, **21**). These systems are non-covalently assembled by a sequential deposition of three or more components. For $n = 1$ (see Fig. 29, **21**) three components are used: first, decanethiol linked to a 4-pyridyl-2,6-dicarboxylic acid ligand is deposited, followed by Cu(II) ions. The last step is the introduction of the pyrenyl-substituted pyridyl-2,6-dicarboxylic acid ligand. The formation of the monolayers was monitored by conductivity and impedance measurements [188].

## 4.3.6 SAMs of Dyad and Triad Chromophore-Arrays

SAMs built from molecules consisting of multiple chromophore units in a linear array (dyads, triads etc.) play an important role in the development of devices for photocurrent generation, artificial photosynthesis or current rectification [26, 27, 61].

Imahori et al. synthesised a series of dyad and triad chromophore arrays in order to mimic the highly efficient multistep electron transfer in natural photosynthesis [26, 27]. In case of a porphyrin-linked $C_{60}$ compound experi-

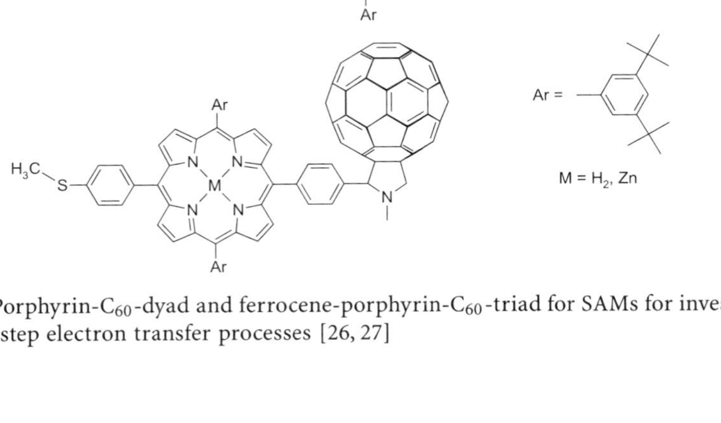

**Fig. 30** Porphyrin-$C_{60}$-dyad and ferrocene-porphyrin-$C_{60}$-triad for SAMs for investigation of multistep electron transfer processes [26, 27]

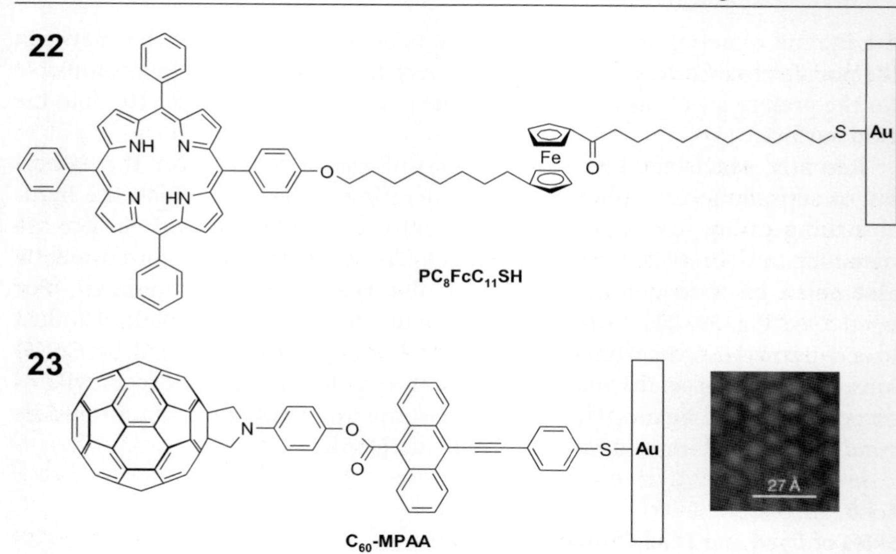


**Fig. 31** Porphyrin– $C_8$ – Fc – $C_{11}$ – SH (**22**) [117, 194] and $C_{60}$-(4-mercaptophenyl)anthrylphenylacetylene ($C_{60}$-MPAA, **23**) SAMs including STM image of a pseudohexagonal ordered assembly of $C_{60}$-MPAA on Au(111) surface. Reproduction with permission of [195]

mental data revealed that the molecules were tilted nearly parallel to the gold surface, leading to the formation of loosely packed structures [26, 189, 190]. Self-assembled monolayer systems of porphyrin-$C_{60}$ dyads have been extended to a linear array of a ferrocene-porphyrin-$C_{60}$ triad with an alkanethiol (Fig. 30) [27, 191–193]. Surface coverage measurements indicate that the tethered oligoalkyl-thiol makes it possible to arrange the triad in a well-packed almost perpendicular orientation to the gold surface, which again emphasises the advantage of alkyl spacers in order to support well-ordered SAMs.

Uosaki et al. investigated a dyad comprising porphyrin, ferrocene, and thiol groups acting as photoactive, electron transport or relay, and surface binding moiety, separated from each other by alkyl chains (Fig. 31, **22**). Angle-resolved X-ray photoelectron spectroscopy (ARXPS) combined with electrochemical coverage determination revealed that the alkyl chains in the ferrocenecarbonylundecanethiol ($FcC_{11}SH$) part of the porphyrin– $C_8$ – Fc– $C_{11}$ – SH dyad is oriented with a tilt angle of ca. 30° normal to the gold surface and that the plane of the porphyrin ring is almost surface-normal [117, 194].

4-Mercaptophenylanthrylphenylacetylene (MPAA) itself forms a highly ordered 2D stacked array SAM owing to strong $\pi$–$\pi$ interactions. This feature was used in an extended $C_{60}$-MPAA dyad to induce an oblique lattice with a pseudohexagonal arrangement on Au(111) (Fig. 31, **23**) [195].

## 4.3.7
## SAMs of Various Kinds of Chromophores

Besides the porphyrins, fullerenes, aromatic azo-compounds and pyrenes widely used as chromophores in SAMs there are some rarely used $\pi$-systems, which are mentioned in this section.

Merocyanines are widely used as solvatochromic dyes. Fujita et al. demonstrated that the Brooker's dye analogue (Fig. 32, **24**) shows a colour change based on protonation and deprotonation in a densely packed SAM (Fig. 32, **24**) [196]. Surface plasmon spectroscopy and X-ray photoelectron spectroscopy indicate that the monolayers are closely packed. The layer thickness is consistent with the molecules having a tilt angle of 30° to the surface normal. IRRAS spectroscopy shows that the chromophores are located in a polar local dielectric situation even in non-polar solvents. Nonetheless, the deprotonated, zwitterionic form of the dye shows a distinct negative solvatochromism depending on the solvent polarity.

Carotenoids constitute an important class of dyes in biological systems where they act as singlet oxygen deactivators, light harvesting chromophores and photoreceptors [197]. The extended conjugated $\pi$-system of carotenoids stimulated many attempts to use them as electrically conducting wires. In fact, a mixed SAM with the thiol-substituted carotenoid **25** depicted in Fig. 32 was over one million times more conductive than the surrounding alkyl chains (measured by conducting AFM) suggesting that the carotenoid indeed acts as a molecular wire. The thiol-substituted carotenoid SAMs are less ordered than alkanethiols due to the steric hindrance along the polyene

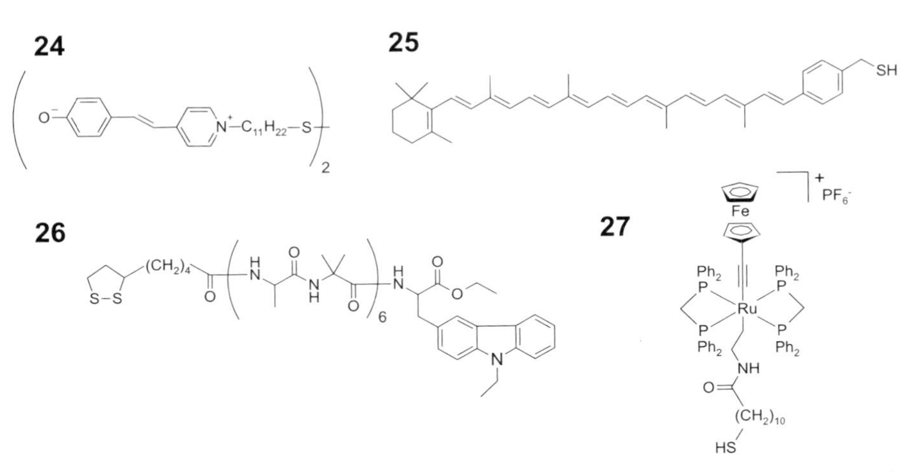


**Fig. 32** Various chromophores used in SAMs: **24**: merocyanines [196]; **25**: carotenoids [199]; **26**: helical peptides with *N*-ethylcarbazolyl groups [118]; **27**: mixed-valence Fe–Ru complexes [200]

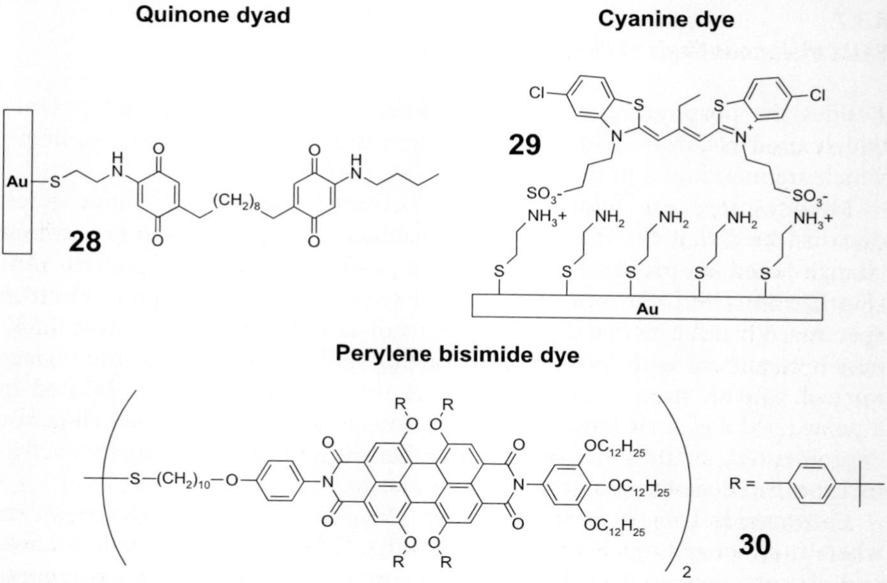

**Fig. 33** Various chromophores used in SAMs

backbone and the bulky terminal $\beta$-ionone ring group, but still present a hydrophobic interface. Analysis of the C – H stretching modes indicates that the polyene chains are disordered (Fig. 32, **25**) [198, 199].

Other chromophores are helical peptides with *N*-ethylcarbazolyl groups for photocurrent generation with tilt angles of the helices to the surface normal of about 40° (Fig. 32, **26**) [118] or mixed-valence iron–ruthenium complexes for quantum dot cells (QCA) which possess an isolated, tethered structure on the gold surface, as indicated by CV measurements (Fig. 32, **27**) [200].

Just briefly mentioned should be quinone dyes **28** [201], cyanine SAMs **29** where the cyanine dye forms highly ordered 2D J-aggregates on a cysteamine SAM with a $7 \times \sqrt{3}$ superlattice [145, 146] and a perylene bisimide dye **30**, which shows J-type aggregation as demonstrated by the bathochromic shift of the fluorescence spectra [202] (Fig. 33).

# 5
# Basic Photophysical Processes in SAMs

## 5.1
## UV/vis Absorption

In general UV/vis absorption spectra of SAMs are difficult to measure because of the extremely low surface concentration of the chromophore, which leads to very low absorbances. Therefore, only dyes with rather high molar absorptivities, such as porphyrins, have been accurately characterised in this way. Virtually all studies were performed in transmission mode using thin, optically transparent gold layers on glass etc. because of the better background compensation compared to reflectance mode measurements.

Absorption spectra have been used to determine the relative orientation of porphyrin molecules by analysing shifts of the Soret band. It is well established that a stacked face-to-face porphyrin $\pi$-aggregation (sandwich-type H-aggregation) leads to a spectral blue shift relative to the monomer, while side-by-side porphyrin $\pi$-aggregation (J-aggregate) leads to a red shift compared to the spectra in solution (Fig. 34) [203]. The shift itself can be explained by exciton coupling theory [204] and depends on the interaction between the chromophores and, consequently, on the distance separating the porphyrin rings [142, 144, 205]. The shape of the absorption bands is also an indicator for the homogeneity of SAMs. Thus, mixtures of side-by-side and face-to-face orientations lead to a broader Soret band, which exhibits both a red and a blue shift [144]. Narrower peaks in UV/vis-spectra of SAMs

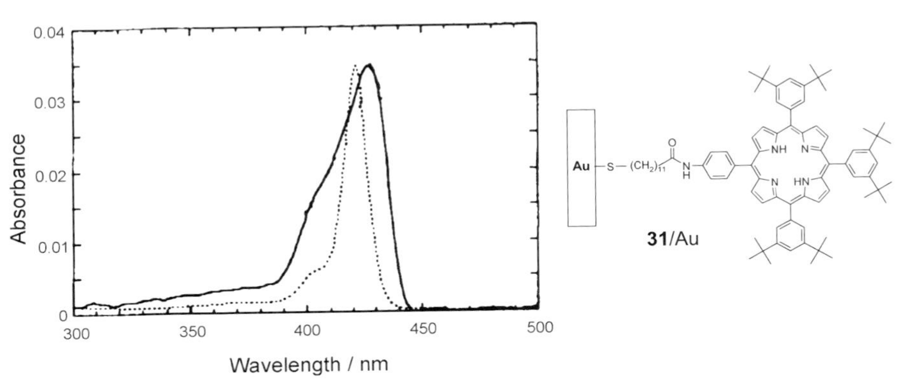


**Fig. 34** UV/vis absorption spectra of a porphyrin chromophor **31** on Au/glass in transmission mode (*solid line*, red shift) and in dichloromethane (*dotted line*). Reprinted with permission from [142]. Copyright (2000) American Chemical Society

than in solution suggest a more uniform orientation of the chromophore molecules in the monolayer [78].

Absorption band shifts of chromophore SAMs were also reported for the excitation of azonaphthalenes [185], azobenzenes [183, 184], carotenoids [199], heme derivatives [119] and cyanine dyes [145, 146].

## 5.2 Fluorescence Measurements

Light emission of chromophores attached to gold is usually quenched by energy transfer into the surface plasmon resonance of gold. Because this quenching effect hampers the application of chromophore-modified gold surfaces in devices in which a long-lived excited state is desirable (as in photocurrent generation, see Sect. 6.5), glass or ITO is often preferred as the supporting SAM surface [1, 3, 55–61, 67, 68]. However, in certain cases fluorescence in SAMs on gold could be observed: Whitesell et al. investigated the distance-dependent fluorescence behaviour of fluoren-9-yl-alkanethiols on SAMs [206]. Comparison of the solution and the monolayer spectrum of fluoren-9-yl-alkanethiol **32** shows that the emission maximum is red-shifted from 305 to 385 nm and that the band is much broader in the SAM. Lifetimes for the terminal fluorenyl groups when bound as a SAM to the Au surface are biexponential, consisting of a long-lived component corresponding to that observed in solution ($\tau \sim 3.2$ ns), and a second shorter lifetime species ($\tau \sim 260$ ps) corresponding to a metal-mediated quenching process. By increasing the alkyl spacer length in a monotonous fashion with the number of methylene groups $n = 6, 7, 8, \ldots, 12$ the fluorescence intensity increases (Fig. 35). Thus, the rate of fluorescence quenching by the metal surface is correlated inversely to the distance separating a covalently bound fluorescent probe molecule from the gold surface.

Chain-length-dependent quenching processes between the excited chromophore (porphyrin) and the gold surface were also observed by Imahori et al. [142, 150] indicating that the excited singlet state of the porphyrin moiety in the SAMs is efficiently quenched by an energy transfer (EN) to the gold surface. In contrast to fluoren-9-yl-alkanethiol SAMs **32**/Au the peak positions as well as the shape of the emissions of the porphyrin-thiols in SAMs are quite similar to those in solution. However, the spectra become somewhat broader in the order of **33** in $CH_2Cl_2$ and **33**/Au ($n = 1$–7, 10, 11), which suggests aggregation of porphyrins in the monolayers and/or interaction between the porphyrins on the gold surface (Fig. 36). Time-resolved fluorescence experiments yielded a single-exponential decay of the SAMs, which shows that only one type of chromophore aggregate is present in the SAMs. The fluorescence lifetimes of **33**/Au ($n = 1$–7, 10, 11) ($\tau \sim 3$–40 ps) are much shorter than those of **33** in THF ($\tau = 9.8$ ns), indicating that the excited porphyrin state is quenched by the gold surface and that the non-radiative decay is enhanced by

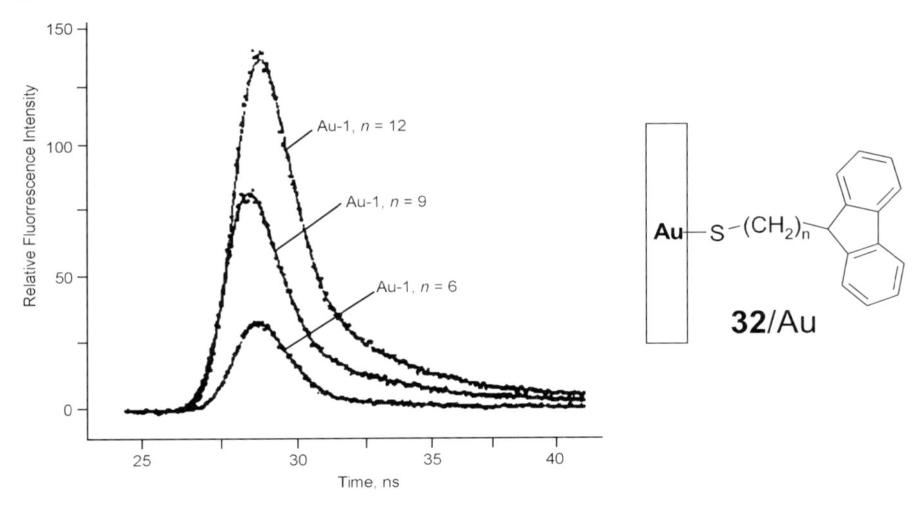


**Fig. 35** Fluorescence decay profiles of SAMs of fluoren-9-yl-alkanethiol **32** (number of methylene groups = 6, 9 and 12) on gold at $\lambda_{exc} = 375$ nm. Reprinted with permission from [206]. Copyright (2001) American Chemical Society

aggregation. The quantum yield of **33**/Au increases from 0.095% for $n = 1$ up to 0.33% for $n = 5$ and stays nearly constant at further increase of the chain length. The authors suggest energy transfer as the predominating quenching process.

Karpovich and Blanchard used 1-pyrenebutanethiol as a fluorescence chromophore embedded in a SAM of octadecanethiol in order to probe the dynamical behaviour. The dilution (3% pyrene) leads to non-aggregate pyrene emission at 400 nm. The fluorescence decay is non-exponential with

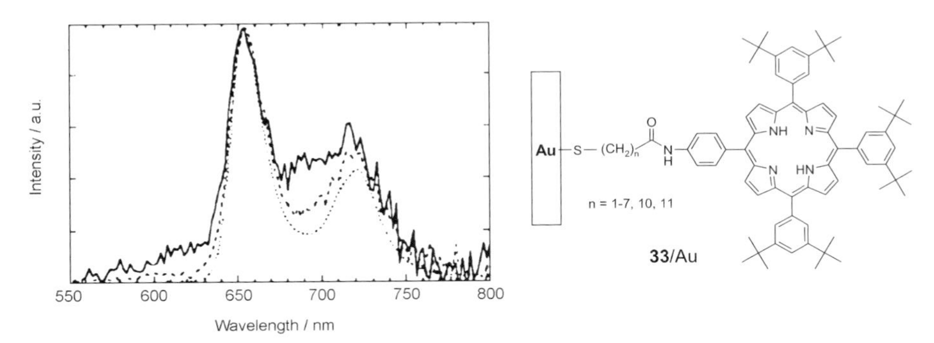


**Fig. 36** Corrected fluorescence spectra of **33**/Au ($n = 1$, *solid line*), **33**/Au ($n = 11$, *dashed line*) and **33** with methyl group replacing the alkanethiol (*dotted line*) in $CH_2Cl_2$. The spectra are normalised for comparison. Reprinted with permission from [142]. Copyright (2000) American Chemical Society

a short component of $\tau_{1/2} \sim 500$ ps, which is much shorter than in solution. Polarisation-dependent fluorescence investigations demonstrate that the alkanethiol monolayers are very rigid between the substrate and the top of the SAM [207].

### 5.3 Infrared Measurements

Infrared spectroscopy is a powerful tool for investigating the local structure of SAMs on gold [1, 3, 208]. For example, comparison of the transmission spectrum of a KBr pellet of a chromophore with the reflectance spectrum of the corresponding SAM can help to confirm the identity and to assess the orientation of the adsorbed molecules on the surface [199]. IR spectroscopy can also yield information about the intermolecular environment of SAMs on gold, as well as about the orientation of specific groups relative to the surface normal [209]. Selection rules for reflectance IR spectroscopy of anisotropic films on a conducting surface allow only those transition moments that are perpendicular to the plane of the substrate surface to exhibit appreciable absorbance [208, 209]. Thus, the peak intensity for a given transition is proportional to the magnitude of its projection along the surface normal. Structural information may be obtained by comparison of relative peak intensities in isotropic bulk phases with those on surfaces [210].

Two examples are mentioned here to illustrate how IR spectroscopy can be used to obtain structural information about SAMs: The molecular structure of a thiol-substituted carotenoid (see Sect. 4.3.7) was investigated by IR spectroscopy. Comparison of the IR spectrum of the carotenoid in the KBr matrix with that in the SAM shows the peak frequencies to be slightly higher ($\sim 5$–$10\ \mathrm{cm}^{-1}$) in the monolayer film (Fig. 37; see also Sect. 4.3.7). The relative

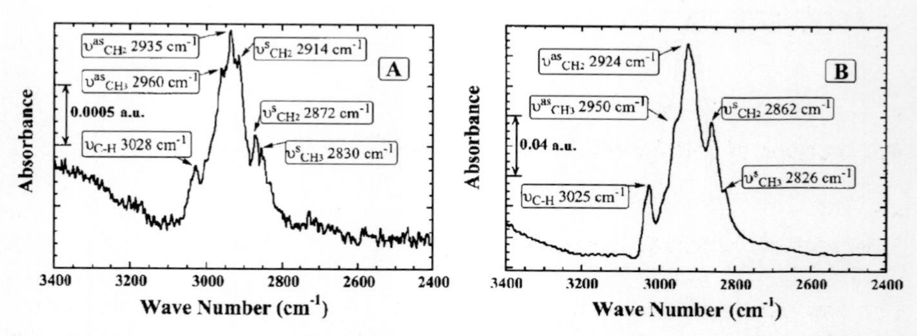


**Fig. 37** Comparison of (**a**) reflection FT-IR spectrum of a carotenoid-thiol modified Au surface with (**b**) transmission FT-IR spectrum of the carotenoid-thiol in KBr matrix in the C – H stretching region. Reprinted with permission from [199]. Copyright (2002) American Chemical Society

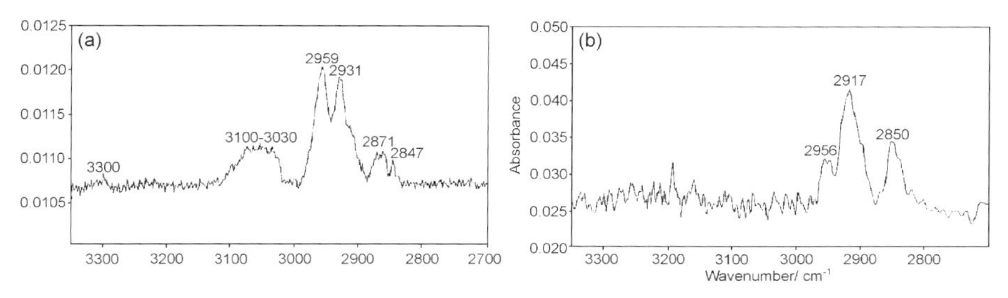


**Fig. 38** Transmission IR spectrum of phthalocyaninealkanethiol on a gold-coated silicon wafer (*left*) and RAIR spectrum of the same SAM on a gold-coated glass slide (*right*) [171]. Reproduced by permission of The Royal Society of Chemistry

intensity of asymmetric and symmetric methyl stretching modes suggests that the carotenoid polyene backbone lies normal to the surface because the intensity ratio of the asymmetric and symmetric methyl stretch is largest in the monolayer. The local $C_{3v}$ symmetry axis of the methyl groups is oriented approximately parallel to the surface in such a bonding geometry. This results in an increase of the intensity of the asymmetric methyl stretching mode and in a decrease of the symmetric stretching mode as observed experimentally (see Fig. 37) at 2960 $cm^{-1}$ (asymmetric $CH_3$ stretch) and at 2830 $cm^{-1}$ (symmetric $CH_3$ stretch) [199].

Figure 38 shows the IR transmission (left) and the RAIR spectrum (right) obtained for the SAM of a phthalocyaninealkanethiol (Pc) on gold. In context with the surface selection rules, the presence of the aromatic C – H stretch (3030–3100 $cm^{-1}$), the N – H stretch (3300 $cm^{-1}$) and the ring vibrations in the transmission spectrum and their absence in the RAIR spectrum suggest that the Pc ring is oriented parallel to the gold surface, because in this case all aromatic C – H and N – H transition moments projected to the surface normal yield zero intensities. In contrast, the alkyl C – H vibrations are still visible in the RAIR spectrum [171].

## 6 Applications of Self-Assembled Dyes on Gold

### 6.1 SAMs of Photoswitchable Materials

Photoswitching of chromophores in general is one attractive method to gate optical or electrical information [211]. A number of chromophores that can be isomerised photochemically have also been investigated in SAMs.

Willner et al. introduced a system that enables the photoswitchable activation and deactivation of Cytochrome c using a mixed monolayer consisting of mercaptopyridine and a photoactive protonated nitromerocyanine, which were both assembled onto a gold electrode (Fig. 39) [8, 16, 17]. At neutral pH, Cyt. c is positively charged and, thus, is electrostatically repelled from the electrode surface and the redox activity is switched off. Photoisomerisation of the nitromerocyanine into spiropyrane leads to a neutral monolayer interface and the redox functions of the hemoprotein are switched on. In the same way, photoisomerisable nitrospiropyrane flavoenzyme glucose oxidase SAMs were prepared to carry out the bioelectrocatalysed oxidation of glucose [8, 16, 212].

Willner et al. also investigated the reversible photoisomerisation of *trans*-quinone (6-hydroxynaphthacene-5,12-dione) to the *ana*-quinone isomer state in a mixed monolayer [213]. The resulting densely packed monolayer includes the *trans*-quinone in a rigidified configuration (Fig. 40). When irradiated with light between 320 and 380 nm the *trans*-quinone isomerises to *ana*-quinone, which is electrochemically inactive but which can be removed by reaction with primary amines.

Other applications of photoswitchable materials on SAMs are the UV-controlled complexation of metal ions to the merocyanine isomer of spironaphthoxazine-thioethers (Fig. 41a) [214]. The long saturation times for the pure spironaphthoxazine-SAM at complexation with zinc ions is an indication for a densely packed spironaphthoxazine-monolayer, with only a small fraction of open merocyanine isomer due to steric constraints.

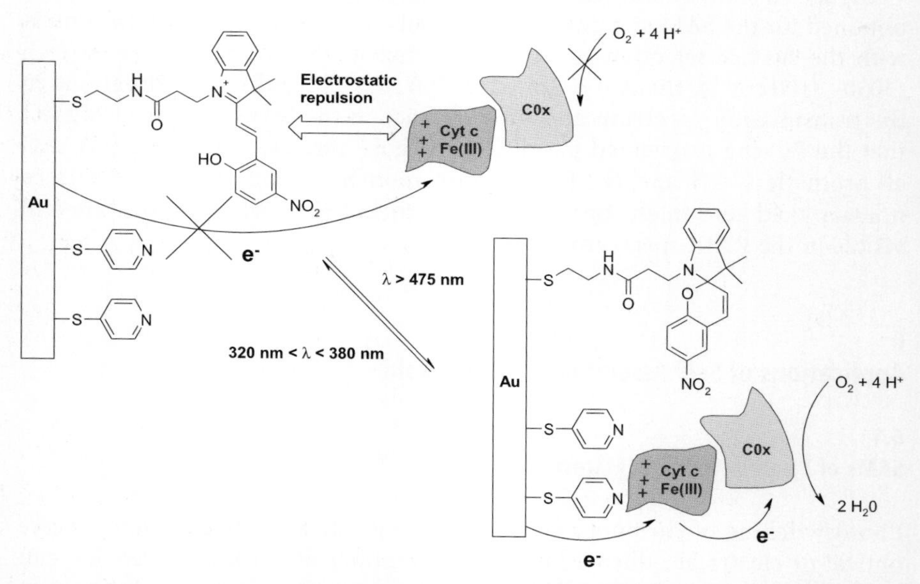


**Fig. 39** Photoswitchable activation and deactivation of the electrical contact of Cyt. c and an electrode, and secondary Cyt. c-mediated biocatalysed reduction of $O_2$ by COx, using a pyridine-nitrospiropyrane photoisomerisable mixed monolayer electrode [17]

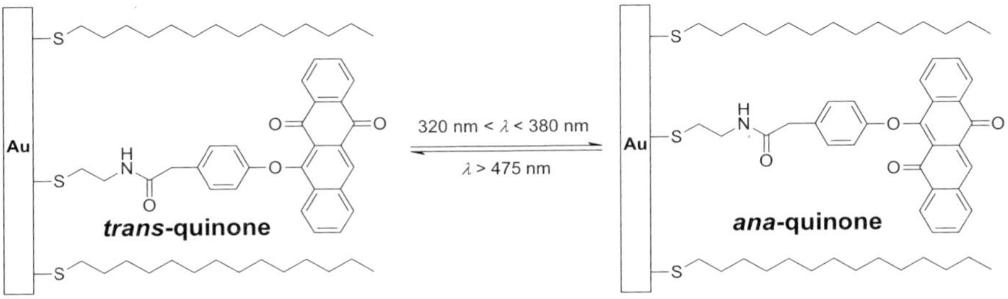

**Fig. 40** *Trans*-quinone/$C_{14}H_{29}SH$ mixed monolayer and photorearrangement [213]

To confirm this assumption, mixed monolayers with alkanethiols and the spironaphthoxazine-thioethers were prepared, which indeed show a significantly shorter saturation time. The results indicate that isolated spironaphthoxazine-thioethers lead to a rapid complexation with zinc ions after UV irradiation.

For wettability studies, non-covalently assembled monolayers with 2,2′-dipyridylethylene ligands for photoinduced *cis–trans* isomerisation have been prepared (Fig. 41b) [215]. The two different SAMs (*cis* and *trans*) display

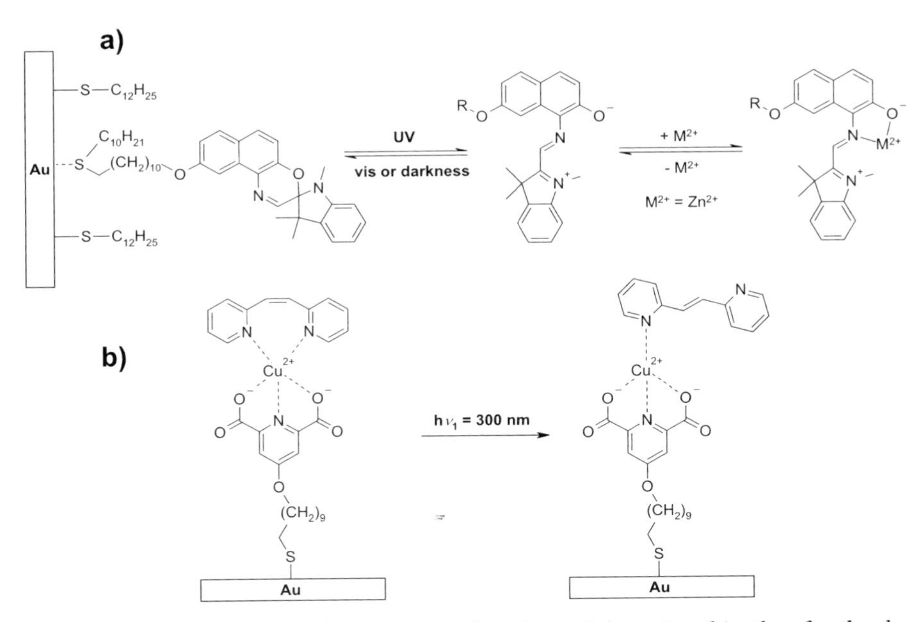


**Fig. 41** (**a**) SAM containing photoisomerisable spironaphthoxazine-thioether for the detection of metal ions [214]; (**b**) *cis–trans* switching of 2,2′-dipyridylethylene ligands in a copper complex containing SAM [215]

distinctly different contact angles. Unfortunately, the isomerisation is irreversible probably because of a very well-ordered *trans*-SAM.

## 6.2 SAMs of Phthalocyanines as Sensors and as Hole-Injection Layers in Organic Light-Emitting Diodes

Phthalocyanines can be used as the chemically active component in both conductometric and optical gas sensors [169, 216]. The technique of self-assembly has the merit of simplicity and, from the viewpoint of an application within a sensor, the advantage of providing a monomolecular layer that should exhibit near instantaneous response to a gaseous analyte (see Sect. 4.3.2, Fig. 23 for structure) [217]. Fluorescence quenching by external molecules can, in principal, provide the basis for an optical sensing device. Exposure of a phthalocyanine alkanethiol SAM with a longer chain tether to 200 ppm $NO_2$ gas showed that the fluorescence is indeed reversibly quenched [218]. In further experiments, changes in the surface plasmon resonance reflectivity signal proportional to the concentration were observed when a phthalocyanine SAM was exposed to $NO_2$, pointing to a further optical method for the detection of this gas [171].

Another interesting application is to use phthalocyanines as a hole-injection layer in organic light-emitting diodes (OLEDs). Zhu et al. demonstrated that a SAM of phthalocyanine thiol (HS-Pc) may act as a hole injection material in OLEDs (Fig. 42). The insertion of a SAM of HS-Pc between the Au anode and the hole-transport layers enhances the hole injection, which in-

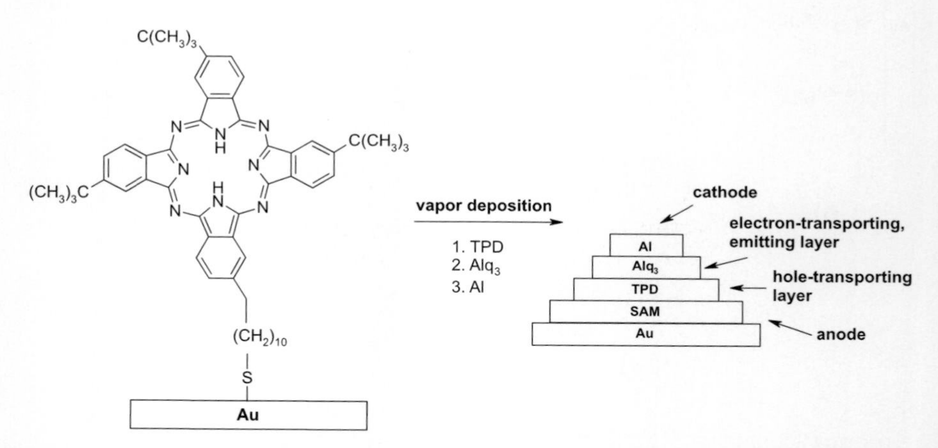


**Fig. 42** A phthalocyanine SAM as the hole injection layer in a two-layer OLED with $N,N'$-bis(3-methylphenyl)-1,1′-biphenyl-4,4′-diamine (TPD) as the hole transport layer and tris(8-hydroxyquinolinato)aluminium ($Alq_3$) as the electron transport and emitting layer [219]

creases the external quantum efficiency by a factor of ∼ 27 compared to the OLED without SAM and decreases the operating voltage from 13 to 8 V [219].

## 6.3 SAMs as Catalysts

Catalysts play an important role in many industrial processes. In terms of selectivity and activity, homogeneous catalysts are superior to their heterogeneous counterparts under mild reaction conditions. Unfortunately, the problem of separating the single-site catalysts from the reaction media is still an important drawback, which makes large scale applications in industry unfavourable. This major problem may be solved by using heterogeneous catalysts built, for example, of SAMs that include the catalytic moiety.

One outstanding example is bis(cobalt)diporphyrin. Films of bis(cobalt)diporphyrin with cofacial porphyrin units adsorbed on edge plane pyrolytic graphite electrodes are one of the most efficient molecular electrocatalysts that are able to reduce oxygen to water in a four-electron process [220]. However, when chemisorbed on a gold electrode surface, the system acts as a two-electron catalyst, producing hydrogen peroxide instead (Fig. 43) [154]. Molecular films of cobalt tetraphenylporphyrin derivatives in which the porphyrin is cofacial to the gold surface even reduce oxygen to hydrogen peroxide with a turnover number of more than $10^5$. This SAM catalyst is more ac-

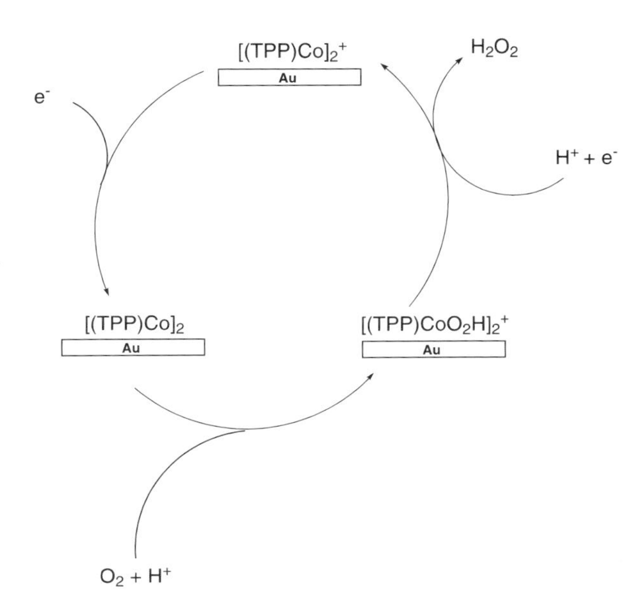


**Fig. 43** Proposed two-electron reduction of oxygen with bis(cobalt)diporphyrin on gold [154]

tive and durable than non-specifically adsorbed cobalt tetraphenylporphyrin (see Sect. 4.3.1, Fig. 18 for structure) [221].

## 6.4 Photoswitchable Chromophores as a Route to Optoelectronic Systems

Photostimulation of molecules, macromolecules and biopolymers can lead to changes of their chemical or physical properties, which provides the basis for the development of future optoelectronic devices. The photostimulated modification of biocatalytic and highly specific recognition and binding functions of biomaterials would enable their application in optoelectronic devices [222].

Photostimulation of redox enzymes could transduce recorded optical signals as an amperometric response by their electrical interaction with electrode interfaces. For example, amperometric transduction of recorded optical signals was accomplished using nitrospiropyran-modified glucose oxidase as photoswitchable material (Fig. 44) [14].

The photoisomerisable nitrospiropyran flavoenzyme glucose oxidase, SP-GOD, was assembled as a monolayer onto a gold surface. The monolayer revealed reversible photoisomerisable properties and illumination of

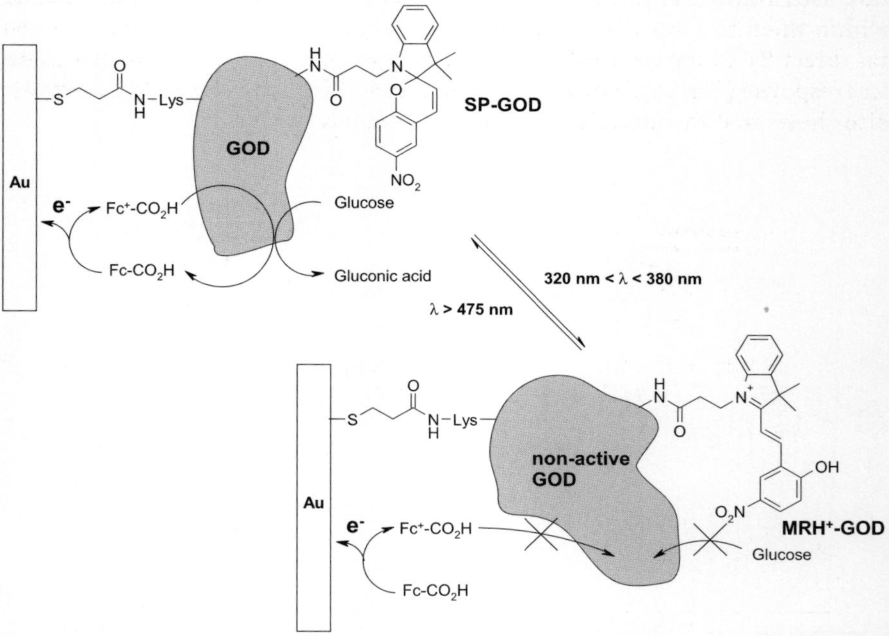

**Fig. 44** Photoisomerisable nitrospiropyran-functionalised glucose oxidase, SP-GOD, in the active (*above*) and inactive (*below*) form in order to mediate the oxidation of glucose [16]

the SP-GOD monolayer between 320 and 380 nm resulted in the protonated merocyanine-substituted enzyme monolayer, $MRH^+$-GOD. In the presence of ferrocene carboxylic acid acting as an electron-transfer mediator, electronic communication between the SP-GOD monolayer and the electrode was achieved. This led to the bioelectrocatalysed oxidation of glucose to gluconic acid and to the observation of an electrocatalytic anodic current. Photoisomerisation yields the $MRH^+$-GOD monolayer state and the enzyme was deactivated by perturbing the active site of the protein towards the electron mediator ($Fe - CO_2H$). Consequently, the bioelectrocatalysed oxidation of glucose was strongly inhibited. Reversible back isomerisation was achieved by irradiation of the merocyanine with light $\lambda > 475$ nm.

Another application for photoswitchable enzymes attached to gold surfaces is the cytochrome c-mediated biocatalysed reduction of $O_2$ by cytochrome oxidase, using a functional pyridine-nitrospiropyran photoisomerisable mixed monolayer electrode (see Sect. 6.1, Fig. 39) [8, 16, 17].

Photoisomerisation of azobenzene was used in a "photoswitchable diode" to regulate the ET events involving a two-component SAM consisting of a 99 : 1 mixture of *cis*-azobenzene (**34**) and *trans*-ferrocenyl azobenzene (**35**) alkanethiols. The ET between the dissolved ferrocyanide and the Au substrate is forced to occur through mediating ferrocenyl sites in the film, which results in a diode-like response. However, the photochemical conversion of bulky *cis*-azobenzene (**34**) to slim *trans*-azobenzene (**34**) increases the free volume within the film. This allows the ferrocyanide to diffuse to the electrode and for direct ET to the electrode surface, which results in a normal electrochemical response (Fig. 45). However, the diffusion of ferrocyanide to the electrode also shows that the monolayers cannot be densely packed [223].

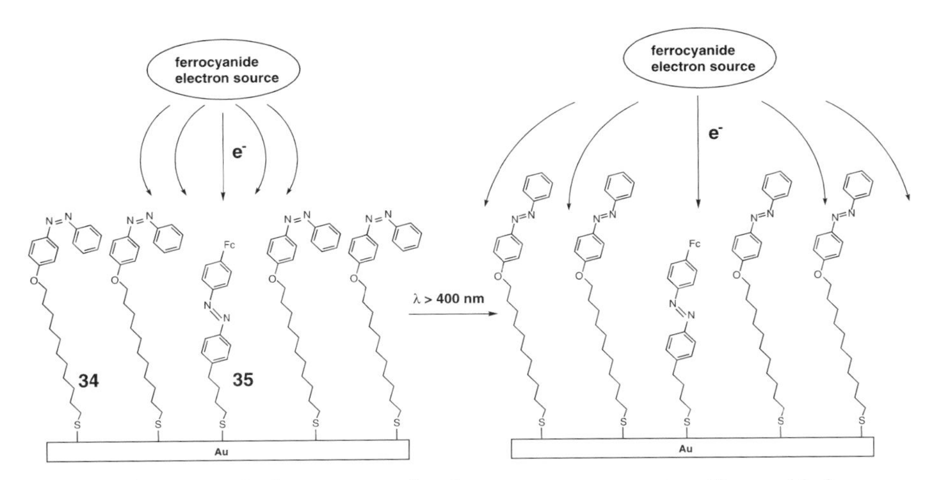


**Fig. 45** Photon-gated electron transfer in a two component self-assembled monolayer [223]

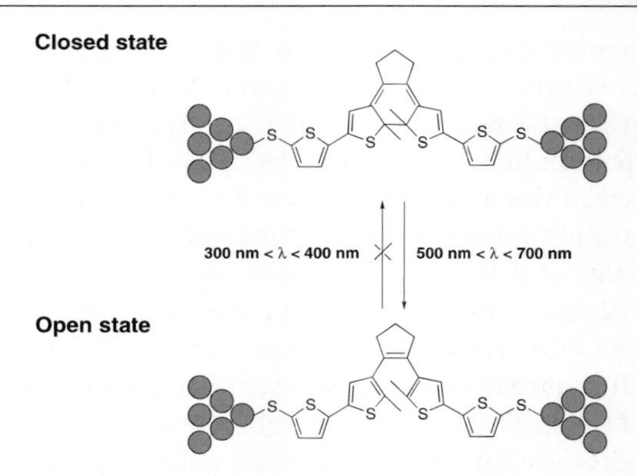


**Fig. 46** Photochromic molecular switch between two Au contacts in closed state and open state [224]

Feringa et al. studied the optoelectronic switching of a photochromic molecule using a mechanical break-junction technique and UV/vis spectroscopy (Fig. 46) [224]. In solution it is possible to switch between the closed and the open state by illuminating the molecule with light of wavelength $\lambda = 546$ nm while it can be switched back by using UV light at $\lambda = 316$ nm. If the chromophore is placed between two Au electrodes only the ring opening reaction occurs using visible light. A switch back from the isolating to the conducting state was not possible, which is attributed to the quenching of the excited state of the molecule in the open form by the gold electrodes [224].

## 6.5 Photocurrent Generation

The fact that all fossil fuel sources are finite makes solar energy conversion and photocurrent generation a topic of utmost importance.

The efficiency and complexity of electron (ET) and energy (EN) transfer reactions in natural photosynthesis have led many chemists to design donor-acceptor-linked systems that mimic these multistep ET processes. In the natural photosynthetic ET reaction centres, porphyrins and metalloporphyrins are the essential chromophore components [27].

One way to convert light energy into electrical energy is to use light-sensitive electrodes. Sensitisation of an electrode can be defined as a process by which a heterogeneous interfacial ET occurs as the consequence of light absorption by a chromophore, which is termed photosensitiser. Light absorption of the chromophore sensitiser leads to an excited state, which is both a stronger reducing agent and a stronger oxidising agent than the chromophore ground state. The excited sensitiser then oxidises (reduces) an electron donor (acceptor) mediator, which serves as a hole (electron) carrier while the sensitiser is concomitantly oxidised (reduced) by the sensitiser-supporting electrode. Depending on the relative energy level of sensitiser, Fermi energy of the electrode, and redox potential of the mediator an anodic (cathodic) photocurrent is observed. Quantification of the energetic level is, thus, of great importance for the development of photoelectrosynthetic or regenerative cells. One important competitive mechanism of photocurrent generation is the deactivation of the excited sensitiser by energy transfer to the metal electrode, as discussed in Sect. 5.2 [225].

The ET and EN processes in chromophore SAMs have been widely studied with different systems such as porphyrins [142, 149, 150, 160, 161], fullerenes [28, 177–179], carbazols [118], pyrenes [147] and dyads or triads containing these units [26, 27, 117, 189, 191].

Photoelectrochemical studies of SAMs containing porphyrin moieties attached to a gold electrode via methylene spacers of different chain length ($(CH_2)_n$, $n = 1$–7, 10, 11) indicate that the ET takes place from the gold electrode to a counter electrode through the porphyrin monolayer **36**/Au and an electron carrier. The quantum yield of the photocurrent generation in this system increases in a zigzag fashion with increasing spacer length up to $n = 6$ and then decreases slightly with a further increase in spacer length. Such a dependence of the quantum yield on the spacer length can be rationalised by the competition between an ET and an EN of the porphyrin singlet excited state (Fig. 47) [142]. The zigzag behaviour reflects the different orientation of the porphyrin chromophores depending on the even/odd number of methylene spacer groups (see Sect. 4.3.1).

Photoinduced ET in SAMs of porphyrin-linked $C_{60}$ systems shows that the maximum of the photocurrent using free-base porphyrin $C_{60}$ moieties is five times larger than in an analogous free-base porphyrin SAM system without $C_{60}$. In the former system and in a $C_{60}$-terminated oligothiophene mentioned

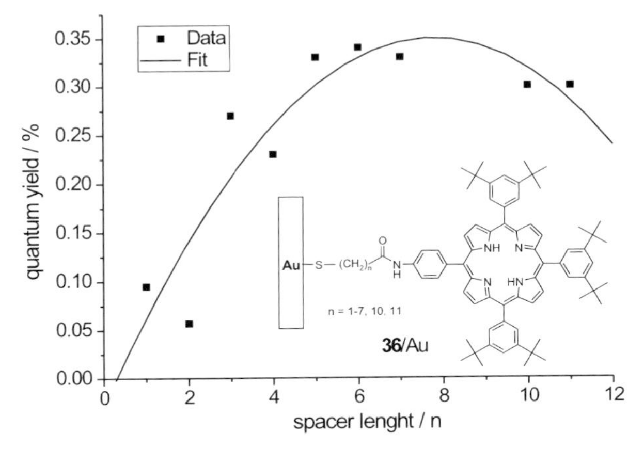


**Fig. 47** Photocurrent quantum yields vs. spacer length ($n$ = number of methylene groups). The *solid curve* is only a guide to the eye. The data were taken from [142]

below the porphyrin and the oligothiophene, respectively, act as photosensitisers which absorb the light. These studies demonstrate that $C_{60}$ serves as an effective mediator in the multistep ET processes, which is probably due to its small reorganisation energy [189]. Uosaki et al. [117] investigated a photoelectrochemical cell in which the photosensitiser (free-base porphyrin) is separated from the gold electrode by a covalently bound redox relay (mediator), which is ferrocene. In this system a quantum yield of photocurrent generation of 11% could be achieved. The surprisingly high quantum yield is with its high electron transfer rate mediated by ferrocene due to the low reorganisation energy, and to the inhibition of back electron transfer as well as of energy transfer quenching of the excited porphyrin state.

Taking all the above-mentioned results together Imahori et al. designed a triad-modified SAM that combines all the features for efficient photocurrent generation: a free-base porphyrin as the photosensitiser terminated by $C_{60}$ as covalently attached redox mediator, both bound to the gold electrode by a second redox mediator/relay (ferrocene). This separates the photosensitiser from the metal electrode in order to minimise back electron transfer and excited state quenching. In this way an exceptionally high quantum yield of 25% could be achieved [191]. The photoelectrochemical experiments used the triad-modified gold electrode, a platinum wire counter electrode and a Ag/AgCl reference electrode in the presence of an electron carrier (e.g. oxygen and methyl viologen) (Fig. 48). The cathodic photocurrent was detected under irradiation at 438.5 nm. An increase in the cathodic photocurrent was observed with increasing negative bias to the gold electrode. This indicates that the direction of the electron flow is from the gold electrode to the counter electrode through the SAM and the electron carrier. Further investigations showed that the photocurrent efficiency is controlled by the heterogeneous ET between the gold electrode and the ferrocene and that the porphyrin is indeed the major photoactive species for the photocurrent generation.

In the photoinduced ET from the porphyrin singlet excited state to the $C_{60}$ moiety high quantum yields of up to 20–25% could be achieved, followed by the charge shift from the ferrocene to the resulting porphyrin radical cation to yield a charge-separated state, $Fc^+$-P-$C_{60}^-$ in the monolayer (Fig. 48). This charge-separated state, $Fc^+$-P-$C_{60}^-$, reduces electron carriers such as oxygen ($E^0_{red}$ =– 0.48 V for $O_2/O_2^-$ vs. Ag/AgCl) and/or methyl viologen ($E^0_{red}$ =– 0.62 V for $MV^{2+}/MV^+$ vs. Ag/AgCl), which eventually transport an electron to the counter electrode. The ET rate from the gold electrode to $Fc^+$ in $Fc^+$-P-$C_{60}^-$ is controlled by the potential applied to the gold electrode, i.e. the ET rate from the gold electrode to $Fc^+$ increases with decreasing potential and, thus, the photocurrent increases [191, 192].

Quite recently, Aso and Imahori et al. developed a different route to achieve even higher photocurrent quantum yields than those in the triad SAM. Both $C_{60}$ and oligothiophenes are molecular units that are often used in optoelectronic devices. Therefore, $C_{60}$-terminated oligothiophenes were employed

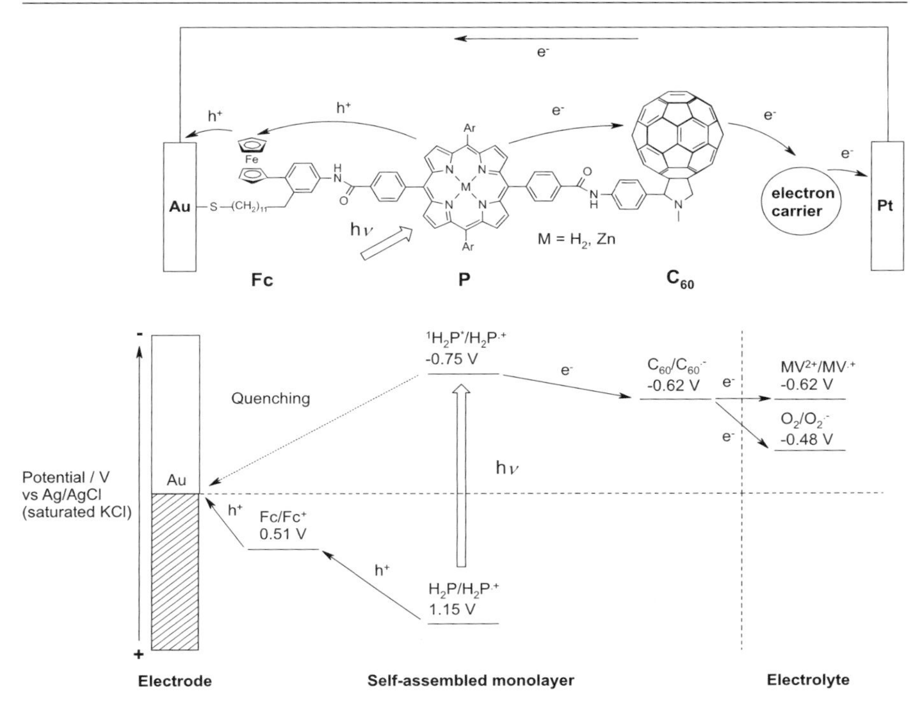


**Fig. 48** Photoinduced multistep ET at gold electrodes modified with SAMs of ferrocene-porphyrin-$C_{60}$ alkanethiols, electron carrier: oxygen or viologen [27]

as sensitisers for photocurrent generation. The ET and, consequently, the photocurrent efficiency in oligothiophene-$C_{60}$ molecules depends on the position of the molecule towards the gold electrode (horizontal or perpendicular orientation) and on the interaction between the oligothiophenes. For short oligothiophene chains the molecules are aligned horizontally to the gold surface, resulting in the quenching of the excited state by energy transfer to the gold surface, while for longer chains the initial photoexcited state is quenched due to significant van der Waals interactions of the oligothiophenes [28, 179]. To circumvent these difficulties a three-armed anchor group was chosen so that the molecules have a perpendicular orientation to the gold surface and the oligothiophenes are further apart from each other (see Sect. 4.3.3, Fig. 26). This set-up yields very high quantum yields of 17% for the quarterthiophene-$C_{60}$ dyad and 35% for the octithiophene-$C_{60}$ dyad (see Sect. 4.3.3, Fig. 26) while those of the analogous oligothiophenes without tripodal anchor are 0.1 and 5%, respectively. This demonstrates that avoiding quenching processes in the sensitiser monolayer is of prime importance.

If one takes into account that in natural photosynthesis the light-harvesting systems and the charge-separating system pertain to different molecular entities, the combination of an artificial antenna system with an artificial reaction centre as a mixed SAM on gold electrodes might lead to even more efficient photocurrent generation. Therefore, to examine the photoinduced EN in a SAM, mixed SAMs of pyrene- (antenna system) and porphyrin-thiols (reaction centre) were prepared (see Sect. 4.3.5, Fig. 29, **20**) [147]. Although an efficient singlet–singlet EN takes place from pyrene to the porphyrin moiety (> 62%), pyrene can absorb light only in the UV region, which hampers light harvesting in the visible region (> 400 nm). Therefore, a boron dipyrrin thiol that absorbs light at 502 nm was used as a more suitable light-harvesting molecule to achieve an efficient EN from the boron dipyrrin moiety ($^1B^*$) to the porphyrin chromophore (P) in the blue to green region of visible light. About 100% EN quantum yield was achieved with a mixture of 69 : 31 B : P in the SAM. However, the photocurrent quantum yield of a cell with a mixed B/P SAM was quite low (~ 2%) (Fig. 49) [193].

Therefore, the triad depicted in Fig. 48 was used in combination with the light-harvesting boron dipyrrin. Indeed, a mixed SAM with 37 : 63 B/triad showed a quantum yield of 50% for photocurrent generation. These studies demonstrate that highly efficient photocurrent generation is possible by combination of suitable antenna molecules with multistep ET arrays [193].

Diederich et al. described a new method to increase the photoelectrochemical response and the stability of $C_{60}$-SAMs using a potassium-selective polyurethane membrane cast (Mb/**37**/Au) compared to the same $C_{60}$-SAM (**37**/Au) without the membrane [226]. The photocurrent generation of the stabilised $C_{60}$-SAM is higher than for the bare $C_{60}$-SAM, but by applying a bias under positive conditions the photocurrent becomes lower for the stabilised $C_{60}$-SAM (Fig. 50).

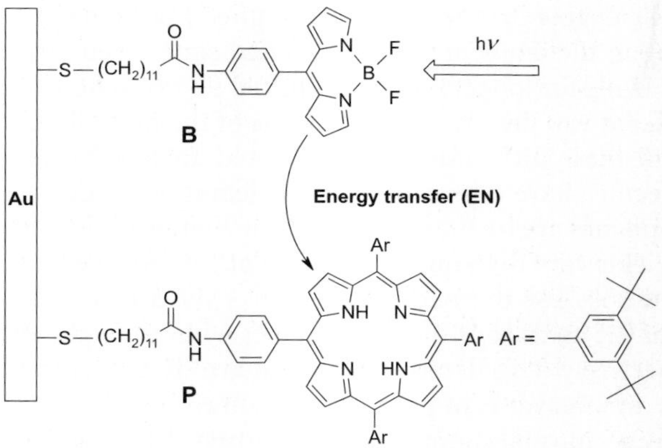

**Fig. 49** Photoinduced energy transfer at gold electrodes modified with SAMs of boron dipyrrin thiol (**B**) and porphyrin alkanethiol (**P**) [27]

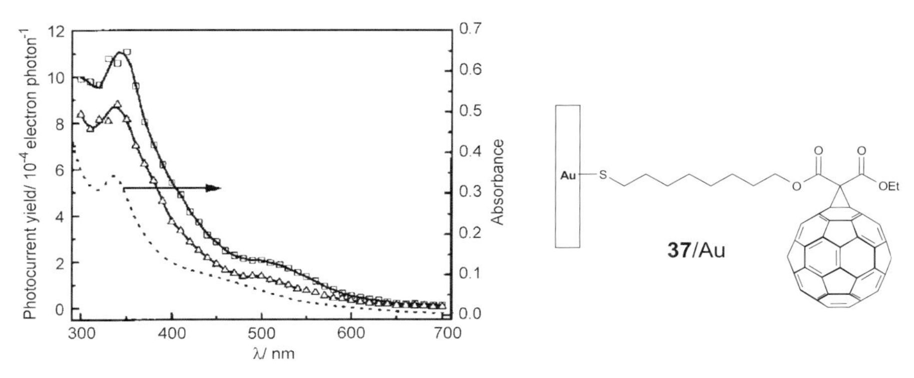


**Fig. 50** Photocurrent action spectra for cells with **37**/Au (△) and Mb/**37**/Au (□) as working electrodes (Ar-saturated $10^{-3}$ M aqueous KCl solution, pH 5.6, no applied voltage) and spectrum of a 49 nm thick spin coated film of **37** on sapphire (*dotted line*) (*arrow:* increasing absorption in the visible region) [226]. Reproduced by permission of The Royal Society of Chemistry

## 6.6 SAMs as Testing Materials for Molecular Wires

A detailed understanding of how electrons are transferred through organic molecules is an important topic in several areas: rationalising ET in organic conducting and semiconducting materials, biological systems such as the photosynthetic reaction centre, fabricating molecular electronic devices such as organic light emitting diodes (OLED), memory devices or field-effect transistors (FETs) and developing quantum computers on a single-molecular scale [227]. Recent work on the electrochemistry of SAMs on electrodes provides a general route for creating surface structures in which redox-active molecules are linked to electrodes via well-defined molecular bridges. One great advantage of this approach is that the distance of the redox centres from the electrode, the chemical surrounding of the redox centres as well as the type of the molecular bridges can be varied systematically. These SAM structures can then serve as excellent model systems for studying bridge-mediated ET.

Creager et al. as well as Chidsey et al. investigated the ET of ferrocene groups attached to gold electrodes with oligo(phenylenethynylene) bridges of variable length and structure in mixed monolayers with hydroxyalkanethiol dummy molecules (Fig. 51) [228–230]. The conjugated bridges allow a strong electronic coupling between the gold electrode and the ferrocene redox centre, which results in a rapid electron transfer over long distances. Using impedance spectroscopy [231], the effects of bridge length on the ET show an exponential distance dependence of the bridge-mediated ET rates from $k_{et} = 350\ s^{-1}$ for the longest bridge (six phenylethynyl units, **42**) to

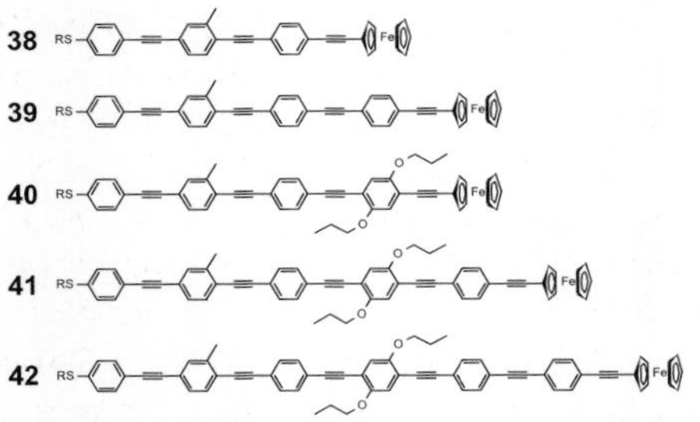

| Monolayer components | $k_{et}$ [$s^{-1}$] |
|---|---|
| **38** + HO-$(CH_2)_{16}$-SH | 5 x $10^5$ |
| **39** + HO-$(CH_2)_{16}$-SH | 6 x $10^4$ |
| **40** + HO-$(CH_2)_{16}$-SH | 6.5 x $10^4$ |
| **41** + HO-$(CH_2)_{16}$-SH | 5 x $10^3$ |
| **42** + HO-$(CH_2)_{16}$-SH | 3.5 x $10^2$ |

**Fig. 51** Structures and rate constants for the heterogeneous ET of ferrocene oligo(phenylenethynylene) thiols in SAMs (R = protection groups) [229]

$k_{et} = 500\,000\ s^{-1}$ for the shortest bridge with three phenylethynyl units **38** (Fig. 51). The effect on the ET rates if two propoxy groups are attached to one of the phenyl rings of the bridge was found to be minimal (cf. **39** versus **40**).

In comparison to the above mentioned ferrocene oligo(phenylenethynylene) SAMs, a group of ferrocene-terminated oligo(phenylenevinylene)methane thiols **43–47** was prepared and the ET behaviour in the corresponding SAMs was investigated [232]. Studies of the monolayers containing oligomeres of the same length with and without ethoxy solubilising groups using cyclic voltammetry revealed that both types of oligomers form well-packed SAMs. The position of the solubilising groups in the oligomer chain does not influence the packing in the monolayer significantly. ET measurements were performed on these systems with the indirect laser-induced temperature jump method. The measured ET rate constants are all greater than $5 \times 10^5\ s^{-1}$ for the five oligomers **43–47** [233].

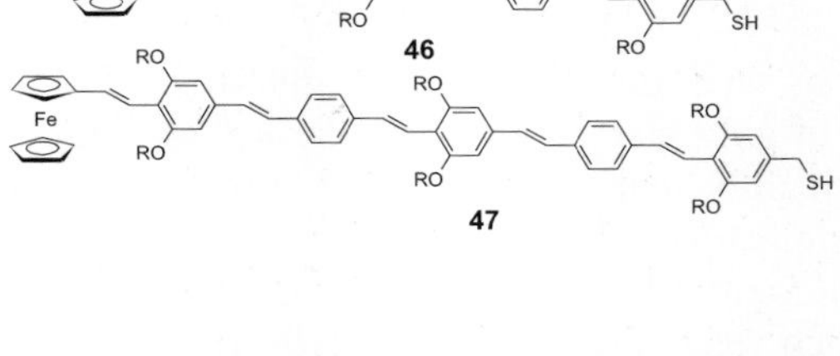

**Structure 1**

Other interesting candidates for bridges in molecular wires are oligophenylenevinylene (OPV) chromophores **48–54** with two thioacetyl groups terminating each end [234]. Contact angle, FT-IRRAS and XPS measurements show that molecules with linear hexyloxy-substituted OPV structures **48–50** form disordered SAMs. The [2.2]paracyclophane containing OPVs (**53**, **54**) form significantly better SAMs than examples with side groups. The best structures for obtaining a monolayer are the unsubstituted OPVs **51** and **52**. Ellipsometry measurements indicate that in the case of molecules **51** and **52** the orientation of the molecule coincides with the surface normal or is tilted up to 30° from the surface normal. Electrochemical investigations show that these derivatives, especially the [2.2]paracyclophane-OPVs **53** and **54** in which through-space and through-bond delocalisation is possible, are promising candidates for molecular wires.

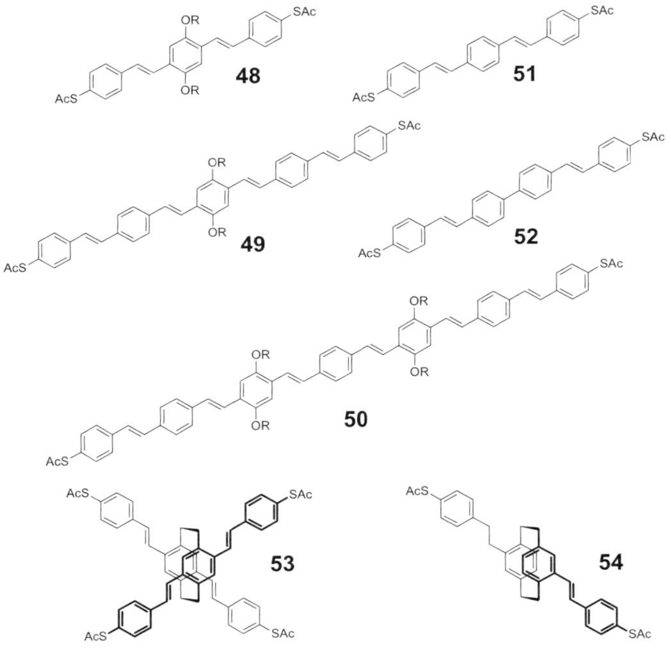

OR SAc 48 AcS OR

SAc 51 AcS

SAc OR OR 49 AcS

SAc 52 AcS

SAc OR OR OR OR 50 AcS

AcS SAc 53 AcS SAc

AcS 54 SAc

**Structure 2**

Another new group of chromophores for testing in molecular wires are triarylamines because of their outstanding physical and chemical properties. For ET investigations triarylamines and phenothiazines with different substituted phenylene bridges and acetylene spacers have been prepared in order to probe the influence of electron-donating or electron-withdrawing groups on the ET rates (Fig. 52). The SAMs are built of the chromophores, which are diluted with alkanethiols as dummy molecules [235].

The high affinity of thiols to gold has also lead to the preparation and investigation of single molecular wires using break junction techniques [236–239], but this topic is beyond the scope of this review.

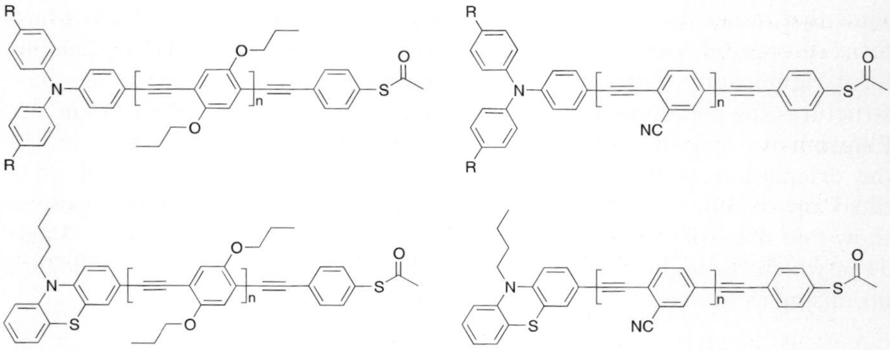

**Fig. 52** Different triarylamines and phenothiazines for testing in molecular wires (R = OMe, Cl; $n = 0$–3) [235]

# 7 Conclusion and Outlook

Undoubtedly, much has been learned about the structure and electronic properties of SAMs containing chromophores as functional subunits. Starting from structural investigations of simple aromatic thiols on gold, the development has led to the construction of complex systems such as gas sensors, catalyst systems, photoswitches and dyad and triad SAM systems for photocurrent generation. The advantages of chromophore SAMs on gold are quite clear: the ease of preparation and their relative stability, together with the strong background information available on gold SAMs, makes them an almost ideal playground for surface research.

However, some possible disadvantages should not be overlooked: the long-term stability of SAMs on gold strongly depends on the organic layer and might be very weak. The molecular order of very complex SAMs and of mixed SAMs has, in most cases, not been assessed in detail and might be quite low with all the consequences for the functional properties of the SAM. The quenching of excited states by the gold surface is an inherent problem. While several successful attempts have been made to circumvent the above mentioned problems in particular cases, generally applicable strategies are not yet known.

Whether SAMs will ever be incorporated in commercial devices is not yet clear, but quite recently a major step was achieved by Baldo et al. who produced a photosynthetic device consisting of a SAM of photosystem I isolated from spinach chloroplasts stabilised with surfactant peptides and coated with a protective organic semiconductor. This device, which has an internal quantum efficiency of 12%, was stable for several weeks under ambient conditions [240]. Therefore, we conclude that functional SAMs have a good prospect of being implemented in a variety of electronic devices in the future.

**Acknowledgements** We thank the Fonds der Chemischen Industrie and the Volkswagen Foundation for financial support and PD Dr A. Terfort/Hamburg for helpful discussion.

## References

1. Ulman A (1991) An introduction to ultrathin organic films: from Langmuir–Blodgett to self-assembly. Academic, Boston
2. Finklea HO (1996) In: Bard AJ, Rubinstein I (eds) J Electroanal Chem 19. Dekker, New York
3. Ulman A (1996) Chem Rev 96:1533
4. Crooks RM, Ricco AJ (1998) Acc Chem Res 31:219
5. Mallouk TE, Harrison JD (1994) Interfacial design and chemical sensing. American Chemical Society, Washington, DC
6. Flink S, van Weggel CJM, Reinhoudt DN (2000) Adv Mater 12:1315
7. Haussling L, Knoll W, Ringsdorf H, Schmitt FJ, Yang J (1991) Macromol Chem Macromol Symp 46:145
8. Willner I, Rubin S (1996) Angew Chem, Int Ed 35:367
9. Dhirani A, Lin P-H, Guyot-Sionnest P, Zehner RW, Sita LR (1997) J Chem Phys 106:5249
10. Kondo T, Horiuchi S, Yagi I, Yw S, Uosaki K (1999) J Am Chem Soc 121:391
11. Naraoka R, Kaise G, Kajikawa K, Okawa H, Ikezawa H, Hashimoto K (2002) Chem Phys Lett 362:26
12. Mishina E, Miyakita Y, Yu Q-K, Nakabayashi S, Sakaguchi H (2002) J Chem Phys 117:4016
13. Tsuboi K, Seki K, Ouchi Y, Fujita K, Kajikawa K (2003) Jpn J Appl Phys 42:607
14. Willner I, Lion-Dagan M, Marx-Tibbon S, Katz E (1995) J Am Chem Soc 117:6581
15. Willner I, Lion-Dagan M, Katz E (1996) J Chem Soc Chem Commun 623
16. Willner I (1997) Acc Chem Res 30:347
17. Willner I, Willner B (2003) Coord Chem Rev 245:139
18. Kondo T, Kanai T, Uosaki K (2001) Langmuir 17:6317
19. Wang Z, Cook MJ, Nygard A-M, A RD (2003) Langmuir 19:3779
20. Haussling L, Ringsdorf H, Schmitt FJ, Knoll W (1991) Langmuir 7:1837
21. Zamborini FP, Crook RM (1998) Langmuir 14:3279
22. Donhauser ZJ, Mantooth BA, Kelly KF, Bumm LA, Monnell JD, Stapleton JJ, Price Jr DW, Rawlett AM, Allara DL, Tour JM, Weiss PS (2001) Science 292:2303
23. Feldheim DL, Keating CD (1998) Chem Soc Rev 27:1
24. Fendler JH (2001) Chem Mater 13:3196
25. Reed MA, Chen J, Rawlett AM, Price Jr DW, Tour JM (2001) Appl Phys Lett 78:3735
26. Imahori H, Sakata Y (1999) Eur J Org Chem:2445
27. Imahori H, Mori Y, Matano Y (2003) J Photochem Photobiol C 4:51
28. Otsubo T, Aso Y, Takimiya K (2002) J Mater Chem 12:2565
29. Abbott NL, Folkers JP, Whitesides GM (1992) Science 257:1380
30. Lopez GP, Biebuyck HA, Frisbie CD, Whitesides GM (1993) Science 260:647
31. Gorman CB, Biebuyck HA, Whitesides GM (1995) Chem Mater 7:252
32. Rozsnyai LF, Wrighton MS (1995) Langmuir 11:3913
33. Chan KC, Kim T, Schoer JK, Crooks RM (1995) J Am Chem Soc 117:5875
34. Bell CM, Yang HC, Mallouk TE (1995) In: Interrante LV, Caspar LA, Ellis AB (eds) Materials chemistry. American Chemical Society, Washington, DC p 211

35. Fuhrhop J-H, Köning J (1995) In: Stoddart JF (ed) Membranes and molecular assemblies: the synkinetic approach, monographs in supramolecular chemistry. Freie Univerisität, Berlin, p 149
36. Roy D, Fendler J (2004) Adv Mater 16:479
37. Daniel M-C, Astruc D (2004) Chem Rev 104:293
38. Kamat PV (2002) J Phys Chem B 106:7729
39. Dulkeith E, Morteani AC, Niedereichholz T, Klar TA, Feldmann J, Levi SA, van Veggel FCJM, Reinhoudt DN, Möller M, Gittins DI (2002) Phys Rev Lett 89:203002
40. Li X-M, Paraschiv V, Huskens J, Reinhoudt DN (2003) J Am Chem Soc 125:4279
41. Gopidas KR, Whitesell JK, Fox MA (2003) J Am Chem Soc 125:14168
42. Gopidas KR, Whitesell JK, Fox MA (2003) J Am Chem Soc 125:6491
43. Subramanian V, Wolf EE, Kamat PV (2004) J Am Chem Soc 126:4943
44. Ipe BI, George Thomas K, Barazzouk S, Hotchandani S, Kamat PV (2002) J Phys Chem B 106:18
45. Sudeep PK, Ipe BI, George Thomas K, Geroge MV, Barazzouk S, Hotchandani S, Kamat PV (2002) Nano Lett 2:29
46. Frankamp BL, Boal AK, Rotello VM (2002) J Am Chem Soc 124:15146
47. Verma A, Nakade H, Simard JM, Rotello VM (2004) J Am Chem Soc 126:10806
48. Jeoung E, Rotello VM (2002) J Supramol Chem 2:53
49. Hong R, Emrick T, Rotello VM (2004) J Am Chem Soc 126:13572
50. Boal AK, Rotello VM (2000) J Am Chem Soc 122:734
51. Boal AK, Rotello VM (2000) Langmuir 16:9527
52. Levi SA, Mourran A, Spatz JP, van Veggel FCJM, Reinhoudt DN, Möller M (2002) Chem Eur J 8:3808
53. Thomas KG, Ipe BI, Sudeep PK (2002) Pure Appl Chem 74:1731
54. Thomas KG, Kamat PV (2003) Acc Chem Res 36:888
55. Hasobe T, Imahori H, Kamat PV, Fukuzumi S (2003) J Am Chem Soc 125:14962
56. Flink S, van Veggel FCJM, Reinhoudt DN (2001) J Phys Org Chem 14:407
57. Crego-Calama M, Reinhoudt DN (2001) Adv Mater 13:1171
58. Flink S, van Veggel FCJM, Reinhoudt DN (1999) Chem Commun 2229
59. van der Veen NJ, Flink S, Deij MA, Egberink RJM, van Veggel FCJM, Reinhoudt DN (2000) J Am Chem Soc 122:6112
60. Berlin A, Zotti G (2000) Macromol Rapid Commun 21:301
61. Imahori H, Fukuzumi S (2004) Adv Funct Mater 14:525
62. Taniguchi I, Toyoshima K, Yamaguchi H, K Y (1982) J Chem Soc, Chem Commun 1032
63. Nuzzo RG, Allara DL (1983) J Am Chem Soc 105:4481
64. Langmuir I (1917) J Am Chem Soc 39:1848
65. Blodgett KB (1935) J Am Chem Soc 57:1007
66. Metzger RM (2003) Chem Rev 103:3803
67. Lahav M, Gabriel T, Shipway AN, Willner I (1999) J Am Chem Soc 121:258
68. Chrisstoffels LAJ, Adronov A, Frechet JMJ (2000) Angew Chem Int Ed Engl 39:2163
69. Golan Y, Margulis L, Rubinstein I (1992) Surf Sci 264:312
70. Widrig CA, Chung C, Porter MD (1991) J Electroanal Chem 310:335
71. Evans SD, Goppert-Bearducci KE, Uranker E, Gerenser LJ, Ulman A (1991) Langmuir 7:2700
72. Guo L-H, Facci JS, McLendon G, Mosher R (1994) Langmuir 10:4588
73. Creager SE, Hockett LA, Rowe GK (1992) Langmuir 8:854
74. Steinberg S, Tor Y, Sabatani E, Rubinstein I (1991) J Am Chem Soc 113:5176
75. Steinberg S, Rubinstein I (1992) Langmuir 8:1183

76. Uosaki K, Shen Y, Kondo T (1995) J Phys Chem 99:14117
77. DiMilla PA, Folkers JP, Biebuyck HA, Härter R, Lopez GP, Whitesides GM (1994) J Am Chem Soc 116:2225
78. Kalyuzhny G, Vaskevich A, Ashkenasy G, Shanzer A, Rubinstein I (2000) J Phys Chem B 104:8238
79. Link S, El-Sayed MA (1999) J Phys Chem B 103:8410
80. Henglein A, Meisel D (1998) J Phys Chem B 102:8364
81. Ali AH, Luther RJ, Foss CA, Chapman GB (1997) Nanostruct Mater 9:559
82. Doremus R (1966) J Appl Phys 37:2775
83. Allpress JG, Sanders JV (1967) Surf Sci 7:1
84. Doremus R (1998) Thin Solid Films 326:205
85. Levlin M, Laakso A, Niemi HE-M, Hautojärvi P (1997) Appl Surf Sci 115:31
86. Liu ZH, Brown NMD, McKinley A (1997) J Phys:Condens Matter 9:59
87. Chidsey CED, Loiacono DN, Sleator T, Nakahara S (1988) Surf Sci 200:45
88. Golan Y, Margulis L, Matlis S, Rubinstein I (1995) J Electrochem Soc 142:1629
89. Hsu T (1983) Ultramicroscopy 11:167
90. Trevor DJ, Chidsey CED, Loiacono DN (1989) Phys Rev Lett 62:929
91. Shipway AN, Katz E, Willner I (2000) Chem Phys Chem 1:18
92. Elghanian R, Storhoff JJ, Mucic RC, Letsinger RL, Mirkin CA (1997) Science 277:1078
93. Ribrioux S, Kleymann G, Haase W, Heitmann K, Ostermeier C, Michel H (1996) J Histochem Cytochem 44:207
94. Rodriguez JF, Mebrahtu T, Soriaga MP (1987) J Electroanal Chem 233:283
95. Schreiber F (2000) Prog Surf Sci 65:151
96. Delamarche E, Michel B (1996) Thin Solid Films 273:54
97. Delamarche E, Michel B, Biebuyck HA, Gerber C (1996) Adv Mater 8:719
98. Pradeep T, Sandhyarani N (2002) Pure Appl Chem 74:1593
99. Schwartz DK (2001) Ann Rev Phys Chem 52:107
100. Finklea HO (1996) J Electroanal Chem 19:109
101. Templeton AC, Wuelfing WP, Murray RW (2000) Acc Chem Res 33:27
102. Hostetler MJ, Murray RW (1997) Curr Opin Colloid Interface Sci 2:42
103. Ishida T (2003) Chemistry of nanomolecular systems: towards the realization of molecular devices. Springer, Berlin Heidelberg New York
104. Liu GY, Rodriguez JA, Dvorak J, Hrbek J, Jirsak T (2002) Surf Sci 505:295
105. Dubois LH, Nuzzo RG (1992) Ann Phys Chem 43:437
106. Poirier GE (1997) Chem Rev 97:1117
107. Sellers H, Ulman A, Shnidman Y, Eilers JE (1993) J Am Chem Soc 115:9389
108. Zharinkov M, Frey S, Rong H, Yang Y-J, Heister K, Buck M, Grunze M (2000) Phys Chem Chem Phys 2:3359
109. Buckel F, Effenberger F, Yan C, Gölzhäuser A, Grunze M (2000) Adv Mater 12:901
110. Bain CD, Troughton EB, Tao Y-T, Evall J, Whitesides GM, Nuzzo RG (1989) J Am Chem Soc 111:321
111. Dannenberger O, Buck M, Grunze M (1999) J Phys Chem B 103:2202
112. Peterlinz KA, Gerogiadis R (1996) Langmuir 12:4731
113. Karpovich DS, Blanchard GJ (1994) Langmuir 10:3315
114. Xu S, Cruchon-Dupeyrat SJN, Garno JC, Liu GY, Jennings GK (1998) J Chem Phys 108:5002
115. DeBono RF, Loucks GD, Dellamanna D, Krull UJ (1996) Can J Chem 74:677
116. Dannenberger O, Wolff JJ, Buck M (1998) Langmuir 84:5164
117. Uosaki K, Kondo T, Zhang X-Q, Yanagida M (1997) J Am Chem Soc 119:8367
118. Morita T, Kimura S, Kobayashi S (2000) J Am Chem Soc 122:2850

119. Kobayashi K, Imabayashi S, Fujita H, Nonaka K, Kakiuchi T, Sasabe H, Knoll W (2000) Bull Chem Soc Jpn 73:1993
120. Lötzbeyer T, Schuhmann W, Schmidt H-L (1995) J Electroanal Chem 395:341
121. Zimmermann H, Lindgren A, Schuhmann W, Gorton L (2000) Chem Eur J 6:592
122. Fox MA, Whitesell JK, McKerrow AJ (1998) Langmuir 14:816
123. Frey S, Stadler V, Heister K, Eck W, Zharnikov M, Grunze M, Zeysing B, Terfort A (2001) Langmuir 17:2408
124. Azzam W, Wehner BI, Fischer RA, Terfort A, Wöll C (2002) Langmuir 18:7766
125. Himmel H-J, Terfort A, Wöll C (1998) J Am Chem Soc 120:12069
126. Tao YT, Wu CC, Eu JY, Lin WL (1997) Langmuir 13:4018
127. Chang SC, Chao I, Tao YT (1994) J Am Chem Soc 116:6792
128. Heister K, Rong HT, Buck M, Zharnikov M, Grunze M, Johansson LSO (2001) J Phys Chem B 105:6888
129. Rong HT, Frey S, Yang YJ, Zharnikov M, Buck M, Wuhn M, Woll C, Helmchen G (2001) Langmuir 17:1582
130. Shaporenko A, Brunnbauer M, Terfort A, Grunze M, Zharinkov M (2004) J Phys Chem B 108:14462
131. Dhirani A-A, Zehner RW, Hsung RP, Guyot-Sionnest P, Sita LR (1996) J Am Chem Soc 118:3319
132. Stapleton JJ, Harder P, Daniel TA, Reinard MD, Yao Y, Price DW, Tour JM, Allara DL (2003) Langmuir 19:8245
133. Michalitsch R, Nogues C, Najari A, El Kassmi A, Yassar A, Lang P, Garnier F (1999) Synthetic Metals 101:5
134. Michalitsch R, El Kassmi A, Yassar A, Lang P, Garnier F (1998) J Electroanal Chem 457:129
135. Michalitsch R, Nogues C, Najari A, El Kassmi A, Yassar A, Lang P, Rei Vilar M, Garnier F (1999) Synthetic Metals 102:1319
136. Liedberg B, Yang Z, Engquist I, Wirde M, Gelius U, Götz G, Bäuerle P, Hummel R-M, Ziegler C, Göpel W (1997) J Phys Chem B 101:5951
137. Sigal GB, Bamdad C, Barberis A, Strominger J, Whitesides GM (1996) Anal Chem 68:490
138. Patel N, Davies MC, Hartshorne M, Heaton RJ, Roberts CJ, Tendler SJB, Williams PM (1997) Langmuir 13:6485
139. El Kasmi A, Wallace JM, Bowden EF, Binet SM, Linderman RJ (1998) J Am Chem Soc 120:225
140. Kang JF, Liao S, Jordan R, Ulman A (1998) J Am Chem Soc 120:9662
141. Stranick SJ, Parikh AN, Tao Y-T, Allara DL, Weiss PS (1994) J Phys Chem 98:7636
142. Imahori H, Norieda H, Nishimura Y, Yamazaki I, Higuchi K, Kato N, Motohiro T, Yamada H, Tamaki K, Arimura M, Sakata Y (2000) J Phys Chem B 104:1253
143. Fukuzumi S, Imahori H (2000) Electron transfer in chemistry. Wiley-VCH, Weinheim
144. Boeckl MS, Bramblett AL, Hauch KD, Sasaki T, Ratner BD, Rogers Jr JW (2000) Langmuir 16:5644
145. Owens RW, Smith DA (2000) Langmuir 16:562
146. Kawasaki M, Sato T, Yoshimoto T (2000) Langmuir 16:5409
147. Imahori H, Nishimura Y, Norieda H, Karita H, Yamazaki I, Sakata Y, Fukuzumi S (2000) Chem Commun 661
148. Yamada H, Imahori H, Nishimura Y, Yamazaki I, Fukuzumi S (2000) Chem Commun 1921

149. Imahori H, Norieda H, Ozawa S, Ushida K, Yamada H, Azuma T, Tamaki K, Sakata Y (1998) Langmuir 14:5335
150. Imahori H, Hasobe T, Yamada H, Nishimura Y, Yamazaki I, Fukuzumi S (2001) Langmuir 17:4925
151. Shimazu K, Takechi M, Fujii H, Suzuki M, Saiki H, Yoshimura T, Uosaki K (1996) Thin Solid Films 273:250
152. Yuan H, Woo lK (1997) J Porphyrins Phthalocyanines 1:189
153. Yamada T, Nango M, Ohtsuka T (2002) J Electroanal Chem 528:93
154. Hutchinson JE, Postlethwaite TA, Chen C-H, Hathcock KW, Ingram RS, Ou W, Linton RW, Murray RW, Tyvoll DA, Chng LL, Collman JP (1997) Langmuir 13:2143
155. Zak J, Yuan H, Ho M, Woo K, Porter MD (1993) Langmuir 9:2772
156. Postlethwaite TA, Hutchinson JE, Hathcock KW, Murray RW (1995) Langmuir 11:4109
157. Bramblett AL, Boeckl MS, Hauch KD, Ratner BD, Sasaki T, Rogers Jr JW (2002) Surf Interface Anal 33:506
158. Offord DA, Sachs SB, Ennis MS, Eberspacher TA, Griffin JH, Chidsey CED, Collman JP (1998) J Am Chem Soc 120:4478
159. Ashkenasy G, Kalyuzhny G, Libman J, Rubinstein I, Shanzer A (1999) Angew Chem Int Ed Engl 38:1257
160. Kondo T, Ito T, Nomura S-i, Uosaki K (1996) Thin Solid Films 284–285:652
161. Kondo T, Yanagida M, Nomura S-I, Takahashi M, Uosaki K (1997) J Electroanal Chem 438:121
162. Gryko DT, Clausen C, Lindsay JS (1999) J Org Chem 64:8635
163. Gryko DT, Clausen C, Roth KM, Dontha N, Bocian DF, Kuhr WG, Lindsey JS (2000) J Org Chem 65:7345
164. Roth KM, Gryko DT, Clausen C, Li J, Lindsey JS, Kuhr WG, Bocian DF (2002) J Phys Chem B 106:8639
165. Balakumar A, Lysenko AB, Carcel C, Malinovskii VL, Gryko DT, Schweikart K-H, Loewe RS, Yasseri AA, Liu Z, Bocian DF, Lindsey JS (2004) J Org Chem 69:1435
166. Wei L, Padmaja K, Youngblood WJ, Lysenko AB, Lindsey JS, Bocian DF (2004) J Org Chem 69:1461
167. Cai L, Yao Y, Yang J, Price DW, Tour JM (2002) Chem Mater 14:2905
168. Kobayashi K, Shimizu M, Nagamune T, Sasabe H, Fang Y, Knoll W (2002) Bull Chem Soc Jpn 75:1707
169. Wright JD (1989) Prog Surf Sci 31:1
170. Snow AW, Barger WR (1989) Phthalocyanines - properties and applications. VCH, New York
171. Simpson TRE, Cook MJ, Petty MC, Thorpe SC, Russell DA (1996) Analyst 121:1501
172. Cook MJ (1999) Pure Appl Chem 71:2145
173. Chambrier I, Cook MJ, Russell DA (1995) Synthesis 1283
174. Revell DJ, Chambrier I, Cook MJ, Russell DA (2000) J Mater Chem 10:31
175. Schweikart K-H, Malinovskii VL, Yasseri AA, Li J, Lysenko AB, Bocian DF, Lindsay JS (2003) Inorg Chem 42:7431
176. Mirkin CA, Cladwell WB (1996) Tetrahedron 52:5113
177. Imahori H, Azuma T, Ozawa S, Yamada H, Ushida K, Ajavakom A, Norieda H, Sakata Y (1999) Chem Commun 557
178. Imahori H, Azuma T, Ajavakom A, Norieda H, Yamada H, Sakata Y (1999) J Phys Chem B 103:7233
179. Hirayama D, Takimiya K, Aso Y, Otsubo T, Hasobe T, Yamada H, Imahori H, Fukuzumi S, Sakata Y (2002) J Am Chem Soc 124:532

180. Liedberg B, Yang Z, Engquist I, Wirde M, Gelius U, Götz G, Bäuerle P, Rummel R-M, Ziegler C, Göpel W (1997) J Phys Chem B 101:5951
181. Natansohn A, Rochon P (2002) Chem Rev 102:4139
182. Nishihara H (2004) Bull Chem Soc Jpn 77:407
183. Wang R, Iyoda T, Jiang L, Tryk DA, Hashimoto K, Fujishima A (1997) J Electroanal Chem 438:213
184. Wang R, Iyoda T, Tryk DA, Hashimoto K, Fujishima A (1997) Langmuir 13:4644
185. Zhang A, Qin J, Gu J, Lu Z (2000) Thin Solid Films 375:242
186. Förster T (1969) Angew Chem, Int Ed 8:333
187. Fox MA, Li W, Wooten M, McKerrow AJ, Whitesell JK (1998) Thin Solid Films 327–329:477
188. Soto E, MacDonald JC, Cooper CGF, McGimpsey WG (2003) J Am Chem Soc 125:2838
189. Imahori H, Ozawa S, Ushida K, Takahashi M, Azuma T, Ajavakom A, Akiyama T, Hasegawa M, Taniguchi S, Okada T, Sakata Y (1999) Bull Chem Soc Jpn 72:485
190. Akiyama T, Imahori H, Ajawakom A, Sakata Y (1996) Chem Lett 907
191. Imahori H, Yamada H, Ozawa S, Ushida K, Sakata Y (1999) Chem Commun 1165
192. Imahori H, Yamada H, Nishimura Y, Yamazaki I, Sakata Y (2000) J Phys Chem B 104:2099
193. Imahori H, Norieda H, Yamada H, Nishimura Y, Yamazaki I, Sakata Y, Fukuzumi S (2001) J Am Chem Soc 123:100
194. Yanagida M, Kanai T, Zhang X-Q, Kondo T, Uosaki K (1998) Bull Chem Soc Jpn 71:2555
195. Kang SH, Ma H, Kang M-S, Kim K-S, Jen AK-Y, Hadi Zareie M, Sarikaya M (2004) Angew Chem, Int Ed 43:1512
196. Fujita K, Hara M, Sasabe H, Knoll W, Tsuboi K, Kajikawa K, Seki K, Ouchi Y (1998) Langmuir 14:7456
197. Frank HA (1999) The photochemistry of carotenoids. Kluwer, Dordrecht
198. Leatherman G, Durantini EN, Gust D, Moore TA, Stone S, Zhou Z, Rez P, Liu YZ, Lindsay SM (1999) J Phys Chem B 103:4006
199. Liu D, Szulczewski GJ, Kispert LD, Primak A, Moore TA, Gust D (2002) J Phys Chem B 106:2933
200. Li Z, Fehlner TP (2003) Inorg Chem 42:5715
201. Katz E, Willner I (1997) Langmuir 13:3364
202. Haas U, Thalacker C, Adams J, Fuhrmann J, Riethmüller S, Beginn U, Ziener U, Möller M, Dobrawa R, Würthner F (2003) J Mater Chem 13:762
203. Kroon JM, Koehorst RBM, van Dijk M, Sanders GM, Sudhölter JR (1997) J Mater Chem 7:615
204. Kasha M, Rawls HR, El-Bayoumi MA (1965) Pure Appl Chem 11:371
205. Kroon JM, Sudhölter JR, Schenning APHJ, Nolte RJM (1995) Langmuir 11:214
206. Kittredge KW, Fox MA, Whitesell JK (2001) J Phys Chem B 105:10594
207. Karpovich DS, Blanchard GJ (1996) Langmuir 12:5522
208. Porter MD, Bright TB, Allara DL, Chidsey CED (1987) J Am Chem Soc 109:3559
209. Nuzzo RG, Dubois LH, Allara DL (1990) J Am Chem Soc 112:558
210. Parikh AN, Allara DL (1992) J Chem Phys 96:927
211. Feringa BL (2001) Molecular switches. Wiley-VCH, Weinheim
212. Lion-Dagan M, Katz E, Willner I (1994) J Chem Soc Chem Commun 2741
213. Lahav M, Katz E, Doron A, Patolsky F, Willner I (1999) J Am Chem Soc 121:862
214. Chen H, Li Y, Huo F, Wang Z, Zhang X (2003) Chem Lett 32:1094

215. Cooper CGF, MacDonald JC, Soto E, McGimpsey WG (2004) J Am Chem Soc 126:1032
216. Zhu DG, Petty MC, Harris M (1990) Sens Actuators B 2:265
217. Simpson TRE, Russell DA, Chambrier I, Cook MJ, Horn AB, Thorpe SC (1995) Sens Actuators B 29:353
218. Simpson TRE, Revell DJ, Cook MJ, Russell DA (1997) Langmuir 13:460
219. Wang S, Liu Y, Huang X, Yu G, Zhu DG (2003) J Phys Chem B 107:12639
220. Collman JP, Wagenknecht PS, Hutchinson JE (1994) Angew Chem Int Ed 33:1537
221. Hutchinson JE, Postlethwaite TA, Murray RW (1993) Langmuir 9:3277
222. Willner I, Katz E (2000) Angew Chem Int Ed 39:1181
223. Walter DG, Campbell DJ, Mirkin CA (1999) J Phys Chem B 103:402
224. Dulic D, van der Molen SJ, Kudernac T, Jonkman HT, de Jong JJD, Bowden TN, van Esch J, Feringa BL, van Wees BJ (2003) Phys Rev Lett 91:207402
225. Qu P, Meyer GJ (2001) In: Balzani V (ed) Electron transfer in chemistry, vol 4. VCH, Weinheim, p 353
226. Enger O, Nuesch F, Fibbioli M, Echegoyen L, Pretsch E, Diederich F (2000) J Mater Chem 10:2231
227. Tour JM (2000) Acc Chem Res 33:791
228. Yu CJ, Chong Y, Kayyem JF, Gozin M (1999) J Org Chem 64:2070
229. Creager SE, Yu CJ, Bamdad C, O'Connor S, MacLean T, Lam E, Chong Y, Olsen GT, Luo J, Gozin M, Kayyem JF (1999) J Am Chem Soc 121:1059
230. Sachs SB, Dudek SP, Hsung RP, Sita LR, Smalley JF, Newton MD, Feldberg SW, Chidsey CED (1997) J Am Chem Soc 119:10563
231. Creager SE, Wooster TT (1998) Anal Chem 70:4257
232. Dudek SP, Sikes HD, Chidsey CED (2001) J Am Chem Soc 123:8033
233. Sikes HD, Smalley JF, Dudek SP, Cook AR, Newton MD, Chidsey CED, Feldberg SW (2001) Science 291:1519
234. Seferos DS, Banach DA, Alcantar NA, Israelachvili JN, Bazan G (2004) J Org Chem 69:1110
235. Kriegisch V, Lambert C (unpublished results)
236. Mayor M, Weber HB (2003) Chimia 56:494
237. Nitzan A, Ratner MA (2003) Science 300:1384
238. McCreery RL (2004) Chem Mat 16:4477
239. James DK, Tour JM (2004) Chem Mat 16:4423
240. Das R, Kiley PJ, Segal M, Norville J, Amy Yu A, Wang L, Trammell SA, Reddick LE, Kumar R, Stellacci F, Lebedev N, Schnur J, Bruce BD, Zhang S, Baldo M (2004) Nano Lett 4:1079
241. Ishida A, Majima T (1999) Chem Commun 1299
242. Akiyama T, Imahori H, Sakata Y (1994) Chem Lett 1447
243. Ishida A, Sakata Y, Majima T (1998) Chem Commun 57

# Author Index Volumes 251–258

Author Index Vols. 26–50 see Vol. 50
Author Index Vols. 51–100 see Vol. 100
Author Index Vols. 101–150 see Vol. 150
Author Index Vols. 151–200 see Vol. 200
Author Index Vols. 201–250 see Vol. 250

*The volume numbers are printed in italics*

# Subject Index